C0-ALO-946

HANDBOOK OF PLANT ECOPHYSIOLOGY TECHNIQUES

Handbook of Plant Ecophysiology Techniques

Edited by

Manuel J. Reigosa Roger

KLUWER ACADEMIC PUBLISHERS
DORDRECHT / BOSTON / LONDON

A C.I.P. Catalogue record for this book is available from the Library of Congress.

ISBN 0-7923-7053-8

Published by Kluwer Academic Publishers,
P.O. Box 17, 3300 AA Dordrecht, The Netherlands.

Sold and distributed in North, Central and South America
by Kluwer Academic Publishers,
101 Philip Drive, Norwell, MA 02061, U.S.A.

In all other countries, sold and distributed
by Kluwer Academic Publishers,
P.O. Box 322, 3300 AH Dordrecht, The Netherlands.

Cover design: Nuria Pedrol.

Printed on acid-free paper

Printed in the Netherlands.

CONTENTS

LIST OF CONTRIBUTORS

Aliotta, Giovanni
Facoltà di Scienze Matematiche, Fisiche e Naturali. Dipartimento di Scienze della Vita. Seconda Università degli Studi di Napoli. Via Vivaldi, I-81100 Caserta (Napoli), Italy
E-mail: aliotta@cds.unina.it

Bacon, Mark A.
Biological Sciences Department, Institute of Environmental and Natural Sciences. Lancaster University. Lancaster, LAI 4YQ, United Kingdom
E-mail: m.a.bacon@lancaster.ac.uk

Blanco Fernández, Alfonso
Laboratorio de Botánica. Departamento de Bioloxía Vexetal e Ciencia do Solo. Facultade de Ciencias. Universidade de Vigo
Campus Lagoas-Marcosende s/n. E-36200 Vigo, Spain
E-mail: ablanco@uvigo.es

Bolaño González, J. Carlos
Laboratorio de Ecofisioloxía Vexetal. Departamento de Bioloxía Vexetal e Ciencia do Solo. Facultade de Ciencias. Universidade de Vigo
Campus Lagoas-Marcosende s/n. E-36200 Vigo, Spain
E-mail: bolano@uvigo.es

Bolhàr-Nordenkampf, Harald
Institut für Ökologie und Naturschutz. Universität Wien
Althanstrasse 14, A-1090 Vienna, Austria
E-mail: harald.bolhar-nordenkampf@univie.ac.at

Cafiero, Gennaro
Centro Interdipartimentales di Servizio per la Microscopia Elettronica
Via Foria 223, I-80139 Napoli, Italy
E-mail: gecafiero@unina.it

Chiapusio, Geneviève
Laboratoire de Dynamique des Ecosystèmes d'Altitude (LDEA). Université de Savoie. Campus Scientifique
F-73376 Le Bourget-du-Lac (Chambéry), France
E-mail: chiapusio@univ-savoie.fr

Coba de la Peña, Teodoro
Centro de Ciencias Medioambientales
Departamento de Bioquímica y Fisiología Vegetal
Consejo Superior de Investigaciones Científicas.
c/ Serrano 115-bis, E-28006 Madrid, Spain
E-mail: tcoba@ccma.csic.es

González Rodríguez, Luís
Laboratorio de Ecofisioloxía Vexetal. Departamento de Bioloxía Vexetal e Ciencia do Solo. Facultade de Ciencias. Universidade de Vigo
Campus Lagoas-Marcosende s/n. E-36200 Vigo, Spain
E-mail: luis@uvigo.es

González-Vilar, Marco
Laboratorio de Ecofisioloxía Vexetal. Departamento de Bioloxía Vexetal e Ciencia do Solo. Facultade de Ciencias. Universidade de Vigo
Campus Lagoas-Marcosende s/n. E-36200 Vigo, Spain
E-mail: marco@uvigo.es

Martínez Otero, Ana
Laboratorio de Ecofisioloxía Vexetal. Departamento de Bioloxía Vexetal e Ciencia do Solo. Facultade de Ciencias. Universidade de Vigo
Campus Lagoas-Marcosende s/n. E-36200 Vigo, Spain
E-mail: amaranta@uvigo.es

Meister, Margit
Institut für Ökologie und Naturschutz.
Universität Wien
Altahanstrasse 14, A-1090 Vienna, Austria
E-mail: margit.meister@univie.ac.at
meister@pflaphy.pph.univie.ac.at

Pedrol Bonjoch, Nuria
Laboratorio de Ecofisioloxía Vexetal. Departamento de Bioloxía Vexetal e Ciencia do Solo. Facultade de Ciencias. Universidade de Vigo
Campus Lagoas-Marcosende s/n. E-36200 Vigo, Spain
E-mail: pedrol@uvigo.es

Pellissier, François
Laboratoire de Dynamique des Ecosystèmes d'Altitude (LDEA). Université de Savoie. Campus Scientifique
F-73376 Le Bourget-du-Lac (Chambéry), France
E-mail: pellissier@univ-savoie.fr

Ramos Tamayo, Pilar
Laboratorio de Ecofisioloxía Vexetal. Departamento de Bioloxía Vexetal e Ciencia do Solo. Facultade de Ciencias. Universidade de Vigo
Campus Lagoas-Marcosende s/n. E-36200 Vigo, Spain
E-mail: pramos@uvigo.es

Reigosa Roger, Manuel J.
Laboratorio de Ecofisioloxía Vexetal. Departamento de Bioloxía Vexetal e Ciencia do Solo. Facultade de Ciencias. Universidade de Vigo
Campus Lagoas-Marcosende s/n. E-36200 Vigo, Spain
E-mail: mreigosa@uvigo.es

Sánchez-Moreiras, Adela M.
Laboratorio de Ecofisioloxía Vexetal. Departamento de Bioloxía Vexetal e Ciencia do Solo. Facultade de Ciencias. Universidade de Vigo
Campus Lagoas-Marcosende s/n. E-36200 Vigo, Spain
E-mail: adela@uvigo.es

Santos Costa, Xan Xosé
Laboratorio de Ecofisioloxía Vexetal. Departamento de Bioloxía Vexetal e Ciencia do Solo. Facultade de Ciencias. Universidade de Vigo
Campus Lagoas-Marcosende s/n. E-36200 Vigo, Spain
E-mail: xxsc@uvigo.es

Souto Otero, X. Carlos
Departamento de Enxeñería dos Recursos Naturais e Medio Ambiente. Escola Universitaria de Enxeñería Técnica Forestal
Campus A Xunqueira, s/n. 36005 Pontevedra, Spain
E-mail: csouto@uvigo.es

Tiburcio, Antonio F.
Unitat de Fisiologia Vegetal. Facultat de Farmàcia. Universitat de Barcelona. 08028 Barcelona, Spain
E-mail: afernan@farmacia.far.ub.es

Weiss, Oliver A.
Laboratorio de Ecofisioloxía Vexetal. Departamento de Bioloxía Vexetal e Ciencia do Solo. Facultade de Ciencias. Universidade de Vigo
Campus Lagoas-Marcosende s/n. E-36200 Vigo, Spain

Institut für Pflanzenbau. Rheinische Friedrich-Wilhelms-Universität. Katzenburgweg 5. D-53115 Bonn. Germany
E-mail: oliver@uvigo.es

PREFACE

The Handbook of Plant Ecophysiology Techniques you have now in your hands is the result of several combined events and efforts. The birth of this handbook can be traced as far as 1997, when our Plant Ecophysiology lab at the University of Vigo hosted a practical course on Plant Ecophysiology Techniques. That course showed us how much useful a handbook presenting a bunch of techniques would be for the scientists beginning to work on Plant Ecophysiology. In fact, we wrote a short handbook explaining the basics of the techniques taught in that 1997 course:

Flow cytometry to measure ploidy levels,
Use of a Steady-State porometer to measure transpiration,
In vivo measure of fluorescence,
HPLC analysis of low molecular weight phenolics,
Spectrophotometric determinations of free proline and soluble proteins,
TLC polyamines contents measures,
Isoenzymatic electrophoresis,
Use of IRGA and oxygen electrode.

That modest handbook, written in Spanish, was very helpful, both for the people who attended the course and for other who have used it for beginning to work in Plant Ecophysiology. The present Handbook is much more ambitious, and it includes more techniques. But we have also had in mind the young scientists beginning to work on Plant Ecophysiology.

In 1999 François Pellissier leaded a proposal presented to the European Commission in the Fifth Framework Program in the High Level Scientific Conferences, including three EuroLab Courses* about lab and field techniques useful to improve allelopathic research. The EuroLab Courses should be dedicated to:

Plant Anatomy (Second University of Naples, Caserta, 2000)
Plant Physiology (University of Vigo, Vigo 2001)
Ecological Techniques (University of Savoie, Chambery, 2002)

When European Commission approved the proposal to celebrate a series of three EuroLab Courses, we immediately thought that it would be helpful for the people attending the second one, to put together all the

---

* EuroLab Courses are short practical courses (1-2 weeks) addressed to young European scientists. They are funded by the European Commission and included in the High Level Scientific Conferences.

techniques that will be taught in this Second EuroLab Course. But, with some help, we could extend this handbook to cover a greater number of techniques.

The objective of this handbook is to gather some of the most important techniques useful in Plant Ecophysiology. For each one of the selected techniques, some introductory background is explained, followed by the description of the technique and the utilities and some clues about the interpretation of results. Of course, this general scheme could not be strictly followed in some of the chapters, but the authors have done a good interpretation of what was expected.

The handbook is so, I hope, useful for the non-initiated scientist. It will indeed be useful for anyone initiating the work in plant ecophysiology. A whole "panoply" of techniques is here explained and with little effort the reader can understand the basics and the utilities. And, if interested, the technical aspects are also explained. We have included also some examples that we feel that will help to understand the utilities of the technique. Some examples come from the field of allelopathy whilst other come from other fields. In any case, we hope that they will be useful independently of the type of study to be done.

Although each chapter can be read independently, the authors (and the Editor!) have tried to avoid unnecessary repetitions. Also, the chapters are grouped according to the covered topics.

The first chapter is independent. Giovanni Aliotta and Gennaro Cafiero have written a very special chapter about germination. I wanted this chapter to be the first one due to its importance and to the little attention that this extremely important ecophysiological process is attracting.

After this one, there are four chapters more or less related to the use of flow cytometry. Teodoro Coba, Adela Sánchez and Alfonso Blanco have written those chapters; they have done a very good job mixing their different scientific experience, ranging from biochemistry to botany and plant physiology. An additional chapter shows how to assure the data about mitotic index using a complementary and alternative technique to flow cytometry, written by cited authors and Ana Martínez.

The following four chapters are related to photosynthesis, presenting two gas exchange techniques (Infra Red Gas Analyzers and Oxygen Electrode) that are classic in plant ecophysiology and will occupy also in the next future an important place in this science; the other two try to explain the basics of *in vivo* fluorescence techniques. It is advisable for any plant ecophysiologist to learn at least one gas exchange technique and the use of a fluorometer. At the present time, there are many instruments available and its use has become a must. The authors are plant physiologists or ecophysiologists: Luis González, Carlos Bolaño, Pilar Ramos and Adela Sánchez and plant ecologists: François Pellissier and Oliver Weiss.

The next block of chapters includes all the related to water. There is no need to discuss here the physiological and ecophysiological importance of water. Most topics have been covered by Luis González who is a plant physiologist, with some contributions by Adela Sánchez and Marco González-Vilar and there is a new technique to measure stomata that has been written by Margit Meisner and Harald Bolhàr-Nordenkampf.

Next chapters are devoted to biochemical determinations, ranging from the HPLC analysis of secondary metabolites to the functioning of RubisCO or ATPases, including protein, proline, polyamines, ABA and ions determinations, and a special chapter about the stress protein determination by means of two-dimensional electrophoresis. Several authors have contributed: Carlos Souto, Carlos Bolaño, Luis González, Xan Santos, Nuria Pedrol, Pilar Ramos, Adela Sánchez, Antonio Tiburcio, Teodoro Coba, Ana Martínez, Oliver Weiss and Mark Bacon.

Chiapusio and Pellissier devote the final chapter to the use of radiolabelled compounds to follow the movement and metabolism of molecules. Although it is focused in the research about allelochemicals, we feel that it could be useful for other studies.

As a final note, I would like to announce that we have left out of the handbook some of the less standard "recipes" for the practical use of some specific equipment. Those recipes will be freely available in the Web page of the University of Vigo (http://www.uvigo.es), under the Plant Biology and Soil Science Department link.

April, 6, 2001 Manuel J. Reigosa

ACKNOWLEDGMENTS

I want to give my more sincere thanks to all the people and the institutions that made this handbook possible. Thanks to the authors that have done an incredibly fast and good work. Thanks to them for making possible to put together all this valuable, actual and, I hope, very useful information for all of us working on this challenging young discipline called Plant Ecophysiology. Thanks to Nuria Pedrol for her help with the formatting of the documents. Again, it has been an incredibly fast task; without her, the handbook would still stand in the hard disk of the computer. Thanks to my wife, Sara Alonso and my family for their understanding when I stole them a lot of time.

Finally, thanks to the European Commission and to the University of Vigo, for their financial support and to Kluwer for their fast evaluation and acceptance of this handbook, and for the good work producing this book.

CHAPTER 1

SEED BIOASSAY AND MICROSCOPY IN THE STUDY OF ALLELOPATHY: RADISH AND PURSLANE RESPONSES

Giovanni Aliotta[1] and Gennaro Cafiero[2]

[1]*Dipartimento di Scienze della Vita, Seconda Università degli Studi di Napoli, Caserta, Italy.*

[2]*Centro Interdipartimentale di Servizio per la Microscopia Elettronica, Napoli, Italy.*

INTRODUCTION

Since antiquity the formation of seeds and seed germination have intrigued the human mind because of their role in animal nutrition and agriculture. Seed formation is a complex and fascinating process which involves the development of three basic structures – the embryo, the endosperm and the seed coat – The embryo and the endosperm are derived from both maternal and paternal genetic material, while the seed coat is derived from diploid maternal cells. Seed formation starts in the ovule of the flower where male and female nuclei unite to form a zygote, which develops into embryo, three other nuclei are inspired to join in the dance forming the endosperm, and the ovule teguments develop into a seed coat. Then cell division begins, the suspensor develops, sugar molecules come pouring in and polymers are laid down. All seems set for a burst of vigorous growth – and then the whole performance inexplicably comes to a stop. Water evaporates or is drained away, respiration drops practically to nothing, and many of the newly formed cells outside of the embyro die. The embryo, in spite of its generous warehouses of stored food in the endosperm or cotyledons, is halted right in the heyday of its development (Evenari, 1980, 1984). The seed coat is a structure of considerable importance because it forms the barrier between the embryo and its immediate environment. Several features can be distinguished on the outer surface of the seed. The micropyle, the opening in the integuments

M.J. Reigosa Roger, Handbook of Plant Ecophysiology Techniques, 1–20.

through which radicle protrudes, may be completely obliterated or it may remain as a distinct pore (Corner, 1976). In the place where the seed coat was attached to the *funiculus* a scar, termed the *hilum*, is present. Water can penetrate with relative ease through the hilum. Some growths of the funiculus, termed arils may develop on the surface of the seeds. Arils that contain oil, such as those of *Portulaca oleracea* are the elaiosomes connected with seed dispersal by ants.

Not less interesting is seed germination. It is certainly not by chance that the German philosopher Hegel, in order to illustrate his triodic system of logic, uses the seed as an example: 'the seed of the plant is an initial unity of life which when placed in its proper soil suffers disintegration into its constituents, and yet in virtue of its vital unity keeps these divergent elements together, and reappears as the plant with its members in organic union'. We are dealing therefore with a spiral of unification-discordance-synthesis, which is repeated *ad infinitum* (Evenari, 1980).

Scientifically, seed germination of the higher plants can be regarded as those consecutive events which cause a dry quiescent seed, in response to water uptake, to show a rise in its general metabolic activity and to initiate the formation of a seedling from the embryo. Generally germination ends with the radicle protrusion (Mayer and Poljakoff-Mayber, 1988).

In the last decades, seed germination has been the subject of much fruitful study, although certain problems of seed biology such as the dormancy and the interaction among the three parts of the seed, tegument, endosperm and embryo have lost none of their fascination.

Seed germination is the most widely used bioassay in allelopathy and the literature pertaining to the use of this bioassay and its general suitability for the determination of allelopathic activity among species has been reviewed by Leather and Einhellig (1986) and Inderjit (1995). These authors pointed out that there is little standardization governing seed germination bioassays. Most research is centered on a few plants, especially agricultural seeds to discover biochemical and physiological aspects of seed germination. This raises the question as to what extent do the results obtained with these seeds which have been subjected to man's selection can be extrapolated to the wild plants.

The objectives of our research group are to: (i) assay poisonous plants and vegetable wastes, for their allelopathic activities, and (ii) identify the site (s) of action of allelochemicals on seeds of cultivated plants and weeds. We hypothesize that the outermost living cells of the micropylar region of the seeds are responsible for the *primum movens* of germination, therefore, these cells could represent the main target for allelochemicals.

This chapter deals with the morphological and cytological responses of two different seeds such as: radish (*Raphanus sativus* L. cv. Saxa) and purslane (*Portulaca oleracea* L.), sown in presence of furanocoumarins isolated from the ancient medicinal plant Rue (*Ruta graveolens* L.). Moreover, we have made every effort to outline as simply as possible some basic microscopic techniques, bearing in mind the many little pitfalls one encounters as a beginner.

RUE VERSUS RADISH GERMINATION

We have selected rue for allelopathic studies because 1) our previous studies established coumarins as potent allelochemicals (Aliotta et al., 1992), 2) the presence of large amounts of coumarins on the leaf surface of rue and their easy extraction through leaching make them ideal (Zobel and Brown, 1988), and 3) rue and basil never grow together, or near one another (Grieve, 1967).

We tested rue leachate for its allelopathic activity *in vitro* on radish germination in light and darkness by focusing attention on the hilum-micropylar region, where normally a radicle protrudes. (Aliotta et al., 1994). Rue delayed the onset and decreased the rate of germination in light rather than in darkness. Three active pure compounds were isolated and identified as coumarins: 5-methoxypsoralen (5-MOP), 8-methoxypsoralen (8-MOP), and 4-hydroxycoumarin. Their concentrations in the infusion were 10^{-4} M, 2 x 10^{-4} M, and 0.4 x10^{-5} M, respectively.

To ascertain whether the effect of coumarins on radish germination was at the level of the embryo or was mediated by the seed coat and endosperm, these latter were removed and embryos were tested for their germination in light in the presence of each coumarin. It appeared that each coumarin inhibited the seeds with their coat, but did not significantly inhibit seeds without a coat and endosperm. When the most of the seeds soaked in water were germinating, treated seeds were dormant and different uptake of water into these seeds was evident (Figure 1).

Radish seeds are oval (3x2 mm) with a light brown to orange, reticulate surface. Three layers of cells may be recognized in the seed coat of the mature seed: the epidermis (formed by compressed cells), the palisade layer (which represents cells more or less iso-diametrical or radially elongated), and the inner parenchyma (pigmented layer one cell thick) (Vaughan and Whitehouse, 1971). The endosperm persists as a well-formed aleurone layer intimately associated with the seed coat. The hyalin layer covers the embryo (Figure 2), which is folded with cotyledons against the radicle.

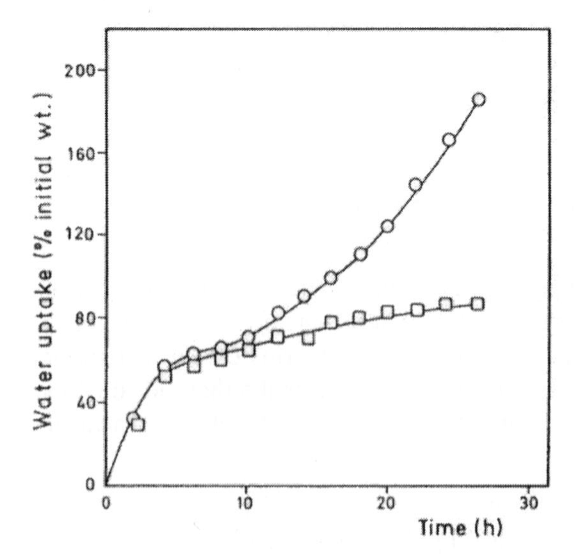

Figure 1. Water uptake by radish seeds in presence (□) and absence (o) of 5-MOP 2 x 10^{-4} M (on a percent of initial weight of seeds). After Aliotta et al. (1994).

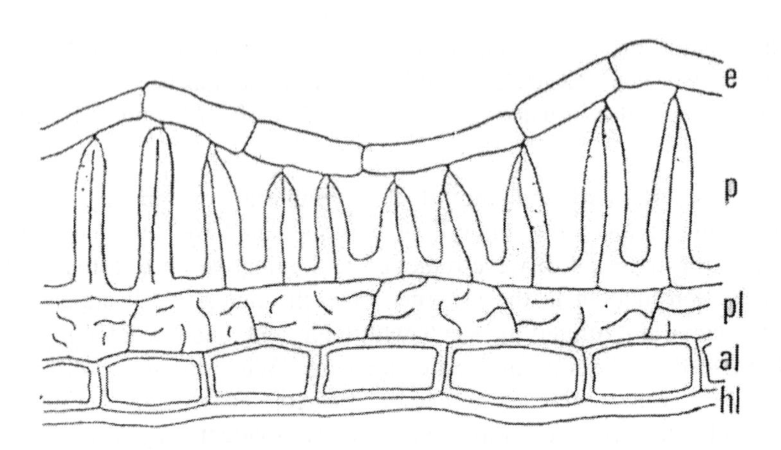

Figure 2. Graphic showing the structural features of a seed coat and endosperm of radish seed, according to Vaughan and Whitehouse, 1971. Abbreviations: e, epidermis; p, palisade; pl, pigment layer; al, aleurone layer; hl, hyalin layer; r, radicle ; hm, hylum-micropylar region.

Figure 3 shows a comparison between control and rue treated radish seeds by stereo, light and SEM microscopy. Different sizes of control and treated seedlings appear evident in Figure 3 a and b. The hilum-micropylar region of the seed was highly specialized and characterized by a flaking epidermis, thickened aleurone cells, and a hyalin layer. Moreover, this area represents a marker for comparison between seeds (Figure 3 c).

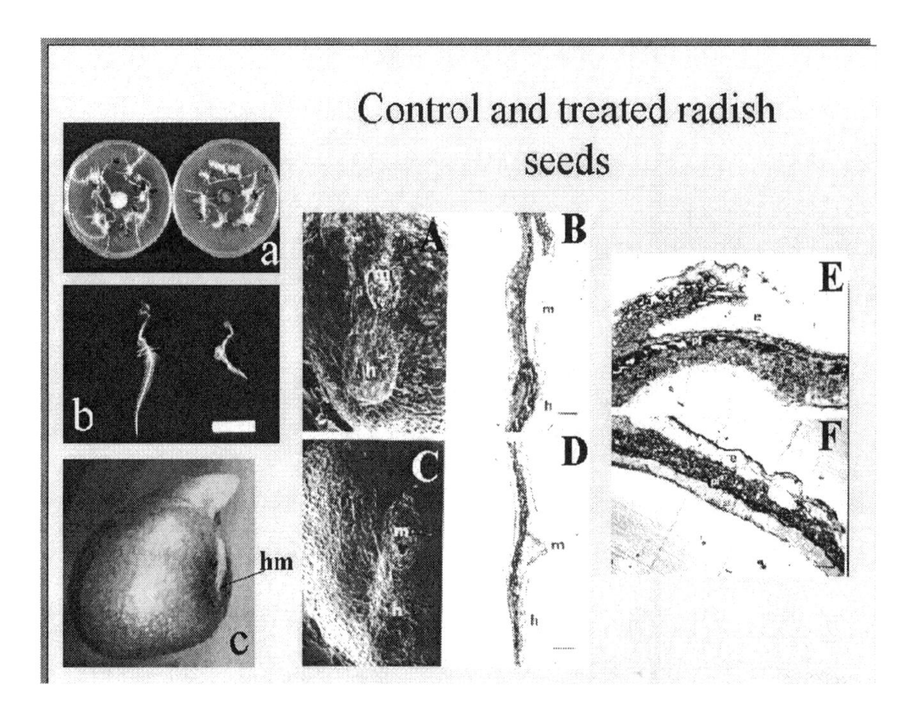

Figure 3. An overview of control and treated radish seeds. Different size of control (left) and treated (right) seedlings (3 a,b); Stereomicrograph of a germinating seed of radish showing the hilum-micropylar region and the radicle (3c). SEM (A, C) and light microscope (B, D) micrographs of the hilum-micropylar region of radish; 18 h after sowing the hilum is more evident in water moistened seed, (A, B), than in 5-MOP- treated seed (C, D). Bar = 0.5 mm. Semithin light microscope section of hilum-micropylar region of water-moistened seed, 18 h after sowing (E) and 5-MOP- treated radish seed (F). e, epidermis; pl, pigment layer; al, aleurone layer. Bar = 150 μm.

SEM and light microscope observations of the hilum-micropylar region showed that the hylum was more evident in the control than in the treated seed (Figure 3, A and C).

This aspect was confirmed by the comparison of the hilum-micropylar semi-thin section of control and treated seeds under the light microscope, where the embryos were removed and the seed coats and endosperms were observed by light microscope (Figure 3, B and D). The hilum of the control seed is thicker and pigmented; the tested pigment layer is more evident, and there are more layers of aleurone cells that are filled by light-dense bodies (Figure 3, E and F).

In this respect it was interesting to compare the TEM ultrastructure of the seed coat and endosperm of control and 5-MOP-treated seeds (Figures 4 and 5), as shown under the light microscope.

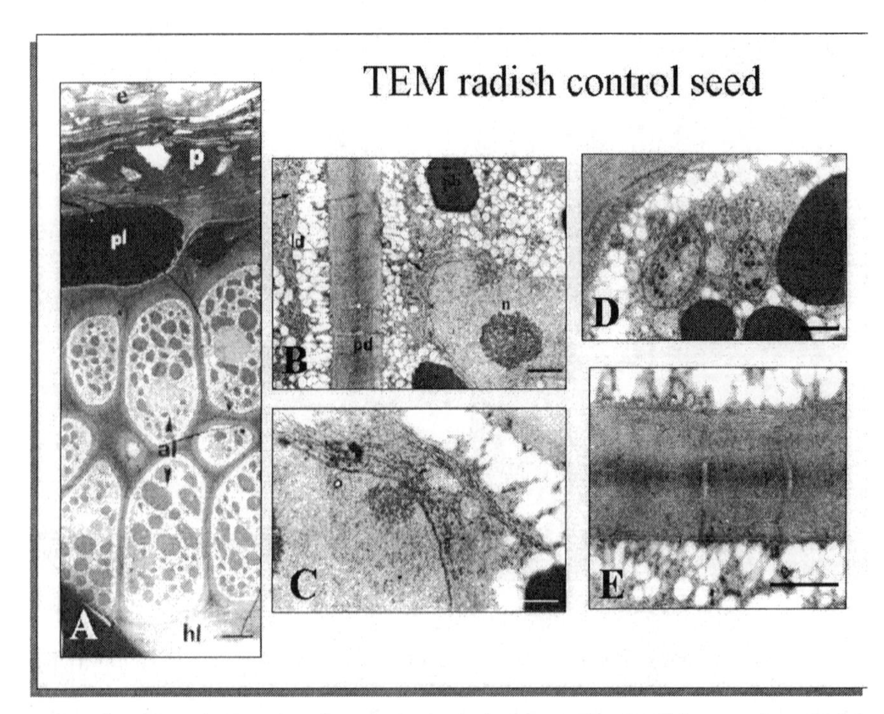

Figure 4. TEM micrographs of seed coat and aleurone cells of radish control seed 18 h after sowing in water; e, epidermis; pl, pigment layer; al, aleurone layer. (A) Bar = 30 μm; Particulars of the aleurone cell showing some organelles: nucleus (n), plasmodesmata (pd), protein bodies (pb) and lipid droplets (ld). Bar = 5 μm.

As can be seen, the palisade layer of treated seed appears thicker than in the control, while comparison between aleuronic cells of the control (Figure 4, A) and treated cells (Figure 5, B), reveals that the cells of the control are healthy with some evident organelles such as the nucleous and the rough endoplasmic reticulum, and other structures, the plastid, the plasmodesmata, conspicous constrictions, protein bodies, and lipid droplets. By contrast, cells of the treated seed resemble those in the dried quiescent seed (Figure 5, C). The differences observed between moistened and 5-MOP-treated radish seeds represent useful signals to establish whether the water uptake of the seed will culminate in radicle emergence.

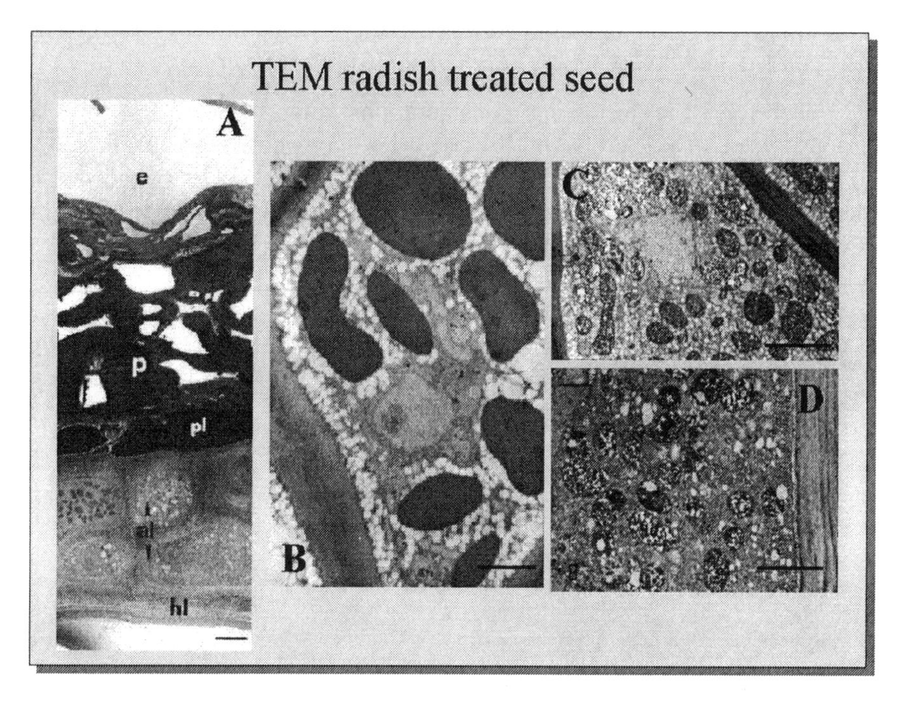

Figure 5. TEM micrographs of seed coat and aleurone cells of radish treated seed 18 h after sowing in presence of 5-MOP e, epidermis; pl, pigment layer; al, aleurone layer Bar = 30 μm; Particulars of aleurone cell (B,C,D,E) showing different profiles respect to those of control (Figure 4) Bar = 5 μm.

We have studied the anatomical and ultrastructural aspects of *in vitro* germination of the purslane seeds using stereo and SEM microscopy (Figures 6 and 7).

RUE VERSUS PURSLANE GERMINATION

On the basis of results obtained from radish seeds, we began to use the rue leachates as a possible bio-herbicide against germination of weeds in an agricultural soil (Aliotta et al., 1995). Greenhouse experiments in pots containing a soil differently perfused with aqueous extracts of rue leaves were carried out in the summer of 1994. In treated pots, weeds emerged later than in the control. In all observations, the presence of purslane was always recorded for about 95 percent of total weeds. Purple nutsedge *(Cyperus rotundus* L.) was noticed for less than 5 percent, while a very small number of redroot pigweed *(Amaranthus retroflexus* L.), common lambsquarters *(Chenopodium album* L.), Bermudagrass *(Cynodon dactylon* L.), and dwarf spurge *(Euphorbia exigua* L.) were recorded.

Therefore, the following considerations refer to purslane only, a serious weed of 45 crops in 81 countries, considered the ninth most common weed on Earth (Holm et al., 1977). It has been estimated that a plowed layer of the soil cropped with maize *(Zea mays* L.) contains about 220,000 purslane seeds per m^2 (Leguizamon and Cruz, 1981).

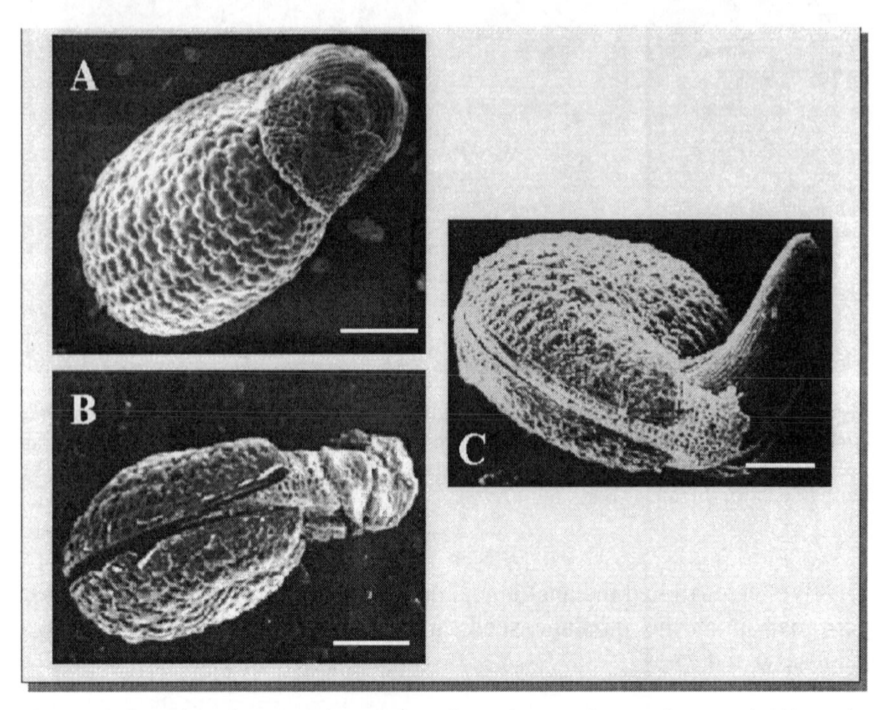

Figure 6. Scanning electron micrographs of a quiescent dry purslane seed (A) an its successive modifications after 25 and 30 h respectively (B, C).

A dry quiescent purslane seed and the successive changes when it is moistened have been reported here. The outer integument (testa) of the seed coat is formed by dead cells sculptured on the surface with stellulae (Figure 6A). The peripheral face of the testa presents an opening; the micropyle and a residue of the funiculus of the placenta which functions as elaiosome. Surprisingly, the first structure that protrudes from the micropyle of the moistened seed, is not the radicle, as in seeds from most species, but the inner integument (tegmen) of the seed coat (Figure 6B). When the germination proceeds further, the radicle breaks the tegmen and protrudes (Figure 6C).

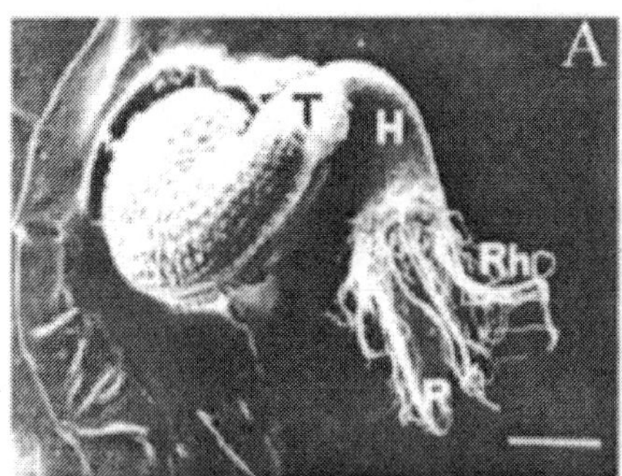

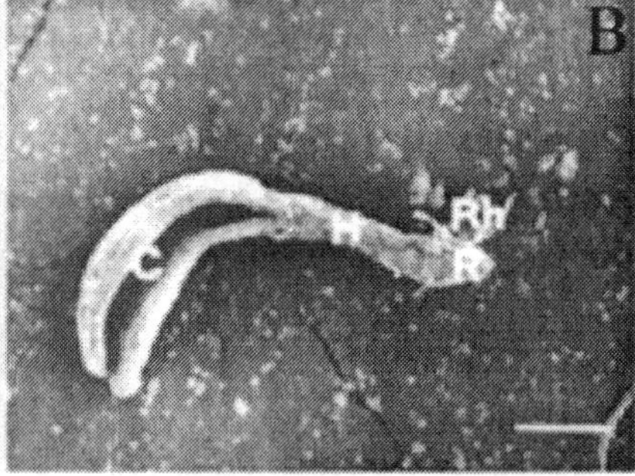

Figure 7. Scanning electron micrographs of developing purslane seedlings of control and with rue infusion. The seedling of control observed after 40 h shows the elongating root covered by root hairs as compared to seedling treated with rue infusion showing a strongly inhibited radicle at 120 h. C = Cotyledons, H = Hypocotyl, R = Radicle, Rh = Root hairs, T = Tegmen. Bar = 0.25mm.

Once the radicle of purslane protrudes in the absence of rue, it rapidly completes the early stages of primary structure development viz., root hairs and secondary roots (Figure 7A). While in the presence of rue infusion, if tegmen regalation fails, the radicle protrudes but is unable to grow and become irreversibly damaged, however, the hypocotyl and cotyledons show normal growth (Figure 7B) (Aliotta et al., 1996).

Furthermore, light and TEM observations were carried out to ascertain the effect of rue infusion on purslane germination (Figure 8). This investigation revealed that the purslane seed is lenticular, the seed coat is bitegmic, and the embryo is curved around the starchy hard perisperm that represents the seed storage reserves. Furthermore, the inner tegument (tegmen) of the control seed appears to be pierced and the radicle protruded (Figure 8, A). By contrast, in treated dormant seed, the tegmen appears intact and not expanded (Figure 8, B). Comparison between a tegmen cell of the control and that of the treated seed (Figure 8, C and D), shows that the cell of the control is healthy, and cytoplasmic structures are well differentiated. Tegmen cells of the treated seed also appears less differentiated. These observations suggest that tegmen, the outermost living structure of the purslane seed, is the primary target of rue leachates. This is in agreement with the similar role as that of the aleuronic layer in a radish seed, where the seed coat is dead.

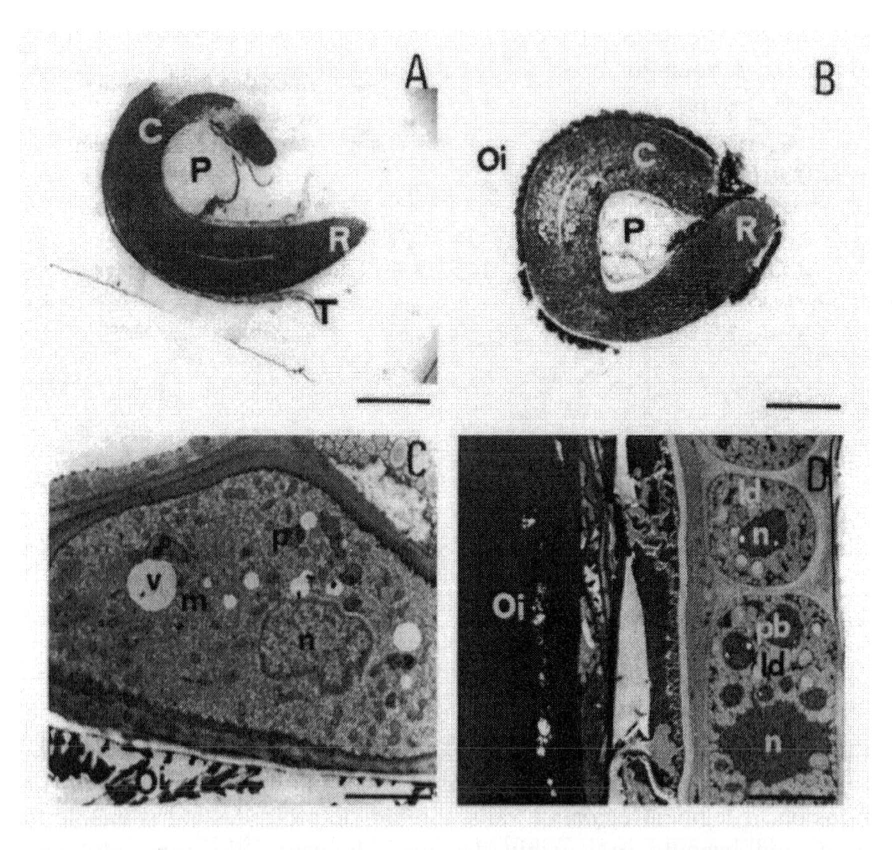

Figure 8. Light microscope micrographs of the control and treated purslane seeds after 30 h of sowing, respectively (A, B), and transmission electron micrographs (C, D) showing the tegmen cells of control and treated seeds at the same time. (A, B) C, Cotyledones; Oi, outer integument; P, perisperm; R, radicle; T, tegmen. Bar = 0.1 mm. (C, D) m, mitochondria; n, nucleus; Oi, outer integument; p, plastids; v, vacuoles. Bar = 5 μm.

Considerable research remains to be done to ascertain the potential of the simple rue infusion, as a better cost-effective tool for weed managements in crops (e.g., its biodegradability and selectivity). Generally, only about one out of every 10,000 chemicals screened enters the second phase of testing, which means the stage of extensive field evaluation (Han, 1987). Furthermore, it is important to study the effects of rue infusions, if any, on chemical, physical and biological soil properties. Nevertheless, the properties of rue open up a promising avenue of research.

HANDS ON

Preliminary approach

Before facing seed allelopathic aspects, it is important to have a knowledge of macro and micro anatomy of the seed basic structures. Therefore, we have made a series of cuts to elucidate the organization of the embryo, the endosperm and the seed coat of both radish and purslane seeds.

These attempts are shown in Figure 9 where radish seeds were excised orthogonally (Figure 9a) or longitudinally (Figure 9c). The whole seed shows the cotyledon prints (Figure 9e). Radicles inside excised and germinated seeds are shown in Figure 9b,f,d. Moreover, scanning micrographs in Figure 10 show the tridimensional structure of the seed coat according the scheme of Vaugan and Whitehouse.

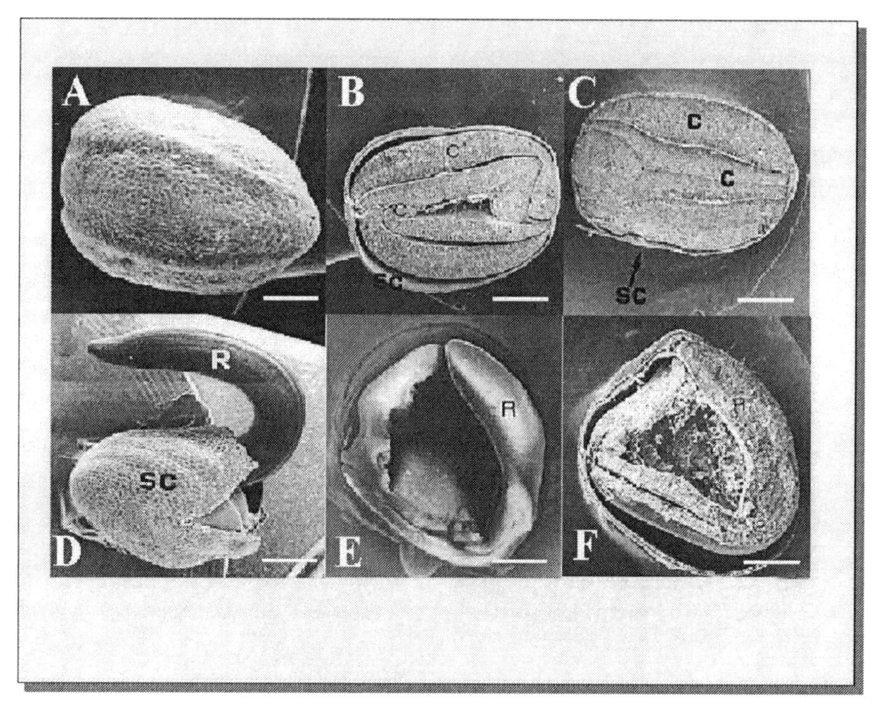

Figure 9. SEM micrographs of the whole radish seed (A), orthogonally excised seed (B), longitudinally excised seed (C), germinating seed, radicles in different cuts of the seed (E, F). c = cotyledons, sc = seed coat, R = radicle. Bar = 1.8 mm.

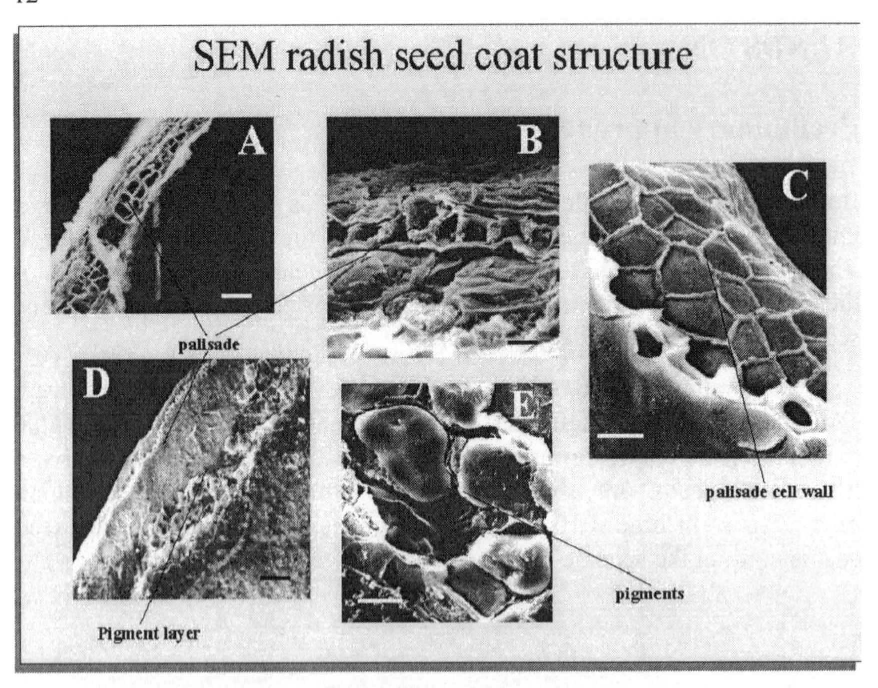

Figure 10. See text.

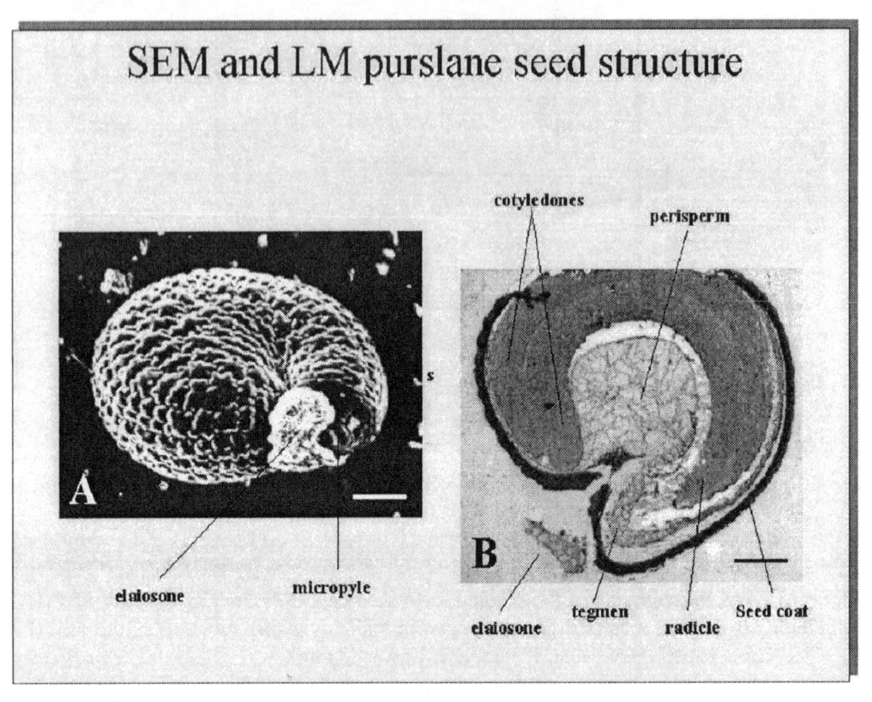

Figure 11. See text.

For purslane seeds we can see the whole seed by SEM (Figure 11 A) and light microscopy diametrical section (Figure 11 B). Light microscope photograph of perisperm is shown in Figure 12 A and SEM micrographs of perisperm starch in Figures 12 B and C.

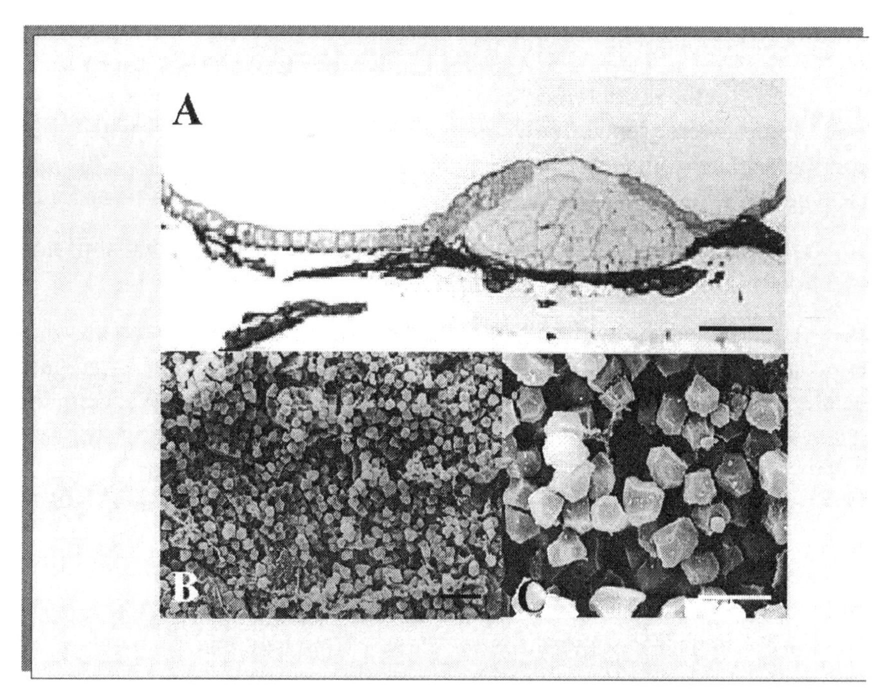

Figure 12. Light microscope photograph of purslane perisperm (A) and SEM micrographs of perisperm starch (B, C).

An outline of Stereo, Light, and Electron Microscopy

Eighteen hours after sowing, control and 5-MOP-treated radish seeds were differently moistened in the presence of light, their weight increased 110 and 80 %, respectively. The water uptake was evaluated as the difference in weight between moistened and unmoistened seeds as a percentage of the weight of unmoistened seeds (percent initial weight).

Whole and excised moistened seeds were observed directly with a Stereomicroscope Wild M3Z.

Moistened seeds were excised (cut in half along the two orthogonal planes of their major axis), the embryos were removed, and the seed coats and endosperms were fixed in 3 % glutaraldehyde in 0.065 M phosphate buffer (pH 7.4) for 2 hr at room temperature. The specimens were then placed into 2 % Os04 in 0.1 M phosphate buffer (pH 6.8) at 4 °C before being dehydrated with ethanol and propylene oxide, and embedded in Spurr's resin.

Thick sections (ca. 1 µm each) were stained with 0. 1 % toluidine blue and observed with a Zeiss light photomicroscope.

Thin sections, obtained with a diamond knife on a Supernova microtome, were sequentially stained at room temperature with 2 % uranyl acetate (aqueous) for 5 min and by lead citrate for 10 min (Reynolds, 1963).

Ultrastructural studies were made using a Philips CM12 transmission electron microscope (TEM) operated at 80 kV.

After fixation and ethanol dehydration, the seed coats and endosperms of control and treated seeds were critical-point dried and finally coated with carbon and gold in a sputter-coater. These specimens were observed at 20 kV with a Cambridge 250 Mark3 scanning electron microscope (SEM).

Some basic considerations

Here we emphasize some different steps of microscopic preparation of radish and purslane seeds related to practical microscopy.

....Moistened seeds were excised (cut in half along the two orthogonal planes of their major axis), the embryos were removed, and the seed coats and endosperms were fixed in 3% glutaraldehyde....

Fixation is an attempt to quickly arrest biological activity and to stabilize cellular components with minimal distortion of conformational and spatial relationships between cellular constituents. The purpose of fixation is to preserve tissue structure with minimal alteration during dehydration, embedding, cutting, staining and viewing the microscope. The most common reason for poor fixation is large specimen size. For example, glutaraldehyde, the fixative most often used in electron microscopy, penetrates to a depth of less than 1mm. To minimise autolytic changes, slices or ribbon of tissue 0.5 mm thick should be placed in fixative promptly (Slayter and Slayter, 1993).

Chemical fixation for transmission electron microscopy prepares cells for the preservation of damage due to subsequent washing with aqueous solvents, dehydration with organic solvents such as ethanol or acetone, embedding in plastic resins, polymerisation of the resins by heat, exothermic catalysts, or ultraviolet radiation, and imaging with high-energy electron beams in an electron microscope.

Materials prepared for scanning electron microscopy are not embedded with resins, but while in solvents they are subjected to high pressures during critical point drying. An ideal fixative would transform the viscous colloidal protoplasm of a cell into cross-linked and stabilized cellular components. The spatial relationship between all organelles and cellular structures are not altered, the cellular components are not solubilized, and the biological activity of complex proteins like antigens and enzymes remain undiminished (Dykstra, 1993).

.....in 0.065 M phosphate buffer (pH 7.4).......

The buffering system is used typically to stabilize the pH of the tissue somewhere near physiological levels during fixation and to maintain nearisotonic conditions. For structural studies, most cells fix well within a pH range from 7.0 to 7.4. Certain highly hydrated tissues fix better at a more alkaline pH (i.e., 8.0 – 8.4), whereas plant cells, nuclear material, and the fibrils of mitotic spindles fix better at a more acid pH (6.0 – 6.8). Phosphate buffers are typically used at physiological pH. They are specifically avoided in situations where the negatively charged phosphate groups could be expected to interact with positively charged ions in incubation media, causing precipitates to form.

There are three major phosphate buffers used in most laboratories: Sorenson's phosphate buffer formulated from sodium and potassium phosphate salts, and Millonig's sodium phosphate buffer (Glauert, 1975; Millonig, 1964).

....The specimens were then placed into 2% Os04 in 0.1 M phosphate buffer (pH 6.8) at 4'C before being dehydrated with ethanol and propylene oxide, and embedded in Spurr's resin....

Spurr' resin was originally designed for plants with cell walls that were hard to infiltrate with higher-viscosity epoxide resins. It requires only a relatively brief infiltration schedule for most materials and thus can shorten overall specimen processing times.

The embedding resin is compatible with acetone, ethanol, and other commonly used solvents. The resin is completely mixed with ethanol, though blocks polymerized following an ethanol only dehydration series often have a slightly tacky surface not found when acetone is used as a transitional solvent. They also may be slightly more brittle. Because of the low viscosity of Spurr' resin, it is not necessary to rotate or agitate specimens in the resin during infiltration. With proper handling, Spurr' resin provides quick infiltration of difficult samples and can be readily trimmed and sectioned, yielding excellent results (Spurr, 1969).

....Thick sections (ca. 1 μm each) were stained with 0. 1% toluidine blue and observed with a Zeiss light photomicroscope. Thin sections, obtained with a diamond knife on a Supernova microtome....

Ultramicrotomes are designed to cut ultrathin sections, semithin sections, and ultrathin frozen sections if suitably equipped. An ultramicrotome with either a thermal or a mechanical advance mechanism will work well for sectioning plastic-embedded samples. If ultrathin frozen sections are needed for the research objectives anticipated, mechanical advance units offer some advantages in terms of consistency of section thickness. (Fig. 13). Semithin sections up to several square millimeters in size and 0.25 – 0.5 μm thick can be cut easily from epoxy or acrylic resin blocks containing fixed samples with glass knives (Fig. 13 A) and examined by light microscopy. Ultrathin sections approximately 0.5 mm^2 in size and 80 – 90 nm thick can be cut with either glass (Fig. 13 B,C) or diamond knives for examination by TEM (Fig. 13 D,E,F).

....Seed coats and endosperms of control and treated seeds, after fixation and ethanol dehydration....

The fixation of biological samples for scanning electron microscopy (SEM) involves all the same principles for preserving specimen structural integrity. Osmium tetroxide postfixation can sometimes be omitted, though samples that have charging problems leading to image distortion can frequently benefit from osmication.

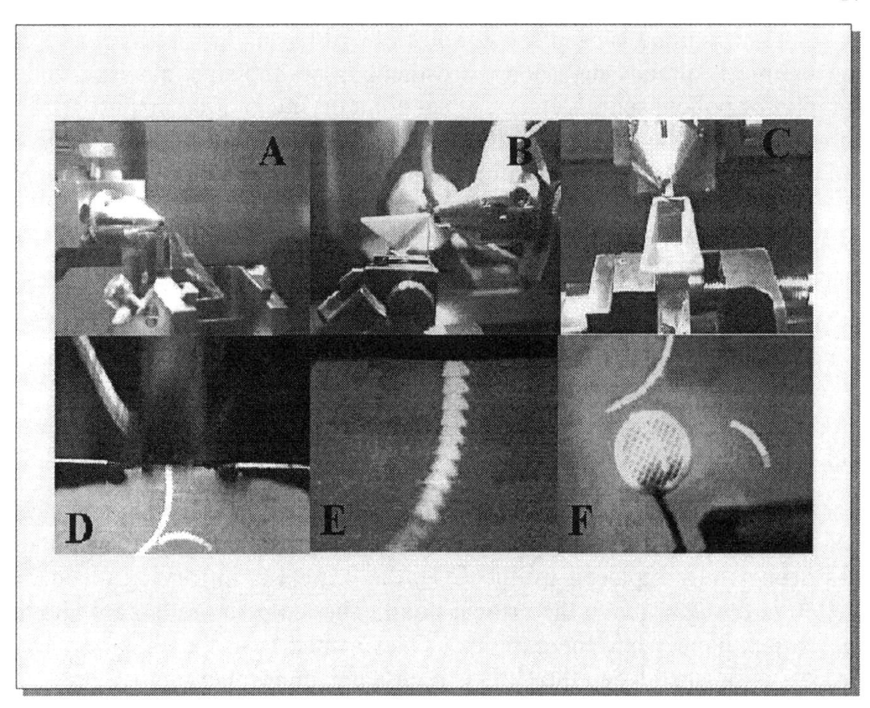

Figure 13. View of the specimen holder and the glass blade while cutting semithin sections (A). Lateral and frontal view of the glass blade mounted with tub for cutting ultrathin sections (B, C). Production of ultrathin sections with diamond knife (D). Particular of ultrathin section film (E). Collection of ultrathin sections by means of copper grid (E) (by G. Cafiero).

Subsequent dehydration in an ethanol or acetone series followed by critical point drying, freeze-drying, or drying with one of the chemical techniques produces a sample that can be introduced into the high vacuum system of the SEM. The dried sample is then attached to support stubs with a variety of materials (colloidal silver, colloidal carbon, double sided tape, or conductive carbon tape, and other) prior to coating with precious metals such as gold-palladium to ensure the electrical conductivity of the specimen surface.

Normally, biological SEM samples are examined in the secondary electron imaging mode (Dykstra, 1993).

…..were critical-point dried…..

This technique is used for samples of soft tissue that are hydrated. If the sample is already dry (bone, prosthetic bone implants, air-dried field samples of pollen, spores, etc.), critical point drying is often inappropriate. In such cases, the dried sample can be immediately mounted on SEM stubs, sputter coated, and examined with the SEM. A hydrated biological sample should be fixed and dehydrated as described for routine SEM preparation. The dehydration fluid (ethanol) must be removed from the specimen prior to sputter coating and viewing with an SEM.

If the sample is allowed to air-dry, pressures up to 2000 $1b/in^2$ can be generated, resulting in major distortion of specimen surfaces. Critical point drying is based on the concept that at a certain temperature and pressure, the vapour and liquid phases of carbon dioxide become indistinguishable.

Liquid CO_2 from a siphon-tube tank is introduced into the chamber and used to replace 100% of the ethanol in the specimen. After the ethanol has been totally replaced by the CO_2, the critical point drying chamber (CPD) is brought above the critical point. The temperature is kept above the critical point while the gaseous CO_2 is vented from the chamber. The process is finished when the CPD is returned to atmospheric pressure.

After critical point drying, the specimen should be totally dry and it is ready to be introduced to the vacuum system of the sputter coater and SEM (Dykstra, 1993; Hayat, 1978).

.....and finally coated with carbon and gold in a sputter-coater....

Sputter coating is a technique for deposing a metal coating on specimen surfaces to be examined with a SEM. The metal is deposited non directionally, allowing even coating surfaces of the specimen, regardless of topography, unlike the directional coating process of vacuum evaporation. Less specimen heating is developed during sputter coating than with vacuum evaporation.

Although it is unlikely that this chapter will provide a complete and final synthesis of the subject, our hope is that the information presented can stimulate young researchers to expand both the depth and breath of microscopy in allelopathy.

REFERENCES

Aiello R., Aliotta G., Molinaro A., Pinto G., Pollio, A. Anti-algal agents isolated from *Arum italicum*. Planta Med 1992; 58:652

Aliotta G., Fuggi A., Strumia S. Coat-imposed dormancy by coumarin in radish seeds: the influence of light. Giorn Bot Ital 1992; 126:631

Aliotta G., Cafiero G., Fiorentino A., Strumia S. Inhibition of radish germination and root growth by coumarin and phenylpropanoids. J Chem Ecol 1993; 19:175

Aliotta G., Cafiero G., De Feo V., Sacchi R. Potential allelochemicals from *Ruta graveolens* L. and their action on radish seeds. J Chem Ecol 1994; 20:2761

Aliotta G., Cafiero G., De Feo V., Palumbo A.D., Strumia S. Inhibition of weeds germination by a simple infusion of rue. Proceedings of the XXII Annual Meeting of the Plant Growth Regulator Society of America. Minneapolis, MN, 1995

Aliotta G., Cafiero G., De Feo V., Palumbo A.D., Strumia S. Infusion of rue for control of purslane weed: biological and chemical aspects. Allelopathy J 1996; 3:207

Corner E.J. *The Seeds of Dicotyledones*. New York, USA: Cambridge University Press, 1976

Dykstra M.J. *A Manual of applied Techniques for Biological Electron Microscopy*. New York, USA: Plenum Press, 1993

Evenari M. The history of germination research and the lesson it contains for today. Israel Bot 1980; 29:4

Evenari M. Seed physiology: from ovule to maturing seed. Bot Rev 1984; 50:143

Glauert A.M. *Fixation, Dehydration and Embedding of Biological Specimens*. North-Holland, Amsterdam, 1975

Grieve M. *A Modern Herbal*. New York, USA: Hafner Publishing, 1967

Han S.K. "Potential Industrial Application of Allelochemicals and their Problems." In *Allelochemicals: Role in Agriculture and Forestry*. G.R. Waller, ed. Washington, D.C.: ACS Symposium Series 330, American Chemical Society, 1987

Hayat M.A. *Principles and Techniques of Scanning Electron Microscopy, Biological Applications*. New York, USA: VNR Company, 1978

Holm L.G., Plucknett D.L., Pancho J.V., Herberger J.P. *The World's Worst Weeds Distribution and Biology*. Honolulu: University Press of Hawaii, 1977

Hunter E. *Practical Electron Microscopy*. New York, USA: Cambridge University Press, 1993

Inderjit. On laboratory bioassay in allelopathy. Bot Rev 1995; 61:28

Leather G.R., Einhellig F.A. "Bioassay in the Study of Allelopathy." In *The Science of Allelopathy*. A.R. Putnam, C. Tang, eds. New York, USA: Wiley-Interscience, 1986

Leguizamon E.S., Cruz P.A., Población de semillas en perfil arable de suelos sometidos a distinto manejo. Revista de Ciencias Agropecuarias 1981; 2:83

Mayer A.M., Poljakoff-Mayber A. *The Germination of Seeds*. Oxford: Pergamon Press, 1988

Millonig G. Study on the factors which influence preservation of fine structure. In *Symposium on Electron Microscopy*. P. Buffa, ed. Rome: Consiglio Nazionale delle Ricerche, 1964

Reynolds E.S. The use of lead citrate at high pH as an electron opaque stain in electron microscopy. J Cell Biol 1963; 17:208-212

Slayter E.M., Slayter H.S. *Light and Electron Microscopy*. New York, USA: Cambridge University Press, 1993

Spurr A.R. A low-viscosity epoxy resin embedding medium for electron microscopy. J Ultrastruct Res 1969; 26:31

Vaughan J.G., Whitehouse J.M. Seed structure and taxonomy of the Cruciferae. Bot F Linn Soc 1971; 64:383

Zobel A.M., Brown S.A. Determination of furanocoumarins on the leaf surface of *Ruta graveolens* with an improved extraction technique. J Nat Prod 1988; 51:941

CHAPTER 2

FLOW CYTOMETRY: PRINCIPLES AND INSTRUMENTATION

Alfonso Blanco Fernández [1], Adela M. Sánchez-Moreiras [1] and Teodoro Coba de la Peña [2]

[1]*Dpto Bioloxía Vexetal e Ciencia do Solo. Universidade de Vigo. Spain*
[2]*Dpto Fisiología y Bioquímica Vegetal. Centro de Ciencias Medioambientales. Consejo Superior de Investigaciones Científicas. Madrid. Spain*

INTRODUCTION

Recent advances in genetic studies have made important the necessity to find an efficient method for the measurement of the total nuclear DNA content of an individual cell in a mixture of different cells. This method must allow also the detection of different subpopulations into the same sample. The small DNA content of some species (< 1 pg), made really difficult to find it (Doležel, 1991).

At the beginning, the use of different microscopic and cytofluorometric techniques allowed to know more about the variations in the DNA content (Ruch, 1966). But it was in the last years when the flow cytometry appeared to be the most efficient and useful technique to obtain information about the cell cycle (Jayat and Ratinaud, 1993). It has been demonstrated as one of the best techniques in both analysis and research laboratories (Haynes, 1988; Michaelson et al., 1991).

Flow cytometry allows to carry out different studies on cell cycle in connection with the effects of drugs and radiations, DNA amount and ploidy determination (in animal and plant tissues, including tumors), different cellular parameters (intracellular pH, Ca^{+2} concentration, membrane potential and fluidity, etc.), the detection of a wide variety of antigens, and it even allows the physical sorting of particles like

M.J. Reigosa Roger, Handbook of Plant Ecophysiology Techniques, 21–34.

organelles and chromosomes. So this technique is very useful in physiology, cytology and immunology (Bergounioux et al., 1992; Jayat and Ratinaud, 1993; Steen, 1990a; Tiersch, 1989).

Principles and Instrumentation

The flow cytometry consists on the measure of particles suspended within a high-velocity fluid stream (Haynes, 1988). Usually, these particles must be labelled with fluorescent molecules, that can be of cellular endogenous nature (for example chlorophyll), or commercial fluorescent dyes, which are different according with the studied material and parameter, to make possible their detection and/or measurement (Galbraith, 1990).

The flow cytometer is based on a sample flow, a sheath fluid that envelopes the sample flow, a light excitation source, a flow chamber where light excites labelled particles, a set of optical filters to collect particle-emitted fluorescence, an acquisition system of the light generated signals (detectors), and a data conversion system able to interpret the results (Steen, 1990a).

The process is as follows: the cell or particle suspension is introduced in a sample flow of the cytometer. A sheath fluid envelopes this sample flow and allows its advance, and the pass of thousands of cells per second in the middle of the sample stream, within a flow chamber, one by one. The velocity of both streams can be regulated using valves. A laser or arc lamp light illuminates the particles into or immediately after their exit of the flow chamber, exciting the cellular dyes and generating one or several emission fluorescence wavelengths. A set of optical filters allows the collection and selection of these optical signals, directing them toward the detectors (photomultiplier tubes or PMT), that convert optical signals (photons) in electronic signals. The cytometer computer software allows the generation of histograms, where the intensity value generated from each measured particle is shown.

Most of the laser-based flow cytometers have essentially the same optics, that is, an orthogonal configuration with three main axis (see Fig. 1): the sample flow, the laser beam and the fluorescence detection optic axis, perpendicular to the other components (Steen, 1990a).

The cell suspension focusing takes place in the flow chamber (see Fig. 2). For this focusing, the sample is injected into a fluid jet of high section and decreasing diameter. This chamber (varying the pumping speed, the sample concentration and the final diameter) allows to run the particles one by one (nuclei, cells) through a luminous sheaf (Haynes, 1988).

The illumination of every particle produces a small pulse of fluorescence. This intensity is proportional to the amount of the fluorescent stained components in the cell. These fluorescent pulses enter in the optical axis, which is constituted by optical filters and light detectors (photodiodes or photomultiplier tubes) to focus it correctly in a sensitive detector. These signals can be used in the cytometer as data sources with an analogical-digital converter to digitise the fluorescent pulses. So they can be treated in a computer unit (Steen, 1990a).

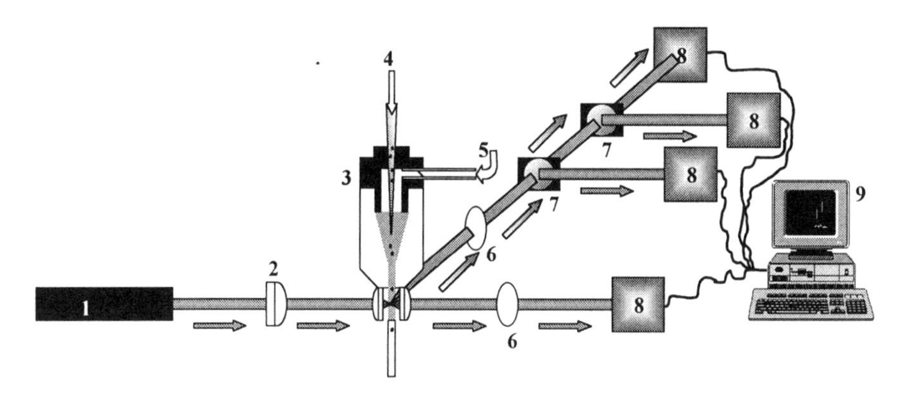

Figure 1. Representative scheme of a flow cytometer. 1. Excitation light ; 2. Focusing lens; 3. Flow chamber; 4. Sample; 5. Sheath fluid; 6. Lens; 7. Dichroic mirrors; 8. Fluorescence detectors; 9. Computer Workstation.

If a cell is placed in the light beam, it will emit fluorescence in all directions, but some light of the laser will be dispersed and able to be detected by an optical sensor denominated diffuse light photodiode or light-scatter detector. The intensity of this light varies in function of different physical (diffusion, cell volume, roundness, granularity/ ruggedness) or biological properties (specific fluorochromes of some cellular component as DNA, proteins, etc.) (Bergounioux et al., 1992).

Cells with the same size can be detected and distinguished if they have different internal refraction properties (Haynes, 1988; Steen, 1990b). All these signals are strong and easily detectable. Therefore, they can be peaked up by an optical system (light-scatter detector) and separated by optical filters placed in the front of the fluorescent detector (photodiode).

The filters are able to stop any diffuse light with an angle of 90° generated by the cell (right angle light-scatter or RALS).

There are also detectors for the measurement of low-angle diffuse light and for measurement of forward-angle diffuse light (forward angle light-scatter, or FALS). The diffuse light in low diffusion angles, below 10°, is dominated by the diffraction of the cell contour, and therefore, it depends mainly on their size (Haynes, 1988; Steen, 1990a).

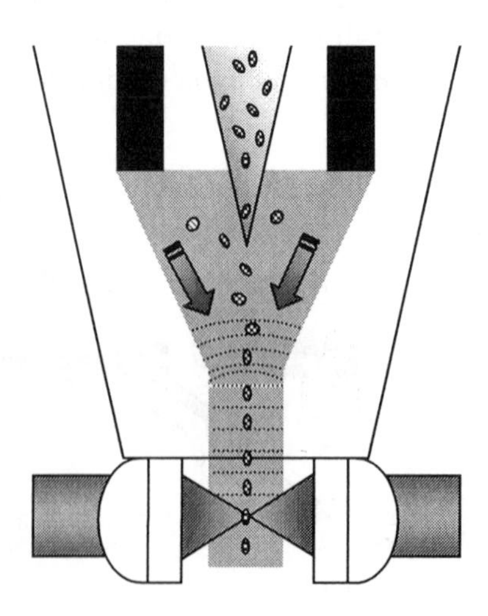

Figure 2. Flow Chamber. Particles are hydrodynamically focused in the central part of the chamber.

Sensitive detectors or photomultiplier tubes (PMT) are used for detecting fluorescent signals and weak side scattered light, and photodiodes for detection of strong forward scattered light; both of them transform the light pulses into electric current pulses (Doležel, 1991). So, the system transforms the light signal to an electric signal with higher value, according with the diffusion or with the fluorescence of the particle (Haynes, 1988; Steen, 1990a). In this way, the fluorescence intensity (that depends on the amount of different cellular components in the cell, labelled with fluorescent dyes), is recorded and quantified for each individual cell (Steen, 1990a).

In contrast with other biochemical techniques that generally give the mean value of a wide number of cells, with flow cytometry a variety of organisms or particles, whole cells, chromosomes, organelles, or protoplasts can be studied (Bergounioux et al., 1992; Michaelson, et al., 1991). These measures are obtained in a sample rate of thousands of particles per second, with a very low variation coefficient (lower than 7%

in plants and even lower in animal cells). In this way, this technique makes possible to distinguish different subpopulations of cells, for example, in the analysis of cell cultures with asynchronous growth, where the different phases of the cell cycle are clearly distinguishable. It is also a very useful technique in immunology, where the flow cytometer distinguishes different subtypes of lymphocytes, expressing different antigens, by the use of fluorescent monoclonal antibodies linked to these antigens. Some instruments, as the cell sorter, have made easier the separation of cells on line, separating them physically according with the values of the measured parameters (Bergounioux et al., 1992; Conia et al., 1987; Steen, 1990a; Veuskens et al., 1992).

The limit of detection of the flow cytometer depends on the number of molecules producing light pulses per each cell, organelle or protoplast. This number must be high enough to discern it from the background noise generated by the rest of the cell and by the medium where the cells are. This limit is determined by the intensity of the fluorescence and the diffuse light that can take place for a cell. The efficiency with which this light can be peaked up and transformed into electric signals (Steen, 1990a) determines also this limit. To balance this measure system, it becomes necessary to standardise it with calibrated particles. For it, some industrial plastic balls (fluorescent beads) are used. These beads are highly stable and they have a well-known and standard size (Haynes, 1988). So, the background decreases, and the portion of noise signal in the measurement is eliminated (Steen, 1990a).

The equipment and the probes, broadly available, allow an easy use of the fluorescence. Therefore, a sensitive, specific, quantitative, quick (by the use of reagents in micro-quantity, avoiding radioactive products), multiparametric (being able to use several different fluorochromes at the same time) and vital analysis (in the case of a non-destructive staining) is provided (Dutch Boltz et al., 1994; Marcus, 1988).

Types of Data Representation

The data received by the computer unit are represented as frequency distribution histograms. The width of the studied parameter is represented in the abscissas axis (from 0 to 255, although each equipment carries out the appropriated transformations), and the number of events by channel (cells, nuclei, etc.) is represented in the OY axis of the histogram. Every time that a cell crosses the sheaf of the laser, it emits a light signal that is transformed, digitised and increased in the corresponding channel (Steen,

1990a). At every time, a great number of elements can be evaluated (Brown and Bergounioux, 1989) generating the histograms. Each histogram is the result of 1,500-5,000 average particles; this allows obtain, for example, a coefficient of variation[1] of 2C peaks lower than 7 % in plants labelled as a cytological preparation and with an equipment stability indicator (Marie and Brown, 1993) when measuring DNA content in plant cells.

Each cell emits several different signals at the same time, but these signals are separated according to their properties. This makes possible to combine two independent parameters in each representation, because each particle has an individual position (Steen, 1990a). In this way, if the nuclear DNA content is measured, the signals emitted by particles of big size and weak fluorescence can be separated by the use of an area of natural selection (gate), which will facilitate the interpretation of the histograms (FALS/yellow, Fig. 3, 4). It is also important to distinguish the particles with a spherical form from the deformed particles, which evidently are not nuclei or are malformed nuclei (peak/integral, see Fig. 5). The signals emitted by red fluorescence (chlorophyll) or by red and yellow at the same time (good nuclei without cytoplasm and, therefore, without chlorophyll (red/yellow, Fig. 6 and 7) can be also separated.

There are four ways for data representation, which correspond to the same data of the same sample, but each with its own value. These values are stored in 'List mode' and they can be recovered to work again with them. Every one of these representations shows the data from a different perspective (Haynes, 1988). In order to facilitate the explication of the interpretation of these representations, figures of DNA ploidy analysis will be used.

Generally, the linear data values are employed, because they are the most precise values (Fig. 8); however, sometimes, it is necessary to use logarithmic values. With the logarithmic amplification we can compress the superior range of signals, and expand the inferior ranges. Often, the result helps to develop, at the beginning, strong gaps between the weak signals and the noises, and to separate, at the end, a wide range of strong signals in recognisable peaks (Haynes, 1988), (see Fig. 9).

In the dot plot representation (Fig. 3), each point (plot) represents one or more cells, which gave a forward or a right angle light-scatter value. Now, the longest cells, which have a bigger FALS and also a bigger

---

[1] The coefficient of variation (CV) in % is the standard deviation of a peak position divided by the mean. Values between 1-10 % are usually considered as good.

RALS, can be seen. This kind of data representation is very used, but a little confusing. It is also difficult to estimate how many cells are in an area, although a quantitative analysis can be carried out.

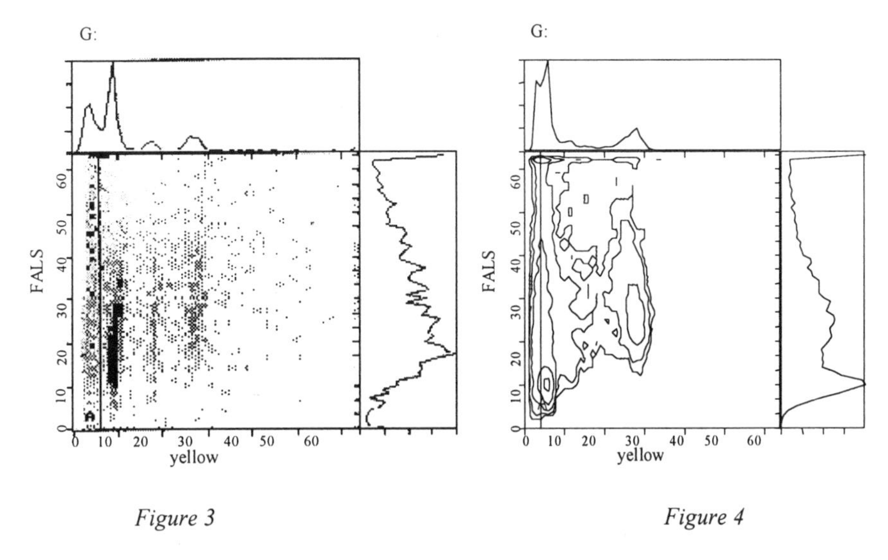

Figure 3

Figure 4

Figure 3. Representation in dot plot. FALS signals are plotted against the yellow fluorescence. The interesting particles are selected with a gate.

Figure 4. Representation in contour. FALS signals are plotted against the yellow fluorescence. There is a gating of the interesting particles.

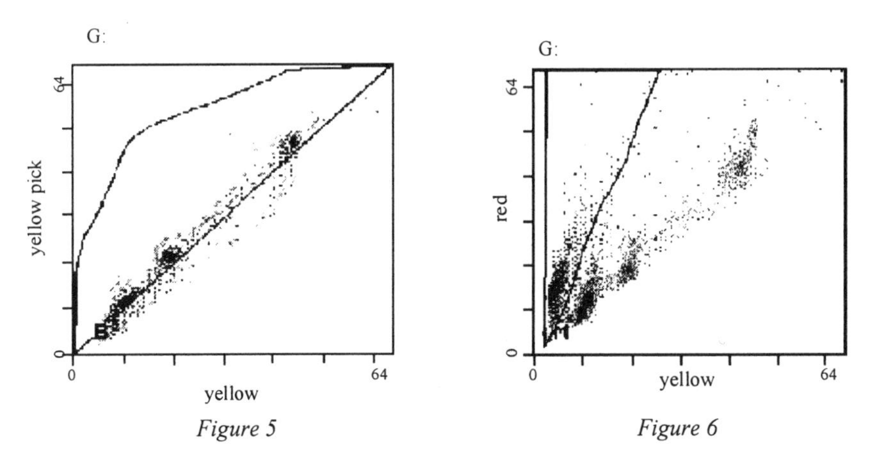

Figure 5

Figure 6

Figure 5. Representation in dot plot of the signals of the yellow fluorescence pulse against the yellow fluorescence pulse integral. Spherical particles are selected.

Figure 6. Representation in dot plot of yellow fluorescence pulse integral signals against red luminescence pulse integral. Chlorophyll and nuclei with rests of cytoplasm are gated.

In the contour plot (see Fig. 4), lines of contour with the same number of cells are drawn. This representation is also quantitative and it defines, in a tangled way, the areas of two cellular types that can be seen as two subgroups enough distinguishable. The use of the contour, as a map of levels, is a two-dimensional representation of a three-dimensional construction, where the mountains are defined as concentric rings of growing elevations. In our case there is an increase in the number of cells.

The isometric representation is the most intuitive form to view the interrelation of the scatter parameters for a sample. It is not as quantitative as the contour representation, but its interpretation is easier, and the cell groups are shown as different mountains (Haynes, 1988), (see Fig. 7).

In the study of ploidy levels, the data can be obtained in two ways: with linear and logarithmic scales. Linear for a higher resolution (see Fig. 8) and logarithmic amplification for an unvaried scale, condensing in a same scale ploidy levels that are really far one from the other (see Fig. 9).

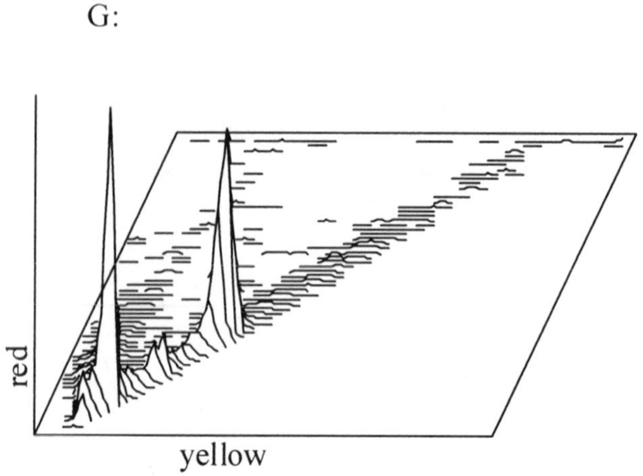

Figure 7. Tridimensional or isometric representation of Fig. 5. The gate is not represented, but it is possible to see the three DNA peaks.

The ploidy (and perhaps the aneuploidy) will be defined with the results of an only plant, or establishing perhaps a formula based in an eventual mixoploidy. This statistical treatment should consider the DNA content (the position of the main peaks in the histogram) and the probability that the nuclei heterogeneity comes exclusively from the proliferation and from the cell differentiation, and not from a mixoploid origin of the plant. This implies the inclusion of an internal pattern for

DNA, and a preliminary study to appreciate the usual heterogeneity in a plant tissue with a well-known ploidy (Laat et al., 1987; Bergounioux et al., 1992; Marie and Brown, 1993).

The plant cells and the protoplasts are frequently rich in starch and pigments that can interfere with the analyses or act as markers. Before starting the flow cytometry with plant materials, it is evident the necessity to characterise them, by using a good method to eliminate the possible pollutants (Brown and Bergounioux, 1989). In our case, the red fluorescence of the chlorophyll is very intense in the most part of the green protoplasts. To filter the emission and to minimise the red spectrum that is overlapped with the other fluorescent markers, a band-pass filter of 685 nm (10 half-wide nm) can be used (Brown and Bergounioux, 1989).

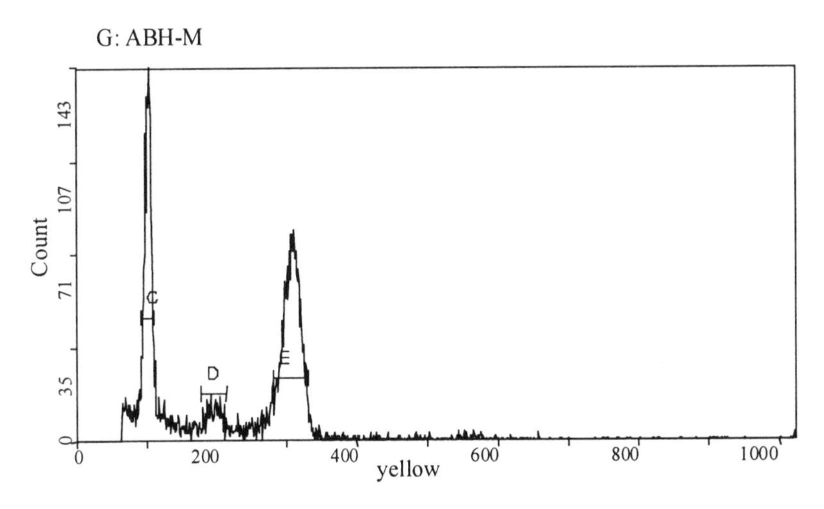

Figure 8. Lineal representation of the 2C (marked with the letter C) and 4C peaks (marked with D) of the inner pattern, and of the 2C peak (marked with E) of the plant sample.

Thus, those parameters that are important to distinguish a population are gated. Before representation, the particles that have passed are selected or edited. The events, selected with a gate according with these parameters, can be taken for the histograms of other parameters, modifying their representation. The use of gates is a very important tool to simplify and to clarify the analysis of subsets of cells, but it can be very deceiving for a casual observer that has not been warned (Haynes, 1988).

These selections can be observed in the Fig. 6. If there is a representation of a previously selected number of cells, the selection carried out by an equation appears after the letter G in the superior part of the graph. Therefore, in the Fig. 3-7, nothing appears, because all the particles are represented. But in the top of the Fig. 8 and 9, the equation is G: ABH-M. That means that the particles that appear are only those particles selected by the gate A in the Fig. 3-4, those selected by the gate B in the Fig. 5, and all the particles that show a constant light emission (gate H). To eliminate the chlorophyll of this selection, it is necessary to type '–M' because the chlorophyll is selected with the gate M in the Fig. 6.

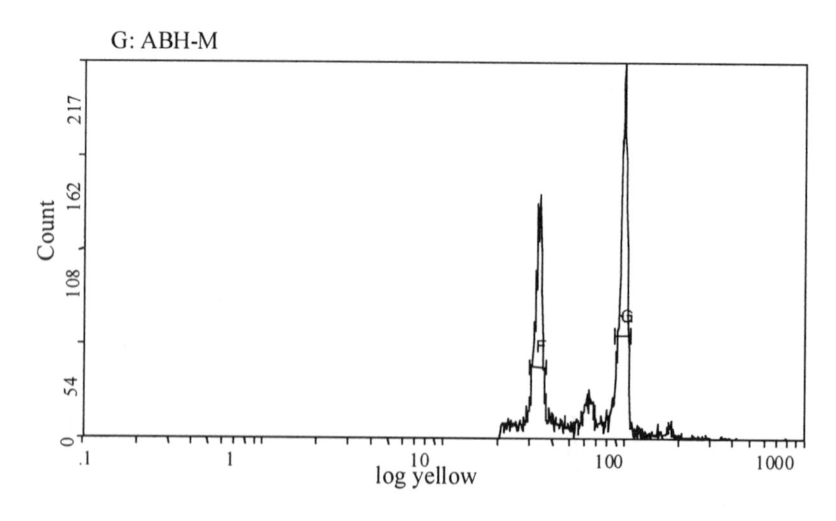

Figure 9. Logarithmic representation of the same data from the Fig. 8, where is marked the 2C peak from the pattern (F) and from the sample (G).

If thousands of interfasic nuclei extracted from a tissue and labelled with a specific DNA fluorochrome are measured one by one, the obtained histogram would have several derived nuclei of cells in the G0/G1 phase of the cell cycle, other nuclei in the S phase (synthesis of DNA), and others in G2 phase.

For a diploid plant, the G0/G1 position represents the amount 2C of DNA. The fluorescence intensity mean of this peak (G0/G1) is half than that of G2 (Brown and Bergounioux, 1989; Brown et al., 1991a; Galbraith, 1990; Jayat and Ratinaud, 1993) (Fig. 8 and 9). These different positions put in evidence a ploidy level, or the presence of a mixture in a seedbed (Petit et al., 1986; Steen, 1990a). It is adequate to know the relative DNA content of lines of cells with different ploidy levels, because the polyploid individuals produce frequently changes in the physiology, which affect the good conditions of growth and competition. These changes favour the occupation of habitats different from the habitats of their diploid predecessors. But these individuals can also be found in the same habitat of their progenitors. In this situation, competition will appear between them (Brown and Bergounioux, 1989; De Laat et al., 1987).

POSSIBLE USES AND FUTURE PERSPECTIVES

The flow cytometry is not only a useful implement to study current hypothesis or strategies, it is also an opportunity to innovate and to develop new concepts in cellular biology. Its development promises fine tools to complete the genetic or somatic manipulation of the plants, and to study the plant cell biology. The list of processes that can be analysed with fluorescent probes is long. Interesting perspectives include:

- RNA analysis. Its use allows the simultaneous study of the nuclear DNA and RNA content and theirs variations (Bergounioux et al., 1988; Brown and Bergounioux, 1989).
- Analysis and physical separation of chromosomes.
- Immuno-labelling. The possibility to perform a multiparametric analysis allows the simultaneous continuous measurement of different antigens (Bergounioux et al., 1992).
- Isolation of fused protoplasts. This technique facilitates the introduction, in plant species, of genetic characteristics with agronomic interest (Brown and Bergounioux, 1989; Galbraith, 1989, and references therein).
- Cell Functions. The study of pH variations (in cytoplasm or vacuolar), calcium activity, modulation of membrane potential, fluidity of the membrane, protein content in the cytoplasm, and diverse enzymatic activities (Bergounioux et al., 1992; Brown and Bergounioux, 1989) can be carried out thanks to flow cytometry.

The accurate protocols for the study of cell functions and signal transduction is recent in plants. Therefore, is expected a future great diversity of works with plant material. Different nuclei extraction methods and a great diversity of species for the cell cycle analysis are developing and improving. In the same way, the techniques for the cytometric study of microorganisms, yeast and micro and macro-algae are now in development, and every time these techniques are applied with more frequency. This kind of work will be facilitated by the use of small portable cytometers that can be carried to the greenhouse or the field (Brown, 1994).

New fluorescent markers have been developed as the green fluorescent protein (GFP) that is a vital marker and its expression can be coupled with that of another gene of interest. This marker does not need substrate or exogenous co-factors. The GFP is very stable and can bind with other interesting proteins. This new tool will allow to monitor by cytometry the genetic expression, signal transduction, co-transfection, transformation, traffic and interactions of proteins, as well as separation and cellular purification (Sheen et al., 1995).

The fluorescent *in situ* hybridisation technique (FISH) allows to show the presence of RNA molecules (or portions of those of DNA) of a gene. It is necessary to add to the tissue or cell preparations a molecule of DNA or synthetic RNA, complementary to the DNA or RNA of interest. Both molecules hybrid because they are complementary. The synthetic molecule is marked with a specific antigen, and the corresponding monoclonal antibody marked with fluorochromes is added. The following analysis by flow cytometry and other techniques of this cell material permits to study the ratio and characteristic of the cell population that express the gene or genes in question. This method allows quantifying also this genetic expression, certain types of genome comparisons, and to study chromosomal crossing (Cytometry 19, 1995, special number on Comparative Genome Hybridisation).

REFERENCES

Bergounioux C., Brown S.C., Petit P.X. Flow cytometry and plant protoplast cell biology. Physiol Plantarum 1992; 85:374-386

Bergounioux C., Perennes C., Brown S.C., Gadal P. Nuclear RNA quantification in protoplast cell-cycle phases. Cytometry 1988; 9:84-87

Brown S.C. Applications of flow cytometry in plant biology and biotechnologies: review and perspectives. Biotech Biotechnol Equipm 1994; 8:75-78

Brown S.C., Bergounioux C. "Plant Flow Cytometry." In *Flow Cytometry: Advanced Research and Clinical Applications*. A. Yen, ed. Boca Raton, Florida: CRC Press, 1989

Brown S.C., Bergounioux C., Tallet S., Marie D. "Flow Cytometry of Nuclei for Ploidy and Cell Cycle Analysis." In *A Laboratory Guide for Cellular and Molecular Plant Biology*. I. Negrutiu, G. Gharti-Cherti, eds. Birkhäauser: Verlag Basel, 1991

Conia J., Bergounioux C., Perennes C., Muller P.H. Flow cytometric analysis and sorting of plant chromosomes from *Petunia hybrida* protoplasts. Cytometry 1987; 8:500-508

Cytometry 19. Special issue on çomparative genomic hybridization, 1995

De Laat A.M.M., Göhde W., Vogelzang M.J.D.C. Determination of ploidy of single plants and plant populations by flow cytometry. Plant Breeding 1987; 99:303-307

Doležel J. Flow cytometric analysis of nuclear DNA content in higher plants. Phytochem Anal 1991; 2:143-154

Dutch Boltz R.C., Fischer P.A., Wicker L.S., Peterson L.B. Single UV excitation of Hoechst 33342 and ethidium bromide for simultaneous cell cycle analysis and viability determinations on *in vitro* cultures of murine b-lymphocytes. Cytometry 1994; 15:28-34

Galbraith D.W. Analysis of higher plants by flow cytometry and cell sorting. Int Rev Cytol 1989; 116:165-228

Galbraith D.W. "Flow Cytometric Analysis of Plant Genomes." In *Methods in Cell Biology*. Z. Darzynkiewicz, H. Crissman, eds. London: Academic Press, 1990

Galbraith D.W., Harkins K.R., Maddox J.M., Ayres N.M., Sharma D.P., Firoozabady E. Rapid flow cytometric analysis of the cell cycle in intact plant tissues. Science 1983; 220:1049-1051

Godelle B., Cartier D., Marie D., Brown S.C., Siljak-Yakovlev S. Heterochromatin study demonstrating the non-linearity of fluorometry useful for calculating genomic base composition. Cytometry 1993; 14:618-626

Haynes J.L. Principles of flow cytometry. Cytometry supp 1988; 3:7-17

Jayat C., Ratinaud M.H. Cell cycle analysis by flow cytometry: principles and applications. Biol Cell 1993; 78:15-25

Kerker M., Van Dilla M.A., Brunsting A. Is the central dogma of flow cytometry true: that fluorescence intensity is proportional to cellular dye content? Cytometry 1982; 3:71-78

Marcus D.A. High-performance optical filters for fluorescence analysis. Cell Motil Cytoskel 1988; 10:62-70

Marie D., Brown S.C. A cytometric exercise in plant DNA histograms, with 2C values for 70 species. Biol Cell 1993; 78:41-51

Michaelson M.J., Price H.J., Ellison J.R., Johnston J.S. Comparison of plant DNA contents determined by Feulgen microspectrophotometry and laser flow cytometry. Am J Bot 1991; 78:183-188

Petit J.M., Denis-Gay M., Ratinaud M.H. Assessment of fluorochromes for cellular structure and function studies by flow cytometry. Biol Cell 1993; 78:1-13

Ruch F. "Determination of DNA Content by Microfluorometry." In *Introduction to Quantitative Cytochemistry*. G. Wied, ed. New York-London: Academic Press, 1966

Sheen J., Hwang S., Niwa Y., Kobayashi H., Galbraith D.W. Green-fluorescent protein as a new vital marker in plant cells. Plant J 1995; 8:777-784

Steen H.B. "Characteristics of Flow Cytometers." In *Flow Cytometry and Sorting*. New York: John Wiley and Sons, 1990a

Steen H.B. Light scattering measurement in an arc lamp-based flow cytometer. Cytometry 1990b; 11:223-230

Tiersch T.R., Chandler R.W., Wachtel S.S., Elias S. Reference standards for flow cytometry and application in comparative studies of nuclear DNA content. Cytometry 1989; 10:706-710

Ulrich I., Fritz B., Ulrich W. Application of DNA fluorochromes for flow cytometric DNA analysis of plant protoplasts. Plant Sci 1988; 55:151-158

Veuskens J., Marie D., Hinnisdaels S., Brown S.C. "Flow Cytometry and Sorting of Plant Chromosomes." In *Flow Cytometry and Cell Sorting*. A. Radbruch, ed. Heidelberg: Springer Berlin, 1992

CHAPTER 3

FLOW CYTOMETRY: DNA QUANTIFICATION

Alfonso Blanco Fernández
Dpto Bioloxía Vexetal e Ciencia do Solo. Universidade de Vigo. Spain

INTRODUCTION

Ploidy[1] level is taken in consideration when studying speciation, systematic, ecology, polysomaty[2], plant conformity following *in vitro* multiplication or somatic embryogenesis, etc. It is recommended to know the relative DNA content in cells lines with different ploidy levels because the polyploid[3] individuals produce frequently changes in the physiology, which affect the good conditions of growth and competition. These changes favour the occupation of new habitats, which can be different than those of their diploid predecessors, although they can also be sharing the same geographical distribution that their progenitors, having to compete with them (De Laat et al., 1987; Brown and Bergounioux, 1989).

In agriculture, ploidy analysis has a considerable importance in commercial seed production, where it is required a certification of the presence of a single ploidy class in the seed like the triploid seedless watermelon or maize. Another application of ploidy analysis is in the field of plant genetic engineering helping to detect quickly undesirable plants or the case of ornamental plants grown in greenhouse (Marie and Brown, 1993; Galbraith et al., 1997).

[1] Ploidy is defined as "the number of complete chromosome sets in a nucleus".
[2] The polysomy is "the occurrence in a nucleus of one or more individual chromosomes in a number higher than that of the rest of nuclei".
[3] Polyploidy, term used by first time by Winkler in 1916, is defined as "the occurrence of related forms possessing chromosome numbers which are three or more times higher than the haploid number".

M.J. Reigosa Roger, Handbook of Plant Ecophysiology Techniques, 35–51.

Classically, plant ploidy level is determined by counting of chromosomes in mitosis or meiosis, but slowness and heaviness of this technique makes that recounts are carried out in a few individuals, generating a statistical problem in the study of populations. Sharma et al. (1983) study the ploidy level of *Nicotiana paniculata* L. and *N. sylvestris* Speg. and Comes and find a direct relationship between the measures of nuclear DNA obtained by flow cytometry and the number of chromosomes. Michaelson et al. (1991) and Bennett and Leitch (1995) compare the data obtained by flow cytometry, microspectrophotometry and microdensitometry and conclude that there is a good correlation among the measures obtained by these techniques in monocotyledonous and in dicotyledonous.

Flow cytometry is a technique that allows to measure nuclear DNA amount quickly and that has demonstrated to be a useful complement to the classic cytogenetic. This technique permits to process each sample in a few minutes. Equipment and probes, broadly available in the trade, permit us to use the fluorescence with great easiness, providing a sensitive, specific, multiparametric, quantitative and quick analysis and, in some cases, vital, allowing approaching the study of the intrapopulational variation. In this sense, Bennett and Leitch (1995) point out that flow cytometry used in the study of plant populations will allow to understand and to evaluate the evolutionary meaning of the intraspecific variation. The use of this method is increasing, principally because it is faster and more accurate than other methods like chromosome counting or microspectrophotometry. The development of flow cytometry promises fine tools to complete the genetic or somatic manipulation of plants as well as to study the plant cellular biology (Bennet and Leitch, 1995; Brown and Bergounioux, 1989; Galbraith, 1990; Galbraith et al., 1983; Marie and Brown, 1993).

The strictly cytogenetic **ploidy** description of a karyotype is the number of copies of the complement[4] for a given species. '**x**' represents the basic chromosome number[5] of a taxa and '**n**' is the number of chromosomes for gametes without replication, constituting the haploid complement. All G0/G1 cells in a organism, with few exceptions, have the same constant value for the nuclear DNA content **(2C)**, this fact is

[4] The complement is defined as "the group of chromosomes derived from a particular nucleus in gamete or zygote, composed of one or more chromosome sets". And the chromosome set is "a minimum complement of chromosomes derived from the gametic complement of a supposed ancestor".

[5] The basic chromosome number is "the supposed number of chromosomes found in the gametes of a diploid ancestor of a polyploid".

referred by the cytogeneticists as the **diploid DNA content**, and it is twice that of gametes (**C**), the **haploid DNA content**, and half that of G2 cells (Marie and Brown, 1993; Rabinovitch, 1992). Mass of the haploid genome (C) and the number of base pairs (bp) are related by the formula:

$$n = \frac{m \times Na}{M}$$

where:

$n =$ number of base pairs
$m =$ mass of DNA
$Na =$ Avogadro's number (6.02* 10^{23} bp/mol)
$M =$ average molecular weight of a base pair (610 g/mol)

In higher plants, C value is between 0.15 pg, approximately, in *Arabidopsis thaliana*, and 127.4 in *Fritillaria assyriaca* (Bharathan et al., 1994). In Pteridophyta, recent studies have shown 2C values from 4.90 pg of DNA in *Sellaginella denticulata* to 51 pg in *Equisetum ramosissimum*. Pteridophyta is a group of plants with an enormous morphologic plasticity, which blurs the taxonomic outlines, with a small number of characters to study them taxonomically, and vivacious, which allows them to live though they are infertile. These facts make flow cytometry a powerful tool to aid to delimit the taxa outlines, allowing to distinguish species, subspecies, varieties or hybrids. For example, this technique has permitted to separate the two subgenera in the genus *Equisetum*: The subgenus *Hippochaetes* has 2C values (38-53 pg) approximately twice larger than those of the subgenus *Equisetum* (22-26 pg), and both have the same chromosome number (2n= 216) (Horjales et al., 1997; Blanco, 1997; Redondo et al., 1998; 1999 a; 1999b).

THE TECHNIQUE

Some Considerations

For the study of mitotic chromosomes is better to select root meristems, and for the study of meiotic chromosomes, selection of the sporophytic and the gametophytic producer tissues are preferable. Thus, in flow cytometry, it is very important to know **which is the best part of the plant for measuring the nuclear DNA content**. Brown et al. (1992)

study the DNA content in all the tissues of *Arabidopsis thaliana.* They observe that young and completely expanded leaves (rosette leaf) are the best tissues to study nuclear DNA amounts, because they have higher percentage of cells at 2C, providing the maximum accuracy.

In order to evaluate the nuclear DNA content, the first step is **to obtain cells or protoplasts in the best conditions** (fresh plant tissues). When the analysis cannot be immediately performed, tissues, cells and protoplasts can be fixed with formaldehyde or with ethanol: acetic acid, or fixed in paraffin. To analysis this fixed material, different protocols have been developed (Rabinovitch, 1992; Galbraith et al., 1997).

To obtain protoplast or isolate nuclei from a fresh plant material, **tissues are chopped with a razor blade in a buffer**. This buffer requires a high osmotic pressure (0,4-0,8 osmolal), because protoplasts are fragile, and they are treated with pectocellulosic enzymes (**mechanical-enzymatic techniques**) or broken with detergents (**mechanical-detergent technique**) (Galbraith, 1984; Brown and Bergounioux, 1989; Bergounioux et al., 1988). In function of the plant material, protective agents as ß-mercaptoethanol, polyethylene-glycol, chelators and others may be added (Steen, 1990a).

Exposition of living cells to enzymes and/or detergents produces breakage of cellular walls and membranes. This breakage allows the isolation of nuclei and other cytoplasmic constituents, which can be labelled with different fluorescent molecules (these can be either molecules of endogenous nature, like chlorophylls, or commercially available fluorescent probes) depending on the studied material and parameter (Galbraith, 1990). Plant cells and protoplasts are frequently rich in starch and pigments that can interfere with the analyses or can be used as markers. Before beginning the study with the flow cytometer, it is obvious that it is necessary to characterise the plant material using a good method **to eliminate all possible pollutants** (Brown and Bergounioux, 1989). In the case that we will see, we will observe that the red fluorescence produced by the chlorophyll is very intense. To filter the emission and to minimise the red spectrum and its overlapping with the fluorescence of fluorescent dyes, a 685 nm band-pass filter (10 nm half-wide) can be used (Brown and Bergounioux, 1989). Other particles, which disturb the sample are apoptotic cells, isolated chromosomes and deformed nuclei. Usually, all of these signals have low fluorescence intensity values and are classified as **debris**. Nevertheless, the use of enzymes and detergents to liberate plant nuclei is better method than others, as the low values of the CV of the mean DNA content of obtained G0/G1 cells show.

Scale of fluorescence intensity is contrasted by the inclusion of an appropriate **biological reference**. It is important that this DNA pattern has a similar DNA content to that of the sample, but it should not overlap with that of the sample; the pattern should be accessible and its DNA amount should remain constant. By this way, the obtained data can be reproduced and compared (Marie and Brown, 1993). Usually the patterns are chicken or human erythrocytes or a plant with a known DNA amount (Brown and Bergounioux, 1989; Michaelson et al., 1991). It is necessary to add this internal pattern before the addition of the dye and of the RNase, and by this way pattern and sample are submitted to the same treatment (Steen, 1990a).

Once obtained the protoplasts, the following step is **to label the nuclei with fluorochromes**. To study DNA, there are different stains as 4',6-Diamidino-2-phenylindole (DAPI), Ethidium Bromide (EtBr), Mithramycin (Mi), etc. DAPI and Hoechst 33342 (Ho) are probes that bind to five serial A-T base pairs. But, owing to the presence of G-C in the DNA, the measure will not be absolute and its value will depend on the ratio of bases of the DNA. Mithramycin binds to three G-C base pairs, then neither it will give an absolute measure of the DNA amount. However, Ethidium Bromide and Propidium Iodide (PI) are inserted among all the base pairs, allowing an absolute measure of the DNA content (they are called 'intercalating dyes'). Working with these three types of fluorochromes on the same material, although not simultaneously, allows the quantification of the DNA amount and of the DNA base-composition with a simple equation (Godelle et al., 1993; Marie and Brown, 1993), which is, sometimes, an useful taxonomic approach.

The number of sites along a DNA molecule that interact with a particular fluorochrome is expected to vary depending on the probe concentration. Therefore, it is necessary to add a minimum concentration of dye. Nevertheless, an excess of the probe makes the solution itself fluorescence (noise). Thus, the optimal dye concentrations are the minimum, but adequate, at which intersample variations do not exist (Darzynkiewicz and Juan, 1997).

Intercalating fluorochromes (Propidium Iodide and Ethidium Bromide) are inserted among the bases and do not provide any type of discrimination between RNA and DNA. This fact makes necessary **to add RNase** in order to carry out only the measures of the cellular DNA. This addition should be made before adding the fluorochrome, to avoid the labelling of the RNA that would suppose a decrease of the resolution capacity in the analysis (Ulrich et al., 1988).

To distinguish between diploid and aneuploid[6] DNA, or to study the ploidy level of different populations requires a high **accuracy** of DNA content measurement. In fixed cells or protoplasts, it is difficult to prevent the adsorption of the dye to cellular membranes because of the opacity of cellular walls and the tendency of dyes to non specific fixation (generally, Triton X-100). This fact decreases the precision of the fluorescent analysis (Brown and Bergounioux, 1989; Ulrich et al., 1988). There are another factors that modify the intensity of fluorescence: the effects taking place by the high concentrations of fluorochromes (the extinction of the fluorescence, the effect of internal filter and the effect of autoabsorption), photodecomposition and variations in the temperature and pH. To estimate the stability and the resolution obtained in cytometry, the coefficient of variation of the G0/G1 population is used, and it will constitute a pattern of gaussian curve, to summarise the quality of the fixation and of the nuclei analysis in a single sample (Steen, 1990a). The value of the **coefficient of variation** (CV) of the mean DNA content of obtained G0/G1 cells is considered as an index of accuracy of DNA amount measurement. Coefficient of variation, expressed in percentage (%), is the standard deviation of the peak position. In the practice, it will be calculated using the width at the half high of the G0/G1 pattern peak. Thus, a result with a CV of 1,5% is considered as excellent, and of more than 7% as not satisfactory. Nevertheless, a series of measures, repeated in the same sample, will give a mean with a standard deviation minor than 2%. Then, the main source of dispersion in the methodology is due to the variability of the cytological preparation and not to its analysis (Steen, 1990a).

Protocol and the Data Analysis

To make the preparations to study the nuclear DNA amount of pteridophytes using the flow cytometer, nuclei must be isolated from young leaves. Two or three sample leaves are putting into a Petri dish together with a piece of the pattern leaf. Usually, the internal plant patterns are young leaves of the diploid *Petunia hybrida* L. PxPC6 (2.85 pg, 41% G-C) of the diploid *Pisum sativum* L. (8.37pg, 40.5% G-C), or of *Triticum aestivum* L. (2C = 30.90 pg; 43.7% G-C).

Using a razor blade, tissues are chopped in 700 µl of a solution buffer, like the Bergounioux buffer, for example (Brown et al., 1991a) into

[6] The aneuploidy is “a deviation from a normal haploid, diploid or polyploid chromosome complement by the presence in excess of, or in defect of, one or more individual chromosomes”.

the Petri dish. Nuclei are liberated into this buffer (Galbraith, 1984; Bergounioux et al., 1988). In function of the plant material, protective agents as ß-mercaptoethanol, polyethylene-glycol, chelators, etc. can be added (Steen, 1990a). In this case, 7 μl of 100% ß-mercaptoethanol/ 1000 μl of Bergounioux buffer are added. The solution is filtered through a nylon filter of 30 μm of pore diameter (according to the method described by Brown et al., 1992). Because in this step there is a lot of material lost and it is difficult to control the quantity, 500 μl of the filtered solution is taken and filtered again, in order to homogenise the samples (see Fig. 1).

Ethidium Bromide (EtBr, Sigma E-8751) is the selected probe to study the nuclear DNA content. The used concentration is 50 μg/ml. As it was explained above, EtBr is an unspecific dye of nucleic acid, and it is necessary to add previously 10 μl/ml of RNase (Boehringer product 109169, from 1% (w/v) stock solution in PBS Tris ClH, pH 7.6). After this step, it is necessary to add the EtBr and to incubate the sample during 30 minutes in darkness at room temperature. If the room temperature is too high, it is better to introduce the samples in ice. In some laboratories, samples are always incubated in ice. The EtBr is excited in blue at 488 nm and its fluorescence is recorded at 610 nm.

When the sample is analysed in the flow cytometer, the histograms are obtained and the interesting signals are selected. The values, which will figure after each histogram, give diverse information about the sample (number of particles, name of the sample...), as well as its statistic parameters (peaks' position, coefficient of variation, ratio of sample particles included in a peak...).

In nuclear DNA amount studies, each cell emits several different signals at the same time, but separated in function of their properties. This fact makes possible to combine in each representation two independent parameters, because each particle has assigned an individualised position (Steen, 1990a). By this way, the signals emitted by the sample reveal particles of big size and weak fluorescence (corresponding to several remains). To facilitate the interpretation of the resulting histograms, these signals can be separated using an area of natural selection, denominated **gate** (FALS/yellow fluorescence, Fig. 2a). It is also important to differentiate those particles that are quite spherical of those that are very deformed, that evidently they will not be nuclei or if they are, at least, they will not be in good state (peak/integral signal, Fig. 2b). Red fluorescence signals can be emitted by chlorophyll, and red/yellow signals correspond to clean nuclei without cytoplasm and, therefore, without chlorophyll. Both types of signals can be separated and discriminated (Figs. 2c and 2d).

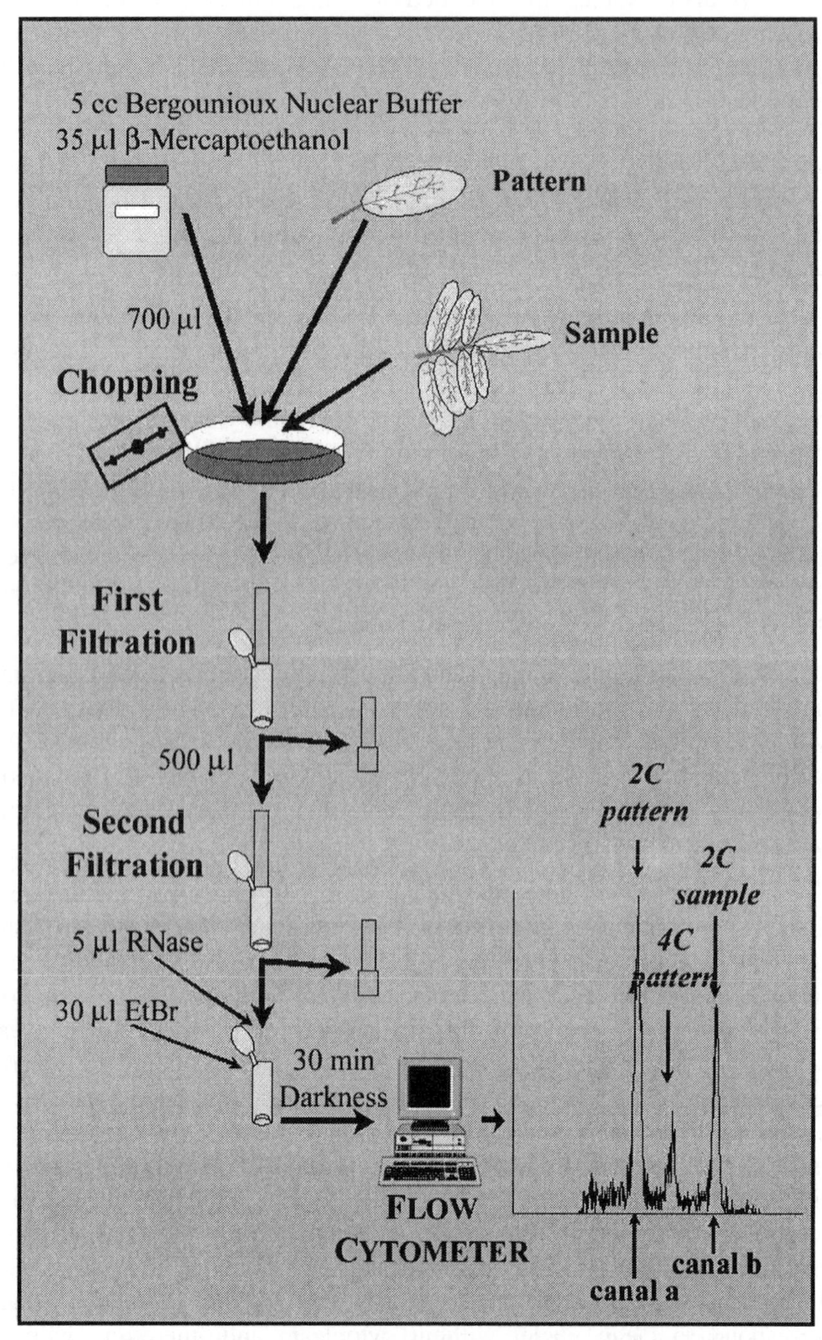

Figure 1. Protocol of the nuclear DNA analysis for pteridophyte samples.

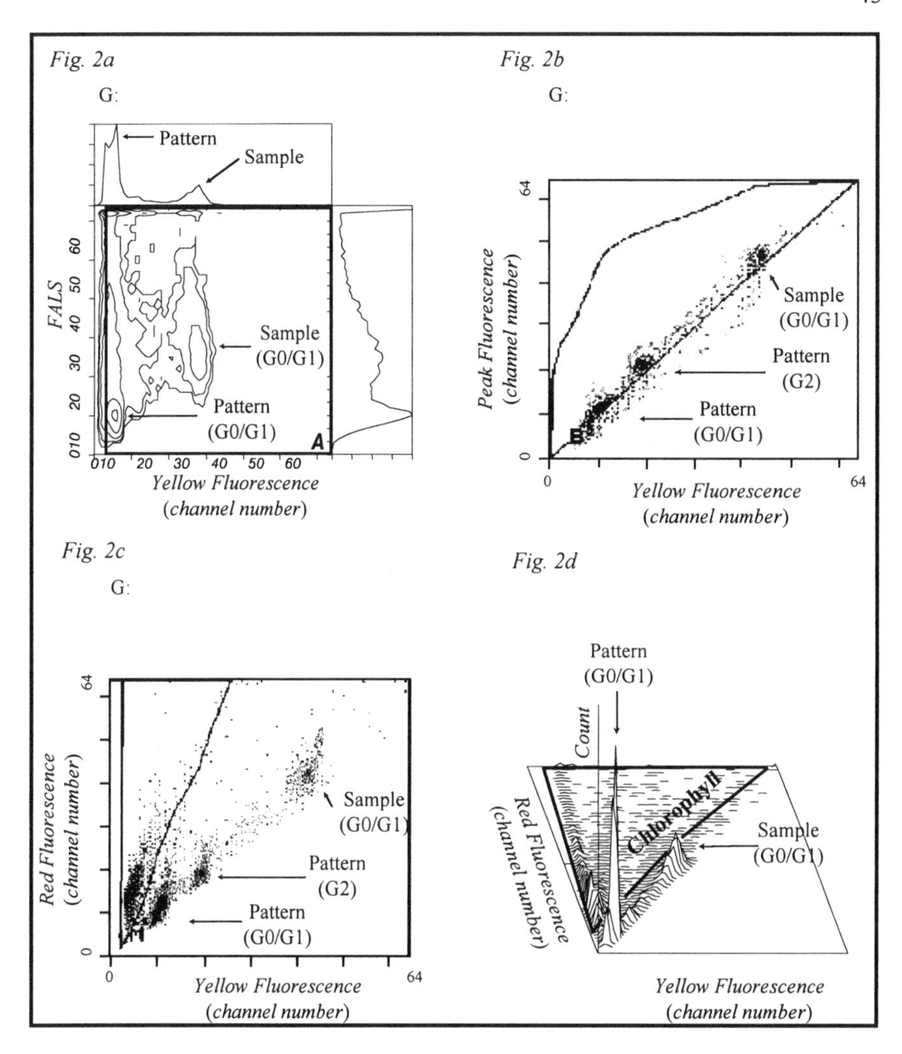

Figure 2a. Contour representations of FALS signals (ordinate axis) and yellow fluorescence (abscissa axis). Selection area (A) can be distinguished. This gate contains the interesting particles. *Figure 2b.* Dot plot representation of yellow fluorescence peak signals (ordinate) and yellow fluorescence integral (abscissa). This gate selects roundness particles. *Figure 2c.* Dot plot representation of yellow (abscissa) and red fluorescence (ordinate). Gate selects chlorophyll and debris. *Figure 2d.* Tridimensional or isometric representation of the figure 2c.

In the study of the ploidy levels, it is necessary to carry out studies of the nuclear DNA amount. The obtained data can be displayed in two ways: through a lineal amplification (for a good resolution, see Fig. 3a) or through a logarithmic amplification (allowing to condense far away-disposed ploidy states in the same scale, Fig. 3b). From the results of a single plant, the ploidy (and perhaps the aneuploidy) will be defined or a formula will be established evoking an eventual mixoploidy[7]. This statistical treatment should consider the DNA content (the position of the main peaks in the histogram) and the probability that the heterogeneity of the nuclei comes exclusively from the proliferation and from the cellular differentiation and not from a mixoploid origin of the studied plant.

These last refinements involve the inclusion of an internal pattern for the nuclear DNA and a preliminary study to appreciate the normal heterogeneity in the tissue of a plant with a well-known ploidy (De Laat et al., 1987; Bergounioux et al., 1992; Marie and Brown, 1993).

Generally, the lineal data values are employed, because they are the most precise values (Fig. 3a). However, sometimes, it is necessary to use their logarithmic values. The global effect of the logarithmic amplification is to compress the superior ranges of signals and to expand the inferior ranges. The result often helps to clearly discriminate valleys between weak signals and noises, in the beginning, and to separate a wide range of strong signals in recognisable peaks, at the end (Haynes, 1988) (Fig. 3b).

If the sample from which the data have been obtained is good, the next step is to calculate the nuclear DNA content of the sample following this equation:

$$\text{Sample DNA amount} = \frac{\text{Canal b}}{\text{Canal a}} x \text{ pattern DNA content}$$

The obtained value is the nuclear DNA amount, expressed in picograms, for each cell of this plant.

Some thousands of interphasic nuclei extracted from a tissue and labelled with a specific DNA fluorochrome are measured one by one. The obtained histogram would have numerous derived nuclei of cells in the G1 phase of the cell cycle, others in the S phase (phase of synthesis of DNA) or still in G2 (phase preceding mitosis). For a diploid plant, the G0/G1 position reflect the quantity 2C of DNA, and the mean fluorescence intensity of this peak (G0/G1) is half that of peak in G2 phase (4C) (Brown and Bergounioux, 1989; Galbraith, 1990; Brown et al., 1991a; Jayat and Ratinaud, 1993) (Figs. 2 and 3). These different positions can put in evidence the ploidy state or the presence of a mixture in a nursery (Petit et al., 1986; Steen, 1990a).

---

[7] The mixoploidy is defined as the presence of cells having different chromosome numbers in the same cell population.

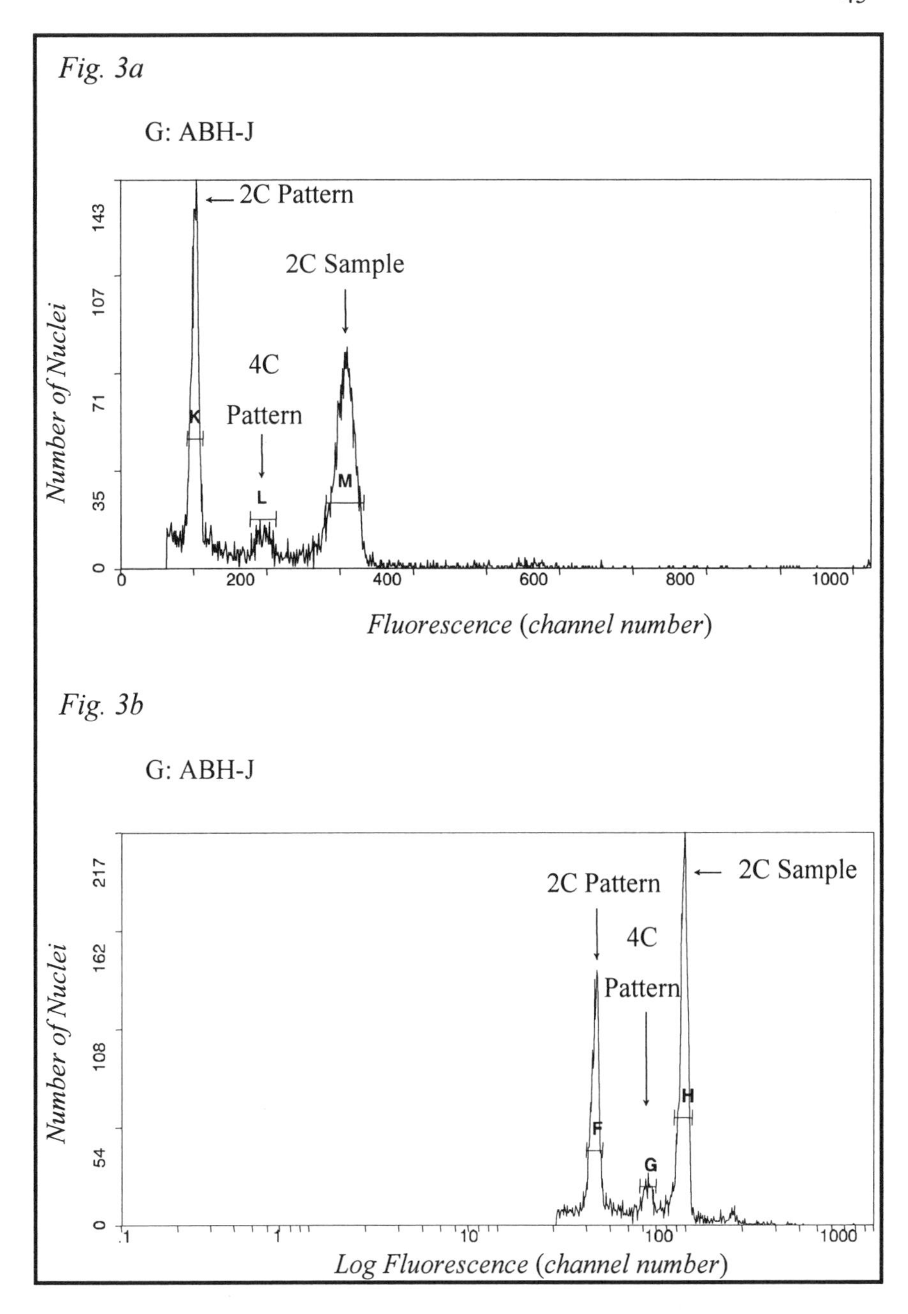

Figure 3a. Lineal representation of 2C (K) and 4C (L) pattern peaks and 2C sample peak (M). *Figure 3b*. Logarithmic representation of the same data. 2C pattern peak is marked with F, 4C pattern peak with G and 2C sample peak with H.

The aspect of 'singular mean' of this technique allows to quantify the variation of a population and to observe subpopulations, which would be impossible for a global mean that could not give nothing else that an intermediary value. It is also available to study the specific interactions, without making a purification of the heterogeneous preparation, of these detected subpopulations (Steen, 1990a).

DNA BASE COMPOSITION

Generally, a reference and a specimen have not the same AT/GC ratio; thus the same relative intensities will not be observed with various base-specific fluorochromes and with intercalating stains. Then, to study the DNA base composition of a given specimen, which sometimes is a useful taxonomic approach, it is necessary to use a combination of three fluorochromes:

- An intercalate stain (Ethidium Bromide or Propidium Iodide), and
- Two types of base-specific dyes:
  - ⇒ A-T base-specific dyes (Mithramycin or Chromomycin)
  - ⇒ G-C base-specific dyes (DAPI or Hoechst)

This combination cannot be simultaneous, then it is necessary to make three preparations for the same sample (using the same leaf, if it is possible) and adding the pattern every time. When the protoplasts are obtained, the following step is to label them with a type of fluorochrome. A possible combination should be Propidium Iodide, Mithramycin and Hoechst.

These fluorochromes absorb and emit light at different wavelengths. This fact makes necessary to use different combinations of emission sources, mirrors and filters to be able to obtain, to deviate and to filter the signal, before read the information.

To calculate the base composition (GC%) of an individual using an internal pattern with this known ratio, Godelle et al. (1993) utilise a curvilinear function to obtain two complementary estimations of base composition of the sample using the AT/GC content of reference:

$$AT\%_{\text{of specimen}} = 100 \text{ x } AT\%_{\text{standard}} \text{ x } (R_{Ho}/R_{EB})^{1/5}$$

and

$$GC\%_{\text{of specimen}} = 100 \times GC\%_{\text{standard}} \times (R_{Mi}/R_{EB})^{1/3}$$

where

R_{EB}= intensity specimen/intensity standard with an intercalating dye

and

R_{Ho} with Hoechst and R_{Mi} with Mithramycin

Marie and Brown (1993) obtain values for the genome composition of about 38% G-C in *Psophocarpus tetragonolobus* (Goa bean), and of about 45% in *Holcus* spp.

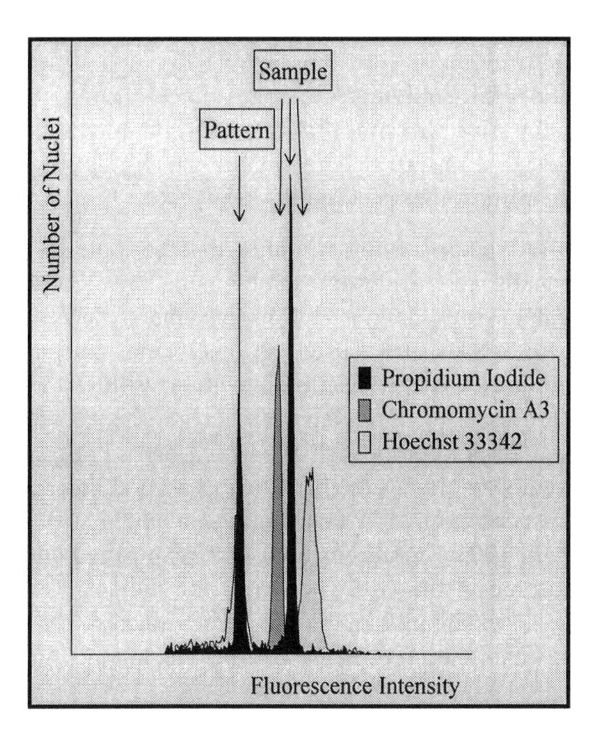

Figure 4. Comparative representation of the three logarithmic histograms obtained using three different fluorochromes. They are aligned by the 2C pattern peak.

RNA ANALYSIS

The fluorochrome Acridine Orange (AO) emits different maximum of fluorescence depending on binding to a double strand nucleic acid (DNA, and in this case AO emits in the green area of the spectrum) or to a simple strand nucleic acid (RNA, and AO emits red fluorescence). The use of this probe allows the simultaneous study of the nuclear DNA and RNA content and their variation (Bergounioux et al., 1988; Brown and Bergounioux, 1989).

ANALYSIS AND PHYSICAL SEPARATION OF CHROMOSOMES

The purification of individual chromosomes allows, using specific chromosome libraries, to simplify the analysis process and the plant genome sequence study. It also facilitates the protoplast transformation with specific chromosomes (Galbraith, 1989).

Using DNA specific fluorochromes, it is possible to analyse and to separate physically certain chromosomes using the flow cytometer. The initial plant cells should have a high mitotic index. In order to obtain it, certain substances are usually used, as colchicine, that permit to obtain high levels of mitotic synchronisation in cellular cultures or in protoplasts. It is also necessary to have appropriate protocols of extraction of chromosomes. The karyotype constituent chromosomes should have an enough different size among them, allowing a good discrimination based on their fluorescence signals when they are labelled with fluorochromes (Métézeau et al., 1993). Veuskens et al. (1995) purified and characterised the Y chromosome in the wild species of the dioicus plant *Melandrium album* and in a sexual mutant of the same species, having previously described the habitual protocols of this technique in other plants (Veuskens et al., 1992).

ACKNOWLEDGEMENTS

I wish to thank to Dr. Teodoro Coba de la Peña for much useful suggestions and for criticism of the manuscript.

REFERENCES

Bennett M.D., Cox A.V., Leitch I.J. Angiosperm DNA C-values database http://www.rbgkew.org.uk/cval/database1.html, 1998

Bennett M.D., Leitch I.J. Nuclear DNA amounts in Angiosperms. Ann Bot 1995; 76:113-176

Bennett M.D., Leitch I.J. Nuclear DNA amounts in Angiosperms – 583 new stimates. Ann Bot 1997; 80:169-196

Bennett M.D., Smith J.B. Nuclear DNA amounts in Angiosperms. Philos T Roy Soc B 1976; 274:227-274

Bennett M.D., Smith J.B., Heslop-Harrison J.S. Nuclear DNA amounts in Angiosperms. Proceed Roy Soc London B 1982; 216:179-199

Bennett M.D., Smith J.B. Nuclear DNA amounts in Angiosperms. Philos T Roy Soc B 1991; 334:309-345

Bergounioux C., Brown S.C., Petit P.X. Flow cytometry and plant protoplast cell biology. Physiol Plantarum 1992; 85:374-386

Bergounioux C., Perennes C., Brown S.C., Sarda C., Gadal P. Relation between protoplast division, cell cycle stage and nuclear chromatin structure. Protoplasma 1988a; 142:127-136

Bergounioux C., Perennes C., Brown S.C., Gadal P. Nuclear RNA quantification in protoplast cell-cycle phases. Cytometry 1988b; 9:84-87

Bharathan G., Lambert G., Galbraith D.W. Nuclear DNA content of monocotyledons and related taxa. Am J Bot 1994; 81:381-386

Blanco A. *Contribución a la Taxonomía de Pteridophyta.* Minor Thesis. Vigo, Spain: Universidade de Vigo, 1997

Brown S.C. Applications of flow cytometry in plant biology and biotechnologies: review and perspectives. Biotech Biotechnol Equipm 1994; 8:75-78

Brown S.C., Bergounioux C. "Plant Flow Cytometry." In *Flow Cytometry: Advanced Research and Clinical Applications. Vol II.* A. Yen, ed. Boca Raton, Florida: CRC Press, 1989

Brown S.C., Bergounioux C., Tallet S., Marie D. "Flow Cytometry of Nuclei for Ploidy and Cell Cycle Analysis." In *A Laboratory Guide for Cellular and Molecular Plant Biology.* I. Negrutiu, G. Gharti-Chherti, eds. Birkhäauser: Verlag Basel, 1991a

Brown S.C., Devaux P., Marie D., Bergounioux C., Petit P. Cytométrie en flux: application à l'analyse de la ploïdie chez les végétaux. Biofutur 1991b

Brown S.C., Kouchkovsky F., Marie D. "Polysomaty in Plants." In *The Porto Workshop on Flow and Image Cytometry. A Collection of Ideas.* Kimler, O'Connor, Sansonetty, Brown, eds. Porto: IPATIMUP. Series 1, Fundaçao Eng. Antonio Almeida, 1992

Brown S.C., Redondo N., Cantrel C., Veuskens A., Horjales M. "Genomesize, a Plant Database." In *Congrès de Cytométrie en Flux et de Cytométrie en Image*. France: Association de Cytométrie en Flux Cercle Français de Microscopie Quantitative, 1993

Conia J., Bergounioux C., Perennes C., Muller P.H. Flow cytometric analysis and sorting of plant chromosomes from *Petunia hybrida* protoplasts. Cytometry 1987; 8:500-508

Darzynkievicz Z., Juan G. "DNA Content Measurement for DNA Ploidy and Cell Cycle Analysis" In *Currents Protocols in Cytometry*. New York, USA: John Wiley and Sons, 1997

De Laat A.M.M., Göhde W., Vogelzang M.J.D.C. Determination of ploidy of single plants and plant populations by flow cytometry. Plant Breeding 1987; 99:303-307

Doležel J. Flow cytometric analysis of nuclear DNA content in higher plants. Phytochem Anal 1991; 2:143-154

Doležel J. Flow cytometry, its Application and Potential for Plant Breeding. In *Current Topic in Plant Cytogenetics Related to Plant Improvement*. T. Lelley, ed. Vienna: Universitätsverlag, 1998

Doležel J., Cíhalíková J., Lucretti S. A high-yield procedure for isolation of metaphase chromosomes from root tips of *Vicia faba* L. Planta 1992; 188:93-98

Doležel J., Lucretti S. High-resolution flow karyotyping and chromosome sorting *in Vicia faba* lines with standard and reconstructed karyotypes. Theor Appl Genet 1995; 90:797-802

Galbraith D.W. Analysis of higher plants by flow cytometry and cell sorting. Int Rev Cytol 1989; 116:165-228

Galbraith D.W. "Flow Cytometric Analysis of Plant Genomes." In *Methods in Cell Biology*. Z. Darzynkiewicz, H. Crissman, eds. London: Academic Press, 1990

Galbraith D.W., Harkins K.R., Maddox J.R., Ayres N.M., Sharma D.P., Firoozabady E. Rapid flow cytometric analysis of the cell cycle in intact plant tissues. Science 1983; 220:1049-1051

Galbraith D.W., Lambert G.M., Macas J., Doležel J. "Analysis of Nuclear DNA Content and Ploidy in Higher Plants." In *Currents Protocols in Cytometry*. New York, USA: John Wiley and Sons, 1997

Godelle B., Cartier D., Marie D., Brown S.C., Siljak-Yakovlev S. Heterochromatin study demonstrating the non-linearity of fluorometry useful for calculating genomic base composition. Cytometry 1993; 14:618-626

Horjales M., Redondo N., Blanco A. Datos sobre la cantidad de DNA en el género *Equisetum* L. NACC 1997; 7: 69-73

Jayat C., Ratinaud M.H. Cell cycle analysis by flow cytometry: principles and applications. Biol Cell 1993; 78:15-25

Kononowicz A.K., Hasegawa P.M., Bressan R.A. Chromosome number and nuclear DNA content of plants regenerated from salt adapted plant cell. Plant Cell Rep 1990; 8:676-679

Lysák M.A., Cíhalíková J., Kubaláková M., Simková H., Künzel G., Doležel J. Flow karyotyping and sorting of mitotic chromosomes of barley (*Hordeum vulgare* L.). Chromosome Res 1999; 7:431-444

Marie D., Brown S.C. A cytometric exercise in plant DNA histograms, with 2C values for 70 species. Biol Cell 1993; 78:41-51

Métézeau P., Schmitz A., Frelat G. Analysis and sorting of chromosomes by flow cytometry: new trends. Biol Cell 1993; 78:31-39

Michaelson M.J., Price H.J., Ellison J.R., Johnston J.S. Comparison of plant DNA contents determined by Feulgen microspectrophotometry and laser flow cytometry. Am J Bot 1991; 78:183-188

Petit J.M., Denis-Gay M., Ratinaud M.H. Assessment of fluorochromes for cellular structure and function studies by flow cytometry. Biol Cell 1993; 78:1-13

Rabinovitch P.S. "Univariate Cell Cycle Analysis: Principles, Mathematical Modelling and Challenges." In *The Porto Workshop on Flow and Image Cytometry. A Collection of Ideas*. Kimler, O'Connor, Sansonetty, Brown, eds. Porto: IPATIMUP. Series 1, Fundaçao Eng. Antonio Almeida, 1992

Redondo N., Blanco A., Horjales M. Estudio del género *Polypodium* L. del Noroeste Ibérico: cantidad de DNA nuclear. NACC 1998; 9:109-116

Redondo N., Horjales M., Blanco A. Cantidades de DNA nuclear y esporas Aspleniaceae: *Asplenium* L., *Phyllitis* Hill y *Ceterach* Willd. NACC 1999a; 9:99-107

Redondo N., Horjales M., Blanco A. Cantidades de DNA nuclear en helechos del Noroeste Ibérico. Real Sociedad Española de Historia Natural 1999b

Steen H.B. "Characteristics of Flow Cytometers." In *Flow Cytometry and Sorting*. M.R. Melamed, T. Lindmo, M.L. Mendelsohn, eds. New York, USA: John Wiley and Sons, 1990a

Steen H.B. Light scattering measurement in an arc lamp-based flow cytometer. Cytometry 1990b; 11:223-230

Tiersch T.R., Chandler R.W., Wachtel S.S., Elias S. Reference standards for flow cytometry and application in comparative studies of nuclear DNA content. Cytometry 1989; 10:706-710

Ulrich I., Fritz B., Ulrich W. Application of DNA fluorochromes for flow cytometric DNA analysis of plant protoplasts. Plant Sci 1988; 55:151-158

Veuskens J., Marie D., Hinnisdaels S., Brown S.C. "Flow Cytometry and Sorting of Plant Chromosomes." In *Flow Cytometry and Cell Sorting*. A. Radbruch, ed. Heidelberg, Germany: Springer Berlin, 1992

CHAPTER 4

USE OF FLOW CYTOMETRY TO MEASURE PHYSIOLOGICAL PARAMETERS

Teodoro Coba de la Peña
Dpto Fisiología y Bioquímica Vegetal. Centro de Ciencias Medioambientales. Consejo Superior de Investigaciones Científicas. Madrid. Spain

INTRODUCTION

In addition to DNA and cell cycle analysis, several parameters involved in signal transduction pathways, enzymatic activities and many other physiological processes can be analysed by flow cytometry. It is the case for intracellular pH, membrane potential, calcium concentration, reactive oxygen species (ROS) generation, and glutathione concentration.

This experimental approach is feasible at present thanks to the development of very specific fluorescent probes. One of the principal advantages of flow cytometry is that it allows the measurement of these parameters in living and small cells, usually in real time during the physiological stimulation.

There is a general method in these studies. First of all, detailed observations for suitable loading and intracellular distribution of the fluorescent probe must be done by fluorescence microscopy and flow cytometry. Secondly, the effects of different ionophores and inhibitors (specific in each case for the physiological parameter under study) on fluorescence is analysed by flow cytometry. These observations allow us to verify that the fluorescence behaviour of the probe is correct and really dependent on the parameter under study, evaluating and discarding possible artifacts. These observations also allow a calibration of the fluorescence and of the range variations of the physiological parameter. Thirdly, physiological effects (and thus, fluorescence variations) induced

M.J. Reigosa Roger, Handbook of Plant Ecophysiology Techniques, 53–64.

by different substances (hormones or other biological substances, chemicals) and physiological conditions (mechanical, water or salt stress, etc.) under study are analysed by flow cytometry. Finally, the recorded data and histograms are analysed.

Most of the fluorescent probes, ionophores, chelators and other substances discussed in this section can be obtained from Molecular Probes (Oregon). Cell loading of the probes referred below is accomplished simply by adding the probe directly to the buffer or cell suspension, at different concentrations and incubation times (other loading techniques, not discussed here, involve electroporation and microinjection).

PHYSIOLOGICAL PARAMETERS

Intracellular pH

Intracellular pH is a critical factor of the intracellular environment. Every cell process can be affected by intracellular pH changes, including metabolism, membrane potential, cell growth and proliferation, movement of substances across membrane, polymerisation of cytoskeleton, etc. In the same way, changes in intracellular pH are a cell response to external applied agents, like hormones, growth factors and others. Different cell organelles, as lysosomes and vacuoles, have a different pH than that of cytosol to accomplish their functions. So cells have developed different mechanisms for regulating intracellular pH. Mechanisms of pH regulation in plants have been reviewed by Smith (1979), Felle (1988) and Kurkdjian and Guern (1989).

Fluorescence properties (fluorescence intensity, emission and excitation spectrum) of several available probes vary depending on the H^+ concentration of their environment, because hydrogen ion binding changes the electronic structure of the probe (Haugland, 1996).

Maximum response of the probe will occur for pH values near its *pKa*. The sensitivity of these probes is around 0.1-0.2 pH units.

Two examples of pH sensitive probes, frequently used in flow cytometry, are BCECF (2',7'-bis-(2 carboxyethyl)-5-(and 6) carboxyfluorescein acetoxymethyl ester) and SNARF1 (Semi naphtho rhoda fluorine acetoxymethyl ester). Covalent binding with an acetoxymethyl (AM) residue allows probes to be permeable across biological membranes. Acetoxymethyl-ester forms of these probes are commercially available. Once into the cell, cellular esterases remove acetoxymethyl residue, and the probes become negatively charged, and thereby membrane impermeable, trapped into the cell or into a cell compartment.

BCECF is optimally excited at 488 nm, and its maximal fluorescence emission is 520 nm. Its *pKa* is 6.98, and thereby it is very suitable for the study of cytosolic pH (6.5-7.5). In case of acidification, BCECF becomes more protonated, and its fluorescence intensity decreases. Cellular alkalinization induces an increase in intensity. BCECF is a fast-response probe, allowing kinetic studies of pH changes in real time. One example of the use of BCECF and flow cytometry in plant material is shown in Giglioli-Givarc'h et al. (1996), where activation of a phosphoenolpyruvate kinase after cytosolic alkalinization in *Digitaria sanguinalis* protoplasts is characterised.

SNARF1 is excited at 488 nm and is a 'ratiometric dye', that is, its emission maximum shifts upon pH changes in the microenvironment. The protonated form of the fluorescent probe has a maximum emission at 540 nm, and the maximum of the deprotonated form is at 630 nm. It is possible to record continuously the fluorescence intensity at both wavelengths by flow cytometry, using two detectors. In fact, the ratio of both fluorescence intensities is a very reliable and specific measure, because it discards fluorescence intensity variations induced by several unspecific factors, like differential individual loading among cells (Haugland, 1996).

Some ionophores and substances used in the validation and calibration of pH fluorescent probes are:

– Nigericine induces a permeabilization of cell membrane to proton ions, so the extracellular and intracellular proton concentrations make equal, if extracellular and intracellular K+ concentrations are the same.

– Propionic is a weak acid that induces intracellular acidification. NH4Cl is a weak base that induces intracellular alkalinization. Both are used for monitoring changes on fluorescence intensity.

The 'null point method' is used for calibrating and converting fluorescence intensity values in pH units. Dye-loaded cells are incubated in a series of buffers at different pHs in presence of Nigericine. Intracellular pH equals extracellular pH, and a direct correspondence between known intracellular pH and fluorescence intensity is established. (Haugland, 1996).

Membrane Potential

Electrical potential differences across membranes of prokaryotic and eukaryotic cells reflect the differential distribution and activity of ions such as Na^+, Cl^-, H^+ and especially K^+ across these biological membranes. Diverse membrane electrogenic pumps generate these ionic gradients, with a contribution from the intrinsic membrane permeability for each ion. This membrane potential plays a major role in the processes involving external stimulation of the cell, photosynthesis, nutrient and ion transport across the membrane and signal transduction. In eukaryotic cells, major examples are cytoplasmic, mitochondrial (inner membrane) and lysosome membrane potential, negative inside the cell (or inside the organelles) relative to the external medium. In chloroplasts, the thylakoid potential is relatively more positive inside, but here the major electrochemical gradient is due to protons, the lumen being acid and the stroma alkaline. The mean potential values in eukaryotic cells are between –10 mV and –100 mV. In the mitochondria, the potential values are around –100 mV, and –50 mV in lysosomal membranes (Shechter, 1984).

Membrane potential changes involve either depolarisation (that is, a decrease in transmembrane potential) or hyperpolarisation (an increase in the potential difference across the membrane).

Many excellent reviews are available concerning the fluorometric methods and probes developed for estimation of membrane potential, especially in organelles or in cells that are too small to allow the use of microelectrodes (Loew, 1982 and 1993; Montana et al., 1989; Gross and Loew, 1989; Haugland, 1996).

Dyes usually used in flow cytometry are molecules with a single negative or positive net charge, are highly hydrophobic and their partition across the membrane is a function of the Nernst equation:

$$[C]_{in}/[C]_{out} = e^{-nFE/RT}$$

where n = net electric charge of the indicator; [C] = intra and extracellular concentration of the indicator; F = Faraday constant; R = gas constant; T = temperature (°K); E = membrane potential.

These dyes are excited at the visible range of the spectrum, and with slow response to environmental changes in membrane potential (Plásek and Sigler, 1996).

Oxonol dyes have one net negative electric charge, and they will accumulate principally in the external volume of a negatively charged membrane, a lesser portion of the dye being retained in the internal compartment. A hyperpolarisation of the membrane produces dye redistribution to the external medium. These dyes are excluded from the negatively charged inner mitochondrial compartment and their fluorescence reflects mainly the plasmic membrane potential. One example is the Oxonol dye $DiBaC_4(3)$, commonly used in cytometry, because it can be excited at 488 nm. These dyes are very sensible to the variation of external ionic concentrations.

Cyanine dyes have one net positive electric charge at physiological pH, so their cellular partitioning is the contrary of oxonol dyes. These dyes are also partially accumulated in some organelles with negative inner membrane potential, like mitochondria and endoplasmic reticulum, and they are relatively toxic to cells. The cellular fluorescence intensity reflects membrane potential from the plasma membrane and also mitochondrial and endoplasmic reticulum membranes. This class of dye is the most used in flow cytometry. Two examples are the Carbocyanines $DiOC_6(3)$ and $DiOC_5(3)$. These dyes can undergo quenching when they are at high local concentration and polymerise.

The Carbocyanine dye JC-1 can be used for the study of mitochondrial potential. At low local dye concentration (low potential), the molecule is in the monomeric state with green fluorescence emission (527 nm) when excited at 490 nm. When the mitochondria are hyperpolarised, the local dye concentration increases and it forms polymer conjugates (J-conjugates) with a shifted red fluorescence (590 nm). This property makes possible ratiometric red/green fluorescence measurements in flow cytometry.

Merocyanine dyes undergo molecular reorientations with membrane depolarisation, forming fluorescent dimers with altered absorption spectra. Merocyanine 540, principally associated with unsaturated lipids, is a common example.

Finally, we can cite Rhodamine 123. Its incorporation depends on the voltage gradient of the mitochondrial inner membrane, and it is less toxic that Carbocyanine dyes. This dye is used in tests for early modifications of energy metabolism.

Validation of the specificity of the dye fluorescence is done using some ionophores that modify membrane potential:

– Valinomycin facilitates the passages of K+ ions down their concentration gradient across the membrane.

– Gramicidine D makes pores in the membrane, facilitating the free passage of mono and divalent ions, and it is generally used for membrane depolarisation.

– Vanadate inhibits the ATPase proteins susceptible to phosphorylation, that is, principally the cytoplasmic membrane ATPases.

– Regarding mitochondrial potential, FCCP (carbonyl cyanide p-trifluoromethoxyphenyl-hydrazone) induces an increase in mitochondrial membrane permeability for H^+ passage, producing depolarisation.

– Nigericine is used to cancel pH gradient, and thereby to hyperpolarise the mitochondrial membrane because there is an interchange that is electrically neutral between K^+ and H^+ ions.

Examples of calibration curves and flow cytometric determinations of absolute membrane potential are shown in Krasznai et al. (1995).

Cytosolic Ca^{2+} concentration

Calcium concentration is a critical factor in the control of many cellular responses. Calcium is a second messenger for a broad variety of stimuli, regulating metabolism and gene expression. Hepler and Wayne (1985) and Bush (1993 and 1995) have reviewed the role of calcium as second messenger in plants.

Grynkiewicz et al. (1985) and Haugland (1996) described several fluorescent probes for measuring cytosolic Ca^{2+} changes. Dissociation constant (K_d for calcium binding) of the probe must be approximately on the same order than resting calcium concentration in the cytosol. Loading of the probe into the vacuole must be avoided, because vacuole acts as a calcium reservoir in plant cells and fluorescent probes would be saturated.

Examples of specific dyes excited in the visible range of the spectrum are Calcium-Green 2 and Fluo-1. An 80-fold increase in the fluorescence intensity of Fluo-1 can be recorded upon binding to Ca^{2+}.

One of the most suitable fluorescent probes for the study of calcium by flow cytometry is indo-1. This is a ratiometric dye, excited in the ultraviolet (338 nm), and its emission spectra shifts following calcium binding (maximum emission for indo-1 in the absence of Ca^{2+} is 490 nm, and 405 nm if bound to Ca^{2+}). Ratio measurements (405/490) allow accurate quantifying of Ca^{2+} concentrations by flow cytometry. Indo-1 is a fast response-dye, and its K_d is 230 nM. Loading can be performed simply by addition of the acetoxymethyl ester form of Indo-1 in the extracellular medium. Darjania et al. (1993) have measured calcium concentration in *Vicia faba* protoplasts using indo-1 and fluorometry. Bush and Jones (1987 and 1990) have developed a methodology for measuring calcium changes in aleurone protoplasts by fluorometry, using this dye.

Some ionophores and other substances used for validation and calibration are:

- Ionomycin forms molecular complex with Ca^{2+} and increases the permeability of biological membranes to calcium, and thereby it allows the concentration-dependent flux of this ion across membranes. Affinity of this ionophore for calcium ion is higher at neutral and alkaline pH.
- Ionophore 4-bromo-A23187 also binds to Ca^{2+}, but its affinity is higher at acidic pH.
- $CaCl_2$ is added to the extracellular medium and, in presence of ionophores, induces a massive entry of Ca2+ into the cell.
- $MnCl_2$ induces fluorescence quenching of all calcium-specific probes.
- Some chelator agents, like BAPTA [1,2-bis(2-aminophenoxy) ethane-N,N,N',N'-tetraacetic acid] and EGTA [ethylene glycol-bis(β-aminoethyl ether)], bind free calcium ion, and they are used, in presence of ionophore, for reducing or regulating extracellular (or even intracellular) free-Ca^{2+} concentrations.

ROS generation

Plant cells can trigger an active defense reaction upon infection by pathogens. One defense-related response is the production of Reactive Oxygen Species (ROS), like H_2O_2, superoxide anion and hydroxyl radical. Extracellular production of H_2O_2 (oxidative burst) has been detected in several plant cell cultures, in response to osmotic and mechanical stress, microbial infection and addition of several microbial substances (Mehdy, 1994; Wojtaszek, 1997). The putative role of oxidative burst as second messenger in signal transduction and activation of defense-related genes has been postulated (Hammond-Kosack and Jones, 1996). Alvarez et al. (1998) have demonstrated that ROS induce localised secondary oxidative bursts in different parts of the plant after leaf infection and primary oxidative burst production, and they seem to regulate systemic acquired resistance signalling. Different roles for ROS at the intracellular level have also been suggested (Allan and Fluhr, 1997). Probably they are involved in the regulation of the redox state of cells, but very little is known about this subject.

Probes whose fluorescence properties depend on ROS production and activity have been developed. Dichlorofluorescin-diacetate (DCFH-DA) and Dihydrorhodamine 123 (DRH123) fluorogenic probes are the most used in flow cytometry. Both are excited at 488 nm, and their emission is comprised between 525 and 560 nm They were first used for detecting production and release of superoxide and hydrogen peroxide in neutrophils, but they have been recently applied to plant cells (Rothe et al., 1988; Haugland, 1996; Allan and Fluhr, 1997).

DCFH-DA is a non-fluorescent and membrane-permeable probe. Once into the cell, it is converted, through the action of cellular esterases and peroxidases, in the fluorescent derivative 2',7'-dichlorofluorescin (DCF). Peroxidase enzymes need H_2O_2 as co-substrate. So labelled cells become increasingly fluorescent depending upon hydrogen peroxide production.

DHR123, as DCFH-DA, is non-fluorescent and membrane permeable. Once into the cell, it is oxidised by H_2O_2 to the fluorescent dye rhodamine 123 (RH123), which is localised in the mitochondrial inner membrane. This reaction seems to be catalysed by peroxidases, but it is independent of esterase activities, which is an advantage in comparison with DCFH-DA. DHR123 also can have a sensitivity to peroxide hydrogen of up to 8 fold that of DCFH-DA (Rothe et al., 1988; Haugland, 1996).

Both fluorogenic probes have two constraints: they can also be significantly oxidised, even in resting cells, by citocrome c, Fe^{2+}, and by the oxidative phosphorylation occurring in the mitochondria (Haugland, 1996). Coba de la Peña (1999) has observed increasing fluorescence intensities in resting (non stimulated) root hairs and plant cell suspensions of *Medicago sativa* labelled with these probes, using confocal microscopy and flow cytometry, respectively. Thereby quantitative estimations of

ROS activity are difficult to perform using this technical approach, but comparison of relative fluorescence intensities is easily feasible.

For validation and calibration in fluorescence and flow cytometry experiments, diphenyleneiodonium chloride (DPI) is used. DPI is an inhibitor of NAD(P)H oxidase, the enzyme that catalyses the conversion of O_2 to superoxide anion. Peroxidase generates H_2O_2 using superoxide anion and H_2O as substrates. So DPI-induced inhibition cancels H_2O_2 production. The enzyme superoxide-dismutase (SOD), catalysing dismutation of superoxide to hydrogen peroxide, is also used.

In fact, several NADPH oxidase-independent ROS generation systems have been detected in plants. Some of them are: the nitric oxide (NO) synthase pathway, the oxalate-and flavine-oxidase systems, and the activation of peroxidases and amino-oxidases by the action of pathogenesis-generated amines and polyamines (Hammond-Kosack and Jones, 1996; Allan and Fluhr, 1997).

Finally, signal transduction components of ROS generation and oxidative burst can be analysed using calcium channel inhibitors, phosphatase and kinase inhibitors, etc.

Allan and Fluhr (1997) have studied intracellular ROS generation in tobacco epidermal cells, using DCFH-DA and fluorescence microscopy. Coba de la Peña (1999) has optimised a methodology for detection and analysis of intracellular and extracellular ROS generation in protoplasts and whole plant cell cultures of *Medicago sativa*, using DCFH-DA and flow cytometry.

Glutathione levels

Glutathione (L-γ-glutamyl-L-cysteinyl glycine; GSH) has an ubiquitous role in physiological processes in both animal and plant cells. GSH mediates neutralisation of oxygen free radicals (GSH is the key antioxidant agent in plant cells), detoxification, amino acid transport and accumulation, synthesis of nucleic acids and proteins and modulation of enzymatic activity. GSH levels also correlate with plant adaptation to environmental stress. An increase in GSH levels is a general response to stress in plants (Alscher, 1989; Noctor et al., 1998; May et al., 1998). Both enzymatic synthesis of GSH (catalysed by γ-glutamylcysteine synthetase) and redox balance between GSH (glutathione reduced form) and GSSG (glutathione disulphide) must be taken into account for a correct analysis and interpretation of changes in glutathione cellular levels.

Ortho-phthaldialdehyde (OPT) is a commonly used probe in studies of GSH carried out by flow cytometry. OPT is not fluorescent initially. After addition and once loaded into the cell, OPT forms a fluorescent and insoluble conjugate with GSH and with thiol (-SH) residues present in proteins. This conjugate is excited at 325 nm, and has two types of emission: specific fluorescence for the OPT-GSH conjugate (450, 525 and 575 nm), and for OPT-thiol (-protein) conjugate (405 nm). Ratiometric measurement of both types of fluorescence serves as a good estimation of GSH content related to thiol residues of cellular proteins (Treumer and Valet, 1986; Hedley and Chow, 1994; Haugland, 1996).

Some chemicals used in GSH validation and calibration are:

- Diethyl maleimide (DEM) induces a decrease in the content of GSH and thiol residues.
- Butionine sulphoximine (BSO) is a specific inhibitor of γ-glutamylcysteine synthetase.
- Dithiothreitol (DTT) is a reducing agent that induces formation of GSH (and thiol residues into the proteins).

Coba de la Peña (1999) has developed an experimental procedure to estimate GSH variations in *Medicago sativa* cell suspensions using OPT and flow cytometry. Coba de la Peña and Brown (2001) have also detected and characterised by flow cytometry a GSH decrease in alfalfa cell suspensions during an oxidative burst induced by pathogenic bacteria *Pseudomonas syringae*. This can be an example of GSH-ROS relationship during defense-related responses in plants.

REFERENCES

Allan A.C., Fluhr R. Two distinct sources of elicited reactive oxygen species in tobacco epidermal cells. Plant Cell 1997; 9:1559-1572

Alscher R.G. Biosynthesis and antioxidant function of glutathione in plants. Physiol Plantarum 1989; 77:457-464

Alvarez M.E., Pennell R..I., Meijer P.J., Ishikawa A., Dixon R.A., Lamb C. Reactive oxygen intermediates mediate a systemic signal network in the establishment of plant immunity. Cell 1998; 92:773-784

Bush D.S. Regulation of cytosolic calcium in plants. Plant Physiol 1993; 103:7-13

Bush D.S. Calcium regulation in plant cells and its role in signaling. Annu Rev Plant Physiol Plant Mol Biol 1995; 46:95-122

Bush D.S., Jones R.L. Measurement of cytoplasmic calcium in aleurone protoplasts using indo-1 and fura-2. Cell Calcium 1987; 8:455-472

Bush D.S., Jones R.L. Measuring intracellular Ca^{2+} levels in plant cells using the fluorescent probes, indo-1 and fura-2. Plant Physiol 1990; 93:841-845

Coba de la Peña T. *Estudio de Acontecimientos Precoces en la Formación de Nódulos en Medicago sativa L.(Study of Early Events in Nodule Formation in Medicago sativa L).* Doctoral Thesis (PhD). Vigo, Spain: University of Vigo, 1999

Coba de la Peña T., Brown S. Flow cytometric characterization of alfalfa whole plant cell suspensions using a time gated amplifier device. (2001, in preparation)

Darjania J., Curvetto N., Delmastro S. Loading of *Vicia faba* guard cell protoplasts with indo-1 to measure cytosolic calcium concentration. Plant Physiol Biochem 1993; 31:793-798

Felle H. Short-term pH regulation in plants. Physiol Plantarum 1988; 74:583-591

Giglioli-Guivarch N., Pierre J.N., Vidal J., Brown S. Flow cytometric analysis of cytosolic pH of mesophyll cell protoplasts from the crabgrass *Digitaria sanguinalis*. Cytometry 1996; 23:241-249

Gross D., Loew L.M. Fluorescent indicators of membrane potential: microspectrofluorometry and imaging. Meth Cell Biol 1989; 30:193-218

Grynkiewicz G., Poenie M., Tsien R.Y. A new generation of Ca^{2+} indicators with greatly improved fluorescence properties. J Biol Chem 1985; 260:3440-3450

Hammond-Kosack K.E., Jones J.D.G. Resistance gene-dependent plant defense responses. Plant Cell 1996; 8:1773-1791

Haugland R.P. *Handbook of Fluorescent Probes and Research Chemicals.* Eugene, Oregon: Molecular Probes, 1996

Hedley D.W., Chow S. Evaluation of methods for measuring cellular glutathione content using flow cytometry. Cytometry 1994; 15:349-358

Hepler P.K., Wayne R.O. Calcium in plant development. Annu Rev Plant Physiol 1985; 36:397-439

Krasznai Z., Márián T., Balkay L., Emri M., Trón L. Flow cytometric determination of absolute membrane potential of cells. J Photoch Photobio 1995; 28:93-99

Kurkdjian A., Guern J. Intracellular pH: measurement and importance in cell activity. Annu Rev Plant Physiol Plant Mol Biol 1989; 40:271-303

Loew L.M. Design and characterization of electronic membrane probes. J Biochem Bioph Meth 1982; 6:243

Loew L.M. "Potentiometric Membrane Dyes." In *Fluorescent and Luminiscent Probes for Biological Activity. A Practical Guide to Technology for Quantitative Real-Time Analysis.* W.T. Masson, ed. London: Academic Press, 1993

May M.J., Vernoux T., Leaver C., Van Montangu M., Inzé D. Glutathione homeostasis in plants: implications for environmental sensing and plant development. J Exp Bot 1998; 49:649-667

Mehdy M.C. Active oxygen species in plant defense against pathogens. Plant Physiol 1994; 105:467-472

Montana V., Farkas D.L., Loew L.M. Dual wavelength radiometric fluorescence measurements of membrane potential. Biochemistry 1989; 28:4536

Noctor G., Arisi A.C.M., Jouanin L., Kunert K.J., Rennenberg H., Foyer C.H. Glutathione: biosynthesis, metabolism and relationship to stress tolerance explored in transformed plants. J Exp Bot 1998; 49:623-647

Plasek J., Sigler K. Slow fluorescent indicators of membrane potential: a survey of different approaches to probe response analysis. J Photoch Photobio 1996; 33:101-124

Rothe G., Oser A., Valet G. Dihydrorhodamine 123: A new flow cytometric indicator for respiratory burst activity in neutrophil granulocytes. Naturwissenchaften 1988; 75:354-355

Shechter E. *Membranes Biologiques*. Masson, ed. Paris: Masson S.A., 1984

Smith F.A. Intracellular pH and its regulation. Annu Rev Plant Physiol 1979; 30:289-311

Treumer J., Valet G. Flow cytometric determination of glutathione alterations in vital cells by o-phthaldialdehyde (OPT) staining. Exp Cell Res 1986; 163:518-524

Wojtaszek P. Oxidative burst: an early plant response to pathogen infection. Biochem J 1997; 322:681-692

CHAPTER 5

FLOW CYTOMETRY: CELL CYCLE

Teodoro Coba de la Peña[1] and Adela M. Sánchez-Moreiras[2]
[1]*Dpto Fisiología y Bioquímica Vegetal. Centro de Ciencias Medioambientales. Consejo Superior de Investigaciones Científicas. Madrid. Spain*
[2]*Dpto Bioloxía Vexetal e Ciencia do Solo. Universidade de Vigo. Spain*

THE CELL CYCLE AND CHARACTERISTICS OF CORRESPONDING FLOW CYTOMETRY HISTOGRAMS

As it was exposed in chapter about DNA quantification (Chapter 3), several available fluorescent probes bind specifically nucleic acids, including intercalating dyes (Ethidium Bromide and Propidium Iodide), AT-binding dyes (Hoechst 33258 and Hoechst 33242, DAPI, DIPI) and GC-binding dyes (Chromomycin, Mithramycin, Olivomycin). The use of this dyes allows to analyse, by flow cytometry, the DNA content, nuclear DNA base content (%AT and %GC), and detection of polyploidy and polysomaty in plant cells and tissues. Another very interesting cellular event that can be analysed is the cell cycle.

Kinetic, dynamic, rate and progression characteristics of the cell cycle in plant cell cultures or tissues are very important parameters involved in multiple physiological events. Flow cytometry makes possible a fine approach to the study of these events, including basic mechanisms of the cell cycle (rates of proliferating and quiescent cells, characterisation of cell subsets and states upon cell cycle length and progression) and also study of effects of different putative modulators and inhibitors (hormones, growth factors, toxins, maybe allelochemicals, etc.) and environmental conditions (including stress) on the cell cycle (Galbraith, 1989; Gray et al., 1990).

M.J. Reigosa Roger, Handbook of Plant Ecophysiology Techniques, 65–80.

Moreover, using simultaneously other fluorescent dyes and fluorescent-labelled monoclonal antibodies, RNA and protein content and synthesis, identification of antigens and molecular markers that are specific and/or critical of a cell cycle subphase can also be analysed at the same time that the cell cycle.

Regarding the cell cycle, cell (and, in particular, its nucleus) can be at different possible states or phases (Jayat and Ratinaud, 1993; Marie and Brown, 1993; Galbraith, 1989): G0, G1, S, G2 and M (Figure 1):

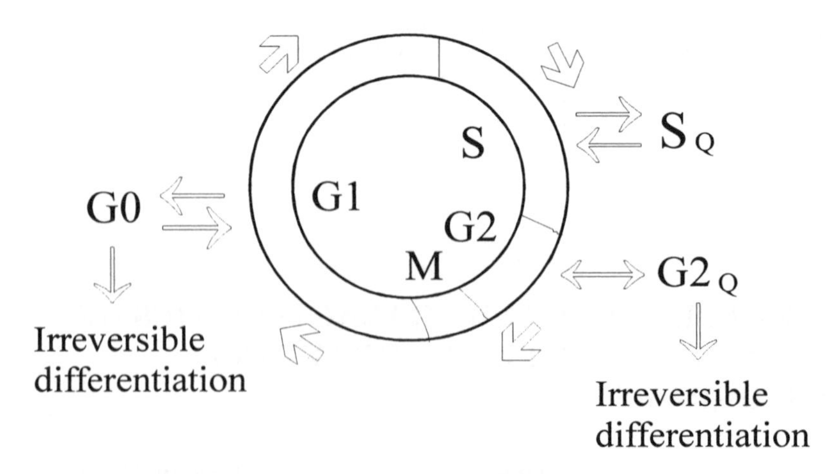

Figure 1. A classical representation of the cell cycle. Different cell cycle phases and terms are explained in the text. The length of the arc indicates the relative duration of a phase, and arrows indicate sense of advance of the cell cycle.

Cells in G0 phase (or Gap 0): cells in quiescent state after mitosis, that is, differentiated or undifferentiated cells that do not divide and are not involved in active (proliferating) cell cycle events. This quiescent state can be reversible, and then cells enter in G1 phase.

Cells in G1 phase: involved in cell growth and in an active cell cycle. They are characterised by a 2C nuclear DNA content (that is, with double DNA amount than that of gametes). This interval precedes nuclear DNA synthesis.

Cells in S (Synthesis) phase: DNA synthesis takes place, and cell can duplicate progressively their nuclear DNA content.

Cells in G2 phase, which is an interval between the end of DNA synthesis and the beginning of mitosis. They are characterised by a 4C nuclear DNA content.

The whole of G1, S and G2 phases is termed 'interphase'.

Cells in Mitosis (M): chromatin condenses, becoming chromosomes. Nuclear envelope disappears. Later on, chromosome segregation occurs, appearing new nuclear membranes, originating two daughter nuclei. This event is usually followed by cell division (cytokinesis). Thereby, this presently 4C cell divides in two 2C daughter cells.

Daughter cells can enter in G0 phase for a time, or they entry directly in the G1 phase of a new cell cycle

In every described phase, cell cycle progression can stop and cell entries in a new quiescent phase. By this way, cells in quiescent G1, S and G2 phases (called $G1_Q$, S_Q and $G2_Q$, respectively) appear. Intermediate phases between quiescent and proliferating cells have also been described, and they are called $G1_T$, S_T and $G2_T$.

G0 and $G2_Q$ can be followed by irreversible differentiation of the tissue cells, that do not divide anymore, although regression to undifferentiated and newly proliferating cells has sometimes been observed in mesophyll cells (Marie and Brown, 1993).

If anomalous mitosis occurs (endomitosis, characterised by no formation of mitotic spindle and no attainment of chromosome segregation), a single nucleus with double number of chromosomes (corresponding to a 4C DNA content) becomes permanent. This event can take place several times, originating cells with a DNA content of 8C, 16C, 32C, 64C, etc., that is, with different ploidy levels. Endopolyploidy is originated by this way. In fact, this phenomenon is common in plants (Brown et al., 1991), and different tissues of a given plant organism can show different ploidy levels (polysomaty).

Length of each phase varies upon species, tissues and cell physiology. In a sample of proliferating cells (as is the case of plant meristems or some plant cell suspensions), most of them are in G1 phase, because this is the longest phase. The higher part of plant tissues is composed by fully differentiated (quiescent) cells, which do not divide anymore.

In this section, we will describe the simplest modality of cell cycle analysis performed by flow cytometry.

Samples to be analysed by flow cytometry are prepared from plant suspensions or meristems as described in chapter of DNA quantification. Briefly, using a Petri dish, intact plant tissue is chopped with a razor blade, into a nuclear buffer. This suspension is filtered through nylon filters (30 μm pore size). If an intercalating dye is used, RNase treatment is necessary, previous to dye addition. Then, a nucleic acid-specific dye, like Ethidium Bromide (EtBr) for instance, is added for nuclei labelling. After an incubation of 30 min, labelled plant nuclei suspension is analysed in a flow cytometer.

Of course, a mixture of nuclei in different cell cycle (active or quiescent) phases is present in this asynchronous suspension. There is a lineal correlation between fluorescence intensity of EtBr-labelled nuclei and DNA content.

One example of monoparametric histogram obtained from a labelled nuclei suspension from root meristems of *Lactuca sativa* by flow cytometry is:

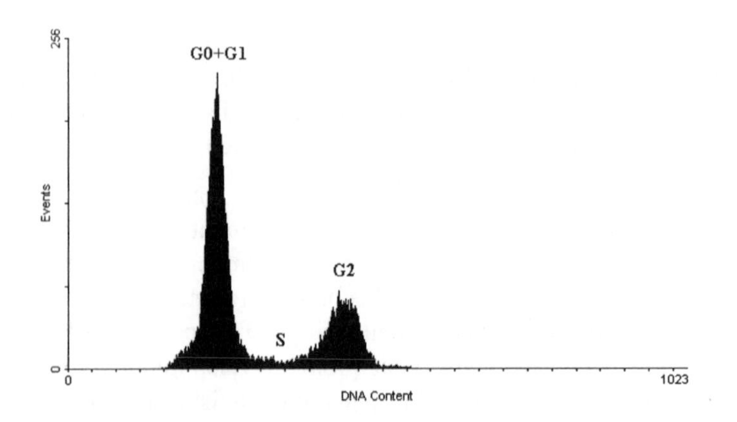

Figure 2. Monoparametric linear histogram for cell cycle analysis of *Lactuca sativa* root meristems.

Both axes are linear scales. Relative fluorescence intensity (proportional to the DNA content, in the X-axis of up to 1024 channels) is represented versus the number of analysed nuclei ('Events' in the Y-axis). In this simple case, three nuclear populations are shown:

- The first one (G0+G1) corresponds to 2C nuclei, and it includes quiescent G0, $G1_Q$ and proliferating G1 undifferentiated cells, and also 2C differentiated cells. We can not distinguish among these different types of nuclei on the only basis of this monoparametric DNA content-depending fluorescence analysis, and all of them are placed in the same peak.

- The second peak (G2) corresponds to 4C nuclei, and it includes G2 cycling nuclei that have finished DNA replication, but also quiescent $G2_Q$

nuclei and differentiated 4C cells, not involved in the proliferating cell cycle. The mean fluorescence intensity of this 4C peak is approximately double than that of 2C peak. Usually, fluorescence intensity ratio 4C/2C is not 2, but 1.8 or 1.9, owing to labelling irregularities due to differences in chromatin condensation state (Galbraith, 1989).

- Finally, the third population (S, between both peaks) is recorded as a strip connecting the first and the second peak population. This population is constituted by S-phase nuclei, in different stages of DNA replication. This is the reason of the strip shape of this population in the flow cytometry histogram. These nuclei include, of course, cycling S and non-cycling S_Q cells.

In fact, the very first analysis of nuclei suspension that must be performed by flow cytometry is a biparametric one: nuclei and debris are identified recording simultaneously EtBr-specific fluorescence intensity and particle size (FALS), as it is shown in Figure 3:

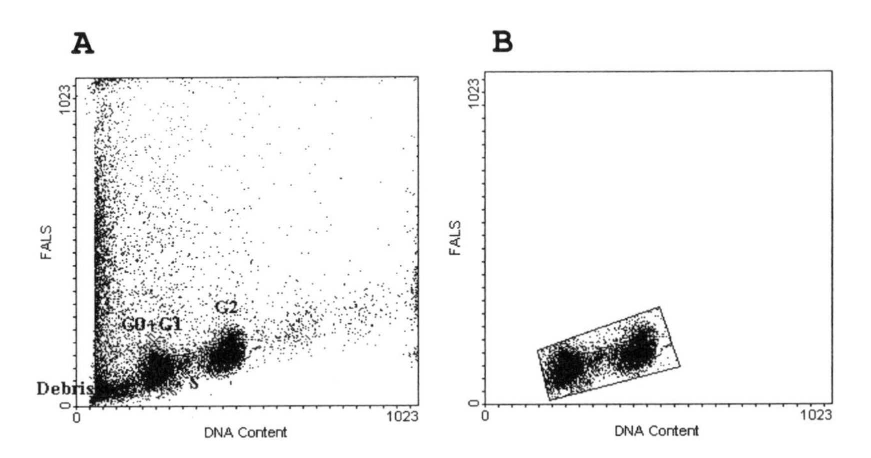

Figure 3. Biparametric cytogram were Forward Angle Light Scatter (FALS, corresponding to particle size) is represented versus EtBr-fluorescence intensity (corresponding to DNA content). **A.** Total events, including the three interesting nuclei populations (G0+G1, S and G2), and population debris. **B.** The three populations referred above are selected using gates, and only these selected nuclei are taken into account for monoparametric histogram display and analysis.

In this cytogram, several principal nuclei populations are clearly identified: G0+G1 corresponds to 2C nuclei. G2 corresponds to 4C nuclei, and their size and fluorescence intensity are approximately the double than in the case of G0+G1 population. A little S population is placed between. Another population has small size and weak fluorescence intensity, and it corresponds to cellular and nuclear debris (broken nuclei, cell and membrane fragments, etc., weakly labelled with EtBr). This debris can be gated and eliminated, using discriminating windows of the cytometer software (Fig. 3B). The histogram showed in Fig. 2 results from gating and projecting the EtBr-fluorescence intensity parameter from Fig. 3, where debris has yet been virtually removed.

When mitosis takes place, nuclear envelope disappears, and the dispersed chromosomes (of different sizes and weak fluorescence intensities) can not be detected or distinguished from debris in this experimental approach. Thereby, mitotic cells are lost and not detected by flow cytometry in these conditions, and this population (M) is not recorded in the histograms. In fact, for a correct evaluation of G2, M and G1 lengths, mitotic indices (percent of mitosis) must be evaluated complementarily to flow cytometry using other techniques, as it is exposed in chapter about mitotic index.

This type of monoparametric analysis by flow cytometry allows simple and useful cell cycle analysis, as it will be shown below.

Usually, it is necessary to apply specific software to the flow cytometry-obtained histograms to perform a suitable cell cycle analysis from the raw data of the initial histograms. Several programs are commercially available, and each one uses different algorithms (Gray et al., 1990). These programs allow a suitable estimation of peak shape, CV of each nuclei population, %G1, %S, %G2, background subtraction, and chi-square (χ^2) estimation of fitting between raw data and estimated data. In our laboratory, the computer program Multicycle (Phoenix Flow Systems, San Diego) is used. Figure 4 shows Multicycle estimation of the monoparametric histogram shown above (Figure 2):

Asynchronous cell populations from different tissues, meristems, and cell suspensions can be analysed, so the percentage of cells in each cell cycle phase can be estimated. But if we are interested in obtaining metaphase chromosomes (for ulterior sorting and characterisation), or in testing the effect of putative cell cycle modulators, a previous synchronisation is required.

There are some commercially available inhibitors that block or stop the cell cycle in a specific phase, as it is the case of aphidicolin and hydroxyurea. Aphidicolin causes a specific and reversible inhibition of the DNA polymerase α, leading to a removable cell cycle block at the G1/S boundary (Cuq et al., 1995 and references therein). Hydroxyurea (HU) reversibly inhibits the enzyme ribonucleotide reductase, and therefore the production of deoxyribonucleotides. Treatment with this inhibitor induces the accumulation of cells in G1 and early S phase (Doležel et al., 1999 and references therein). Starvation and physical methods have also been used for inducing a partial cell cycle synchronisation, principally in cell suspensions, but chemicals are a more specific tool.

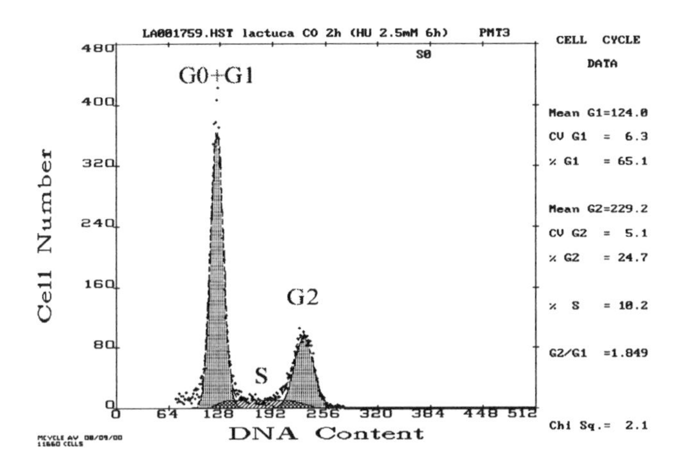

Figure 4. Histogram resulting from application of computer program Multicycle (Flow System, San Diego) on raw data histogram shown in Fig. 2.

Once the commercial inhibitor is added, cycling cells continue the cell cycle progression up to the cycle phase point where that inhibitor has a specific effect, and all the cells will arrest the cell cycle at that phase after an incubation time. After some time, inhibitor is removed from the medium by washing and whole cycling cell population re-start and goes on the cycle simultaneously, and this synchronous cell population progression can be acutely analysed. In the same way, the specific effect of a putative cell cycle modulator under study can be finely analysed. By adding the tested substance at different times after inhibitor removing, cell cycle phase-and subphase-specific effects can be detected. For example, monocerin (benzopyran toxin produced by the fungus *Exserophilum turcicum*) induces a delay in the cell cycle progression of synchronised root meristems of maize, specifically in S and G2 phases, as it was revealed in a study where synchronisation was performed with aphidicolin (Cuq et al., 1995). Lee et al. (1996) used hydroxyurea for root tip synchronisation and subsequent metaphase chromosome isolation from maize.

In our laboratory and for the first time, *Lactuca sativa* root meristems were synchronised using the inhibitor hydroxyurea (HU), blocking the cell cycle in G1 phase. After inhibitor removal, a synchronous cell population of about 25% of total recorded nuclei was detected in progression through S and G2 phases (%S is about 10-12% in asynchronous lettuce meristems; Coba de la Peña et al., 2001). In synchronised plant cell cultures, S nuclei can represent more than 50% of total population. At present, we are testing the putative effects of some allelochemicals on this synchronised cell cycle. Figure 5 shows some steps of synchronous cell cycle progression at different times after HU removal, showing both raw flow cytometry histograms (left) and the corresponding Multicycle-treated data (right):

Immediately after HU inhibitor removal, 79% of detected nuclei are at G0+G1 phase, 8.4% in G2 phase, and 12.4% in S phase (Fig. 5A). HU has induced blockage and accumulation of nuclei in G1 phase. Start and advance of synchronised nuclei in S phase (29.6% of detected nuclei) is observed 30 min after (Fig. 5B). One hour after HU release, synchronised nuclei population begins to incorporate into G2 phase (Fig. 5C). Percentage of synchronised nuclei diminish along time after inhibitor release. Finally, all synchronised nuclei are incorporated in G2 phase of the cell cycle two hours after HU release (Fig. 5D). G2 population, normally about 10% in non-synchronised meristems, amounts to 24.7% in this situation. After this step, synchronised nuclei entries into mitosis and the samples have abundant metaphasic chromosomes. In this particular experimental system, a new synchronised G1 phase is not observed.

BI-OR MULTIPARAMETRIC ANALYSIS OF THE CELL CYCLE

Finally, we can mention that cell cycle analysis by flow cytometry can be performed measuring simultaneously other parameters, like RNA, protein content and a wide range of antigens, using several fluorescent probes and fluorescent-labeled monoclonal antibodies. These measures allow a fine characterisation and discrimination between cycling and non-cycling cells.

Cell cycle can be analysed after 5-bromodeoxyuridine (BrdUrd) incorporation. This is a thymidine analogue that is incorporated in the DNA of S-phase cells. Incorporation of Hoechst 33258 (HO), an AT-binding dye, will be reduced upon the degree of BrdUrd incorporation, owing to Hoechst do not bind DNA if BrdUrd is present instead of thymidine. Propidium Iodide (PI), an intercalating dye that is not affected by BrdUrd incorporation, is also added in this system. Simultaneous analysis of both fluorescent intensities will provide information on relative DNA content (PI) and relative fluorescence quenching (loss of HO intensity) due to DNA synthesis in presence of BrdUrd. Biparametric histograms are analysed. By this way, it is possible to distinguish quiescent from proliferating cells, and to estimate the number of cycles they have progressed. Alternatively, anti-BrdUrd monoclonal antibodies can be used (Jayat and Ratinaud, 1993; Coba de la Peña and Brown, 2001). Biparametric analysis are also shown in Lucretti et al. (1999).

RNA levels can be detected and analysed simultaneously to DNA (Bergounioux et al., 1988) using Acridine Orange. This is a metachromatic dye that stains differentially doubled stranded from single stranded nucleic acids. Acridine Orange fluoresces green in the first case, and red in the second, when excited in blue light (Grunwald, 1993). The resulting DNA-RNA biparametric histograms allow identifying $G1_Q$, S_Q, $G2_Q$, $G1_T$, S_T and $G2_T$ populations.

Total cellular proteins can be estimated by flow cytometry using Sulphorhodamine 101 (SR 101) or Fluorescent Isothiocyanate (FITC). Moreover, a wide variety of fluorescent-labeled monoclonal antibodies against cellular antigens are available, and they can be used simultaneously with DNA-specific fluorescent dyes (Petit et al., 1993). This can be performed in fixed (or even living) cells or protoplasts. DNA-binding vital fluorescent dyes have been recently developed (Haugland, 1996).

Thereby, DNA, RNA, total protein, and even other parameters, can be estimated simultaneously in a flow cytometer, using simultaneously several fluorescent probes, several detectors, up to three lasers for excitation, and multiparametric histograms.

For instead, Onelli et al. (1997) have performed immunocharacterisation of PCNA (Proliferating Cell Nuclear Antigen) in synchronised root meristems of *Pisum sativum* by flow cytometry. An example of combination of this technique with molecular biology is shown in Segers et al. (1996), where it was observed that a cycling-dependent kinase gene is preferentially expressed during S and G phases in meristematic cells of *Arabidopsis thaliana.*

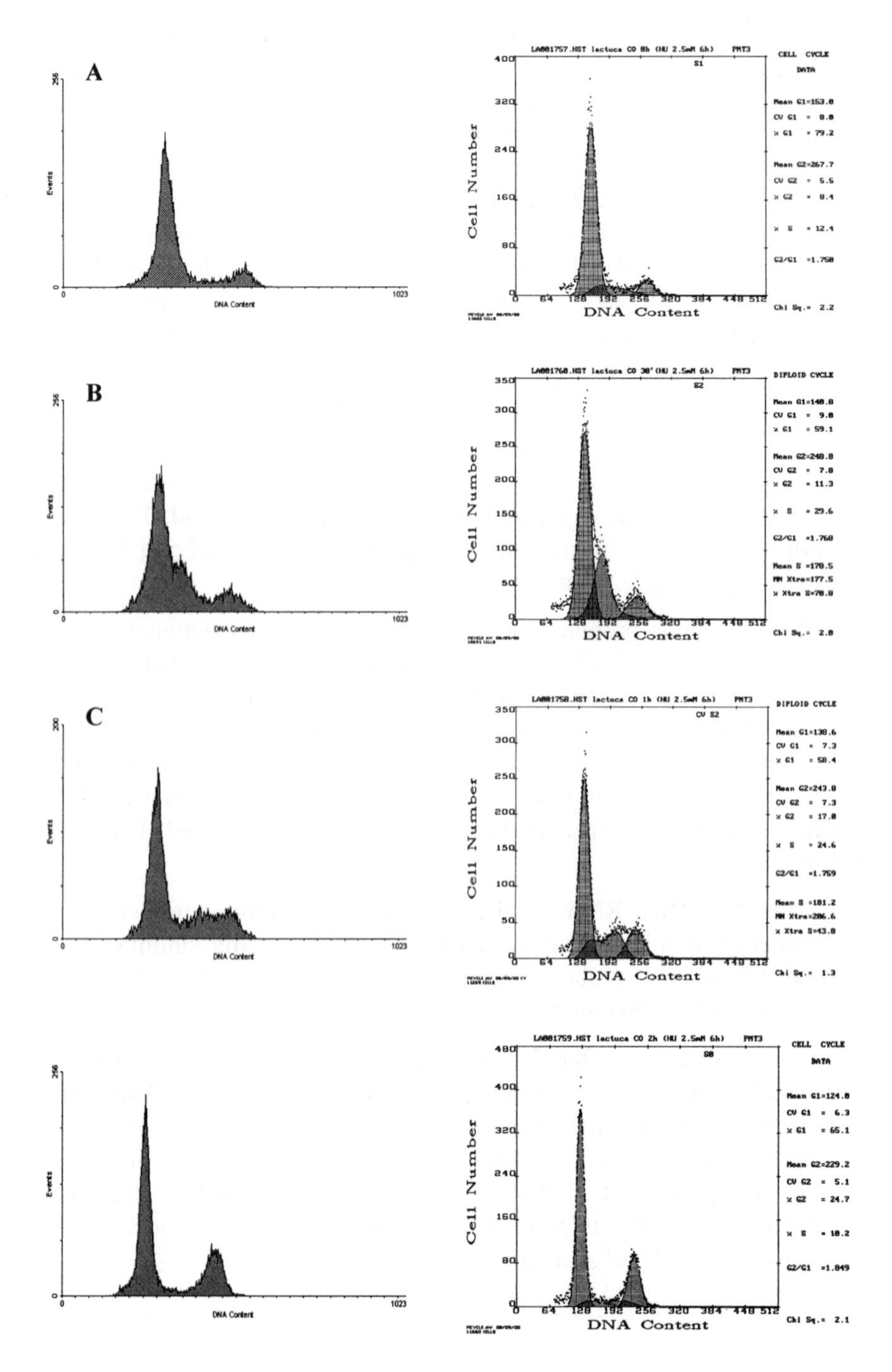

Figure 5. Comparative cell cycle analysis after release of hydroxyurea-synchronised root cell meristems.

Different examples, more information and details about these experimental approaches are exposed in Bergounioux and Brown (1990) and Bergounioux et al. (1992).

PROTOCOL: STUDY OF EFFECT OF BOA ON CELL CYCLE OF SYNCHRONIZED LETTUCE ROOT MERISTEMS BY FLOW CYTOMETRY

A strong effect of the allelochemical BOA (2-benzoxazolinone) on lettuce root growth and membrane permeability has been observed in our laboratory (Sánchez Moreiras, 1996; Sánchez Moreiras et al., 2001). A putative direct or indirect effect of BOA on root meristem cell cycle was investigated.

Coba de la Peña et al. (2001) have developed a methodology for detecting this putative effect on *Lactuca sativa* meristems. Some preliminary experiments suggest a cell cycle delay and inhibition induced by BOA.

Briefly, young lettuce plants are incubated with the cell inhibitor hydroxyurea (HU) for 6 h. After washing with distilled water, plants are immediately incubated with BOA or water (controls). Samples (nuclear suspensions) are prepared from root meristems and analysed by flow cytometry every two hours during 12-14h after HU removal, comparing the synchronised-cell cycle progression of BOA-treated plants with that of corresponding controls. By this way, partial or total inhibition of cell cycle can be detected. In this case, a delay in BOA-treated plants is observed in relation to controls after 6-8h of incubation time, and a clear stop in cell cycle progression after this time is detected.

Equipment and reagents

- Flow cytometer with VIS (visible) excitation source
- Seeds of *Lactuca sativa* cv. Great Lakes, California (Fitó, S.A.)
- Hydroxyurea (Sigma H 8627, 2.5 mM in water, pH 6.0)
- Galbraith nuclear buffer: 45 mM $MgCl_2$, 30 mM sodium citrate, 20 mM MOPS pH 7.0, 0.1% (w/v) Triton X-100, supplemented with 100% beta-mercaptoethanol and Tween 20

- Ehidium Bromide (Sigma product E 8751, stock 10 mg/ml in water)
- RNase A (Boheringer Mannheim 85340024-78, stock 1% solution in Tris-HCl, NaCl and glicerol, pH 7.6)
- BOA (1 mM in water, pH 6.0)
- Heat chamber with a fixed temperature of 26°C
- Petri dishes
- Razor blades
- 30 μm diameter nylon filters
- Micropipettes
- Plastic trays

Method

1. *Lactuca sativa* seeds are placed on moistened filter paper, into plastic trays covered with cooking foil. Seeds are germinated at 27°C and dark for 20h.
2. 1-3 mm-root length plants are transferred to Petri dishes containing filter paper that has been moistened with 5 ml of 2.5 mM hydroxyurea, pH 6.0. Twenty plants are placed in each Petri dish and incubated for 6h at 27°C in the dark.
3. HU is removed by washing twice with distilled water pH 6.0. Immediately after, plants are transferred to other Petri dishes with filter papers that have been moistened with 4 ml of either 1 mM BOA pH 6.0 (treated plants) or distilled water pH 6.0 (control plants). These plants are incubated at 27°C and dark.
4. From this moment, and every 2h, samples of BOA-treated plants and corresponding controls must be processed simultaneously for flow cytometry analysis. The 1 mm-apical tips of root meristems from forty BOA-treated plants (that is, the content of two Petri dishes) are chopping with a razor blade on another Petri dish containing 700 μl of Galbraith buffer, supplemented with 100% Tween 20 (2 drops in 15 ml buffer) and 100% beta-mercaptoethanol (7 μl in 1 ml buffer). The obtained suspension is filtered twice through 30 μm-nylon filters, and 500 μl of filtered nuclei suspension are obtained into Eppendorf tubes. Control plants must be submitted simultaneously to the same process.

The product of forty meristems constitutes one sample for flow cytometry.

5. 5 μl of 1% RNase solution are added to the nuclei suspension and, immediately after, 30 μl of 10 mg/ml Ethidium Bromide (EtBr) are added. Incubation with EtBr is for 30 min at room temperature and dark.

6. Set flow cytometer with the laser turned on 488 nm excitation wavelength. Five types of histograms (or cytograms) must be displayed in the cytometer screen:

   - FALS versus DNA-specific fluorescence (biparametric, see Fig. 3): it allows to gate debris and to eliminate it from analysis
   - Peak signal versus integral signal of DNA-specific fluorescence (biparametric): it allows discarding between single nuclei and doublets
   - Red signal (chlorophyll) versus yellow signal (EtBr-labelled DNA): it allows discarding stained nuclei from pigments and debris with red fluorescence
   - DNA fluorescence in log scale (monoparametric): it allows visualising all peak populations
   - DNA fluorescence in linear scale (monoparametric, see Fig. 2): these are the data for cell cycle analysis

7. Cell cycle histograms are recorded for BOA-treated and control plants every 2h, up to arrive to 12 or 14h of incubation with BOA. At least 10,000 nuclei from each sample must be analysed in the flow cytometer.

8. Data processing begins: clean histograms on a linear scale are obtained by previous gating on the other histograms.

9. Histogram profiles are analysed using the computer program Multicycle (Flow Systems, San Diego), and G0+G1, S and G2 populations are estimated comparatively in control and BOA-treated plants.

Figure 6 shows a schematic representation of protocol of sample preparation and cell cycle analysis by flow cytometry.

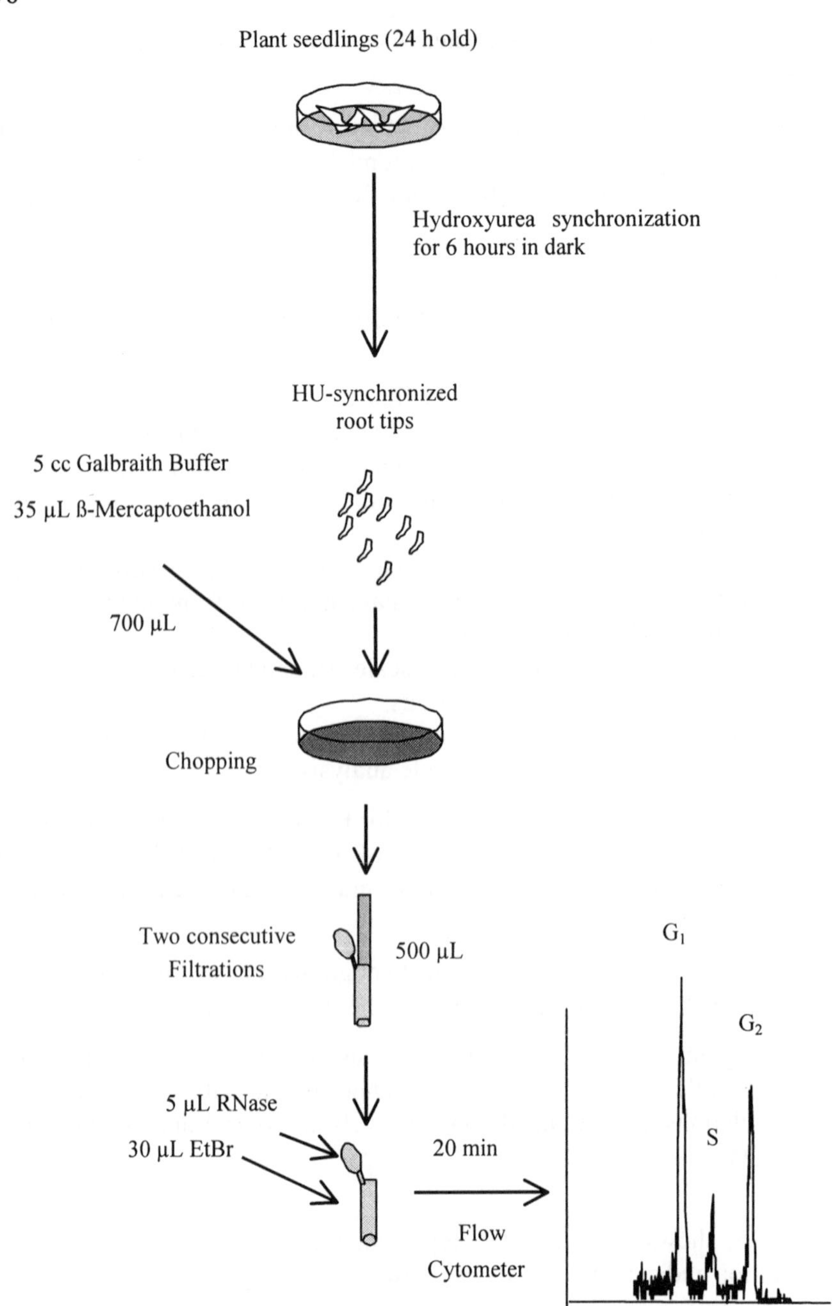

Figure 6. Schematic representation of sample preparation and cell cycle analysis procedure by flow cytometry

REFERENCES

Bergounioux C., Brown S.C. Plant cell cycle analysis with isolated nuclei. Meth Cell Biol 1990; 33:563-573

Bergounioux C., Brown S.C., Petit P. Flow cytometry and plant protoplast cell biology. Physiol Plantarum 1992; 85:374-386

Bergounioux C., Perennes C., Brown S.C., Sarda C., Gadal P. Nuclear RNA quantification in protoplast cell cycle phases. Cytometry 1988; 9:84-87

Brown S.C., Devaux P., Marie D., Bergounioux C., Petit P.X. Analyse de la ploïdie par cytométrie en flux. Biofutur 1991; 105:2-16

Coba de la Peña T., Brown S. "Flow Cytometry." In *Plant Cell Biology: A Practical Approach.* C. Hawes, B. Satiat-Jeunemaitre, eds. Oxford: University Press, 2001 (In Press)

Coba de la Peña T., Sánchez-Moreiras A., Reigosa M. Synchronisation of lettuce root meristems with hydroxyurea: a tool for characterizing cell cycle modifications in *Lactuca sativa* (2001, in phase of redaction)

Cuq F., Brown S.C., Petitprez M., Alibert G. Effects of monocerin on cell cycle progression in maize root meristems synchronised with aphidicolin. Plant Cell Rep 1995; 15:138-142

Doležel J., Cíhalíková J., Weiserová J., Lucretti S. Cell cycle synchronisation in plant root meristems. Method Cell Sci 1999; 21:95-107

Galbraith D.W. Analysis of higher plants by flow cytometry and cell sorting. Int Rev Cytol 1989; 116:165-228

Gray J.W., Dolbeare F., Pallavicini M.G. "Quantitative Cell-Cycle Analysis." In *Flow Cytometry and Sorting.* M.R. Melamed, T. Lindmo, M.L. Mendelshon, eds. New York, USA: John Wiley and Sons, 1990

Grunwald D. Flow cytometry and RNA studies. Biol Cell 1993; 78:27-30

Haugland R.P. *Handbook of Fluorescent Probes and Research Chemicals.* Eugene, Oregon: Molecular Probes Inc., 1996

Jayat C., Ratinaud M.H. Cell cycle analysis by flow cytometry: principles and applications. Biol Cell 1993; 78:15-25

Lee J.H., Arumuganathan K., Kaepler S.M., Kaepler H.F., Papa C.M. Cell synchronisation and isolation of metaphase chromosomes from maize (*Zea mays* L.) root tips for flow cytometry analysis and sorting. Genome 1996; 39:697-703

Lucretti S., Nardi L., Nisini P.T., Moretti F., Gualberti G., Dolezel J. Bivariate flow cytometry DNA/BrdUrd analysis of plant cell cycle. Method Plant Sci 1999; 21:155-166

Marie D., Brown S.C. A cytometric exercise in plant DNA histograms, with 2C values for 70 species. Biol Cell 1993; 78:41-51

Onelli E., Citterio S., O'Connor J.E., Levi M., Sgorbati S. Flow cytometry and immunocharacterization with proliferating cell nuclear antigen of cycling and non-cycling cells in synchronised pea root tips. Planta 1997; 202:188-195

Petit J.M., Denis-Gay M., Ratinaud M.H. Assessment of fluorochromes for cellular structure and function studies by flow cytometry. Biol Cell 1993; 78:1-13

Sánchez-Moreiras A.M. *Efectos Fisiológicos Producidos por la Acción de Aleloquímicos.* Minor Thesis. Vigo, Spain: University of Vigo, 1996

Sánchez Moreiras A.M., Weiss O., Reigosa M.J., Pellissier F. Membrane permeability of plant roots can be affected by some allelochemicals (2001, submitted for publication)

Segers G., Gadisseur I., Bergounioux C., De A.E.J., Jacqmard A., Montagu M.W., Inzé D. The Arabidopsis cyclin-dependent kinase gene cdc2bAt is preferentially expressed during S and G sub (2) phases of the cell cycle. Plant J 1996; 10:601-612

CHAPTER 6

MITOTIC INDEX

Adela Sánchez-Moreiras[1], Teodoro Coba de la Peña[2], Ana Martínez Otero[1] and Alfonso Blanco Fernández[1]

[1]*Dpto Bioloxía Vexetal e Ciencia do Solo. Universidade de Vigo. Spain*
[2]*Dpto Fisiología y Bioquímica Vegetal. Centro de Ciencias Medioambientales. Consejo Superior de Investigaciones Científicas. Madrid. Spain*

INTRODUCTION

The mitotic cell division cycle

Cell division appears in the organisms to be an answer to the necessity of plant growth. This uniform division of all the cell components allows an equilibrated growth of the organism.

As it is exposed in the chapter about cell cycle studies by flow cytometry, proliferating cell cycle is constituted by different events principally divided in four phases: G_1, S, G_2 and M. Each of these phases occurs along a species-dependent time, but normally G_1 is the longest and M is the shortest phase of the cell cycle (see Fig. 1). By other hand, G_2 and S are almost constant, but G_1 can vary according to the cell types (Smith and Martin, 1973).

G_1 is the gap between the cell division and DNA synthesis, where each chromosome is a single chromatide with one DNA molecule; S is the phase of DNA synthesis (a doubling of the DNA content); G_2 is the gap between the DNA synthesis and the cell division, where each chromosome consists of two identical chromatides; and M is the mitosis, that is, the cell division phase (Sharp, 1934; Howard and Pelc, 1951).

M.J. Reigosa Roger, Handbook of Plant Ecophysiology Techniques, 81–95.

Figure1. The mitotic Cell Division Cycle (Quiescent States are not shown)

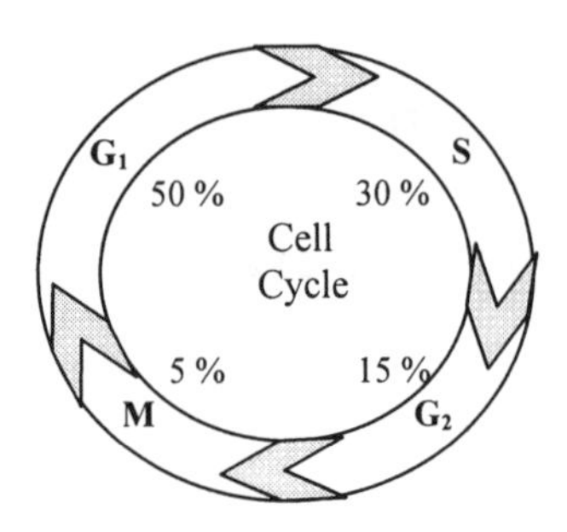

Mitosis occurs after the stage known as interphase (G_1, S and G_2) which prepares the cell to the phenomenon of the mitosis with the synthesis of DNA and of the other complements needed for cell division.

The event of the cell division (M) can be subdivided in prophase, metaphase, anaphase and telophase.

- During prophase, the first stage of the mitosis, the rupture of the nuclear membrane starts, and the chromosomes appear divided in two identical, but still united, chromatides.
- In metaphase a movement of the chromosomes occurs to the centre of the cell along the spindle (a proteinaceous structure formed by microtubules expanding along the cell) and their orientation in a single plane (Bajer and Molé-Bajer, 1972). Chromatides are normally united in one single point, and the rupture of the nuclear membrane has finished (see Fig. 2).
- Anaphase is the shortest period of the mitosis. In this phase the pairs of chromatides are separated and each member of them migrates to one different pole of the spindle (see Fig. 2). So, we have in each pole of the cell the same number and composition of chromosomes.
- With the formation of a new nuclear membrane around each set of chromosomes and the creation of a new cell wall, mitosis has finished. This last phase is known as telophase (see Fig. 2).

After these four stages, the cell starts again with the interphase. Afterwards a new mitotic cell division cycle will start.

But not every cell in an organism is involved in active cell cycle. Most cells of plants are in a quiescent state (G_0 and differentiated cells) and do not divide (Darzynkiewicz, 1983; Zetterberg and Larsson, 1985), as it was exposed above (chapter: Flow Cytometry). The rate of quiescent cells depends on the species (around 80 % in lettuce despite the fact that in other species the number of cells that participate in mitosis can reach up to 50 %). In fact, plants have specific regions or tissues where the mitotic events occur. These areas are termed meristems. We can find leave, shoot, and root meristems along the plant, which are the growth centres of the organism. The size of these areas depends on the species. In any case, it is only in these zones where we will find cells in division. Therefore, when we obtain a plant sample, we must be sure that we have the total meristematic area.

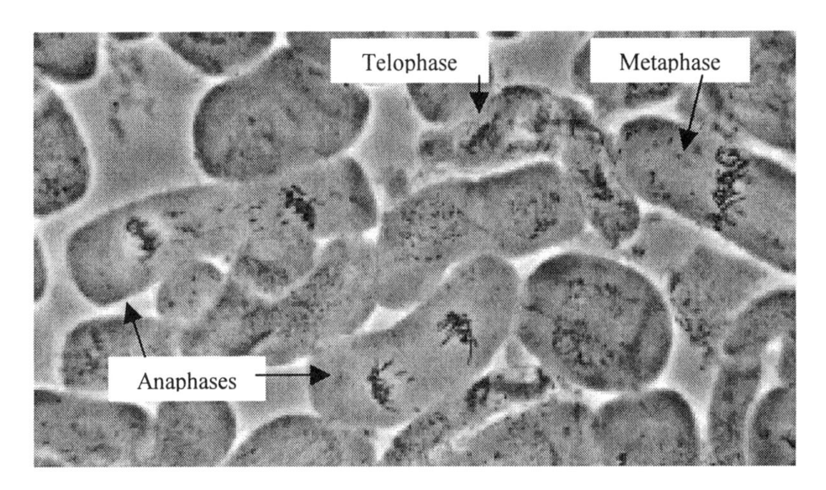

Figure 2. Mitotic cells in root meristems

Cycling cells divide asynchronously in meristems. For that reason, specific cell cycle inhibitors are used for induction of synchronization of the cell cycle as the best form to obtain a sufficient number of cells in mitosis in the studies of cell cycle. Normally, these compounds act blocking and accumulating the cells in a determinate stage of the cell cycle, blocking the formation of the spindle, avoiding a good reorganisation of the chromosomes in the poles of the spindle or inhibiting the synthesis of DNA. When washing eliminates these compounds, the cell cycle continues with an important rate of cells in the same phase, progressing synchronously. By this way, it is possible to detect strong or even weak effects on the cell cycle (Clain and Brulfert, 1980; Nkongolo and Klimaszewska, 1994).

The importance of mitotic activity determination

The study of the mitotic cell division cycle has been used in different works to detect the effects of biotic or abiotic stresses on plant physiology (Evans and Savage, 1959; Murín, 1966; Shehab, 1979; Hess, 1989; Teutonico et al., 1991; Wisniewska and Chelkowski, 1994). For example, the effect of low temperature on barley or in Norway spruce (Harrison et al., 1998; Westin et al., 1999), the reduced mitotic activity in pea seedlings by water stress (Chiatante et al., 1997), or the effects of some contaminants (SO_2 and H_2S) on spruce (Wonisch et al., 1999) could be detected by assessment of mitotic indexes in these plants.

At present, it is well known that allelochemical compounds can interfere with many vital processes of organisms. Stomatal function, photosynthetic activity, respiration, mineral uptake or cell division have been broadly studied to elucidate the mechanism of action of allelochemicals. In this way, information about the mode of action to alter growth is central in allelopathy (Einhellig, 1994).

Some works have been performed using the light and electron microscopy to study the effects of allelopathic substances on plant growth (Aliotta et al., 1993), but in few works direct measurement of mitotic index in the cells exposed to allelochemicals has been done (Shehab, 1979; Alam et al., 1987).

Some works were also performed measuring mitotic activity in plants exposed to secondary metabolites. In 1973, Bhalla et al. could observe mitotic effects of different concentrations of condensed tobacco smoke on the chromosomal number and structure of onion root tips. Some years later, Shehab (1979) saw also in onion an effect of *Pulicaria crispa* extracts on the mitotic index and the percentage of chromosomal abnormalities.

Mitotic effects of other plant extracts were also reported in the literature (Shehab, 1980; Kabarity and Malallah, 1980; Alam et al., 1987). That is the case of Alam et al. (1987), which tested the toxic effects of leaf extracts of *Ipomoea carnea* in the root tip cells of *Allium cepa*. This investigation demonstrated that mitotic index decreased when the concentration of plant extracts increased. Chromosomal anomalies of onion root tip cells were also reported here.

In this way, Wisniewska and Chelkowski (1994), and Packa (1997) studied the potential genotoxicity of *Fusarium* mycotoxins on cells of wheat. In their works, a strong effect from these toxins on mitosis was shown. These secondary metabolites decreased mitotic index, produced excessive condensation of prophasic and metaphasic chromosomes, caused an accumulation of metaphases, and rose significantly the percentage of cells with chromosomal aberrations in the treated seeds and seedlings with respect to the control. One year later, in 1998, Packa could also observe similar effects of these mycotoxins on nuclei and chromosomes from root tip cells of *Vicia faba* and *Pisum sativum*.

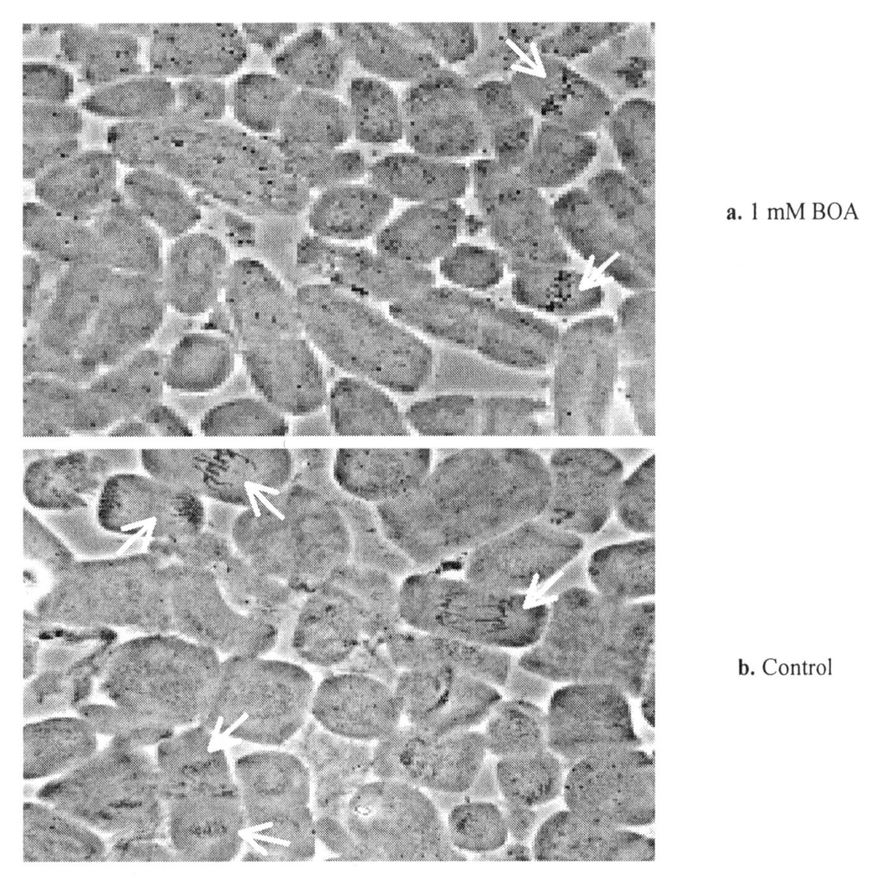

a. 1 mM BOA

b. Control

Figure 3. **a.** Mitotic activity in root tip cells from plants exposed to 1 mM BOA for 4 h after a synchronization of 6 h in HU
b. Mitotic activity in root tip cells from plants washed with destilled water for 4 h after a synchronization of 6 h in HU
* White arrows show cells in mitosis

Investigations carried out in our laboratory (unpublished data) demonstrated the effect of an allelochemical compound, in particular the hydroxamic acid 2-benzoxazolinone (BOA), on the mitotic activity of meristematic root tip cells of lettuce seedlings (see Fig. 3). Lettuce root meristems were previously synchronized with hydroxyurea (HU) for 6 h; this allowed obtaining a significant number of cells starting cell division at the same time (Clain and Brulfert, 1980; Nkongolo and Klimaszewska, 1994). In this work, we detected and counted cells in mitosis in plants exposed to BOA at different times and we compared that with mitotic activity in control plants. The results showed that mitotic activity was affected by this allelochemical, slowing down the cell cycle and reducing mitotic index in the treated plants.

Complementary studies carried out by flow cytometry in our laboratory demonstrated that BOA treatment finally inhibits completely cell cycle progression, according to mitotic index data.

THE MITOTIC INDEX MEASUREMENT

Different Uses for a Same Measurement

Mitotic index describes the sum of cells in prophase, metaphase, anaphase and telophase, expressed as a percentage of the total number of cells observed in a sample.

$$\text{Mitotic Index} = \frac{(\text{Prophase} + \text{Metaphase} + \text{Anaphase} + \text{Telophase})}{\text{Total cells}} \times 100$$

This technique has been used for several studies and for different research teams.

Cuq et al. (1995) quantified mitotic index in maize root meristems to confirm cytometric data. G_1, S and G_2 phases could be observed with a flow Cytometer. With mitotic index, the mitosis event can be studied, allowing a fine estimation of G_2 and M lengths. Therefore, applying these two methods, it is possible to know the accurate stage of every cell cycle phase.

Quastler and Sherman (1959), Wimber and Quastler (1963), and Lavania (1996) used stained chromosomes to estimate the duration of the total mitotic cell cycle in synchronized cells.

Nias and Fox (1971), Clain and Brulfert (1980), Olszewska et al. (1990), and Hawkins and Woods (1995) used the technique to know the synchronization degree obtained with different chemical substances or with physical methods.

A several number of researchers used this technique for different chromosomal studies, as the study of karyotype (account and morphology), ploidy level, mutations, deletions and translocations, or the study of different aberrations produced by toxins (Packa, 1997; 1998).

On the other hand, mitotic index can be also an approached technique to know the role of a substance in the cell cycle parameters. In this way, Hermmerlin and Bach (1998) used mitotic index to study the importance of mevalonic acid for cell cycle progression in tobacco cells.

Fernandez et al. (1999) used it also to know the origin of somatic embryos from an *in vitro* cellular culture of *Triticum durum* Desf.

Finally, the most interesting use for the present work is the knowledge of how many factors can be affecting the cell cycle activity. In this way, many factors were studied:

- **Developmental stage** (O'Really and Owens, 1987; Fielder and Owens, 1989).
- **Stress conditions**, as frost hardiness studied by Colombo et al. (1988), water stress by Chiatante et al. (1997), low temperatures by Harrison et al. (1998), or microgravity by F-Yu et al. (1999).
- **Some chemical substances** were studied too, as contaminants (Wonisch et al., 1999), insecticides (Chauhan, 1999) or secondary metabolites (Packa, 1997; 1998).

All these works show the importance of this technique in the knowledge of cellular division parameters.

How to obtain the measurement?

There are different procedures to assess the mitotic index, but all of them are modifications of two main methods:

Radioactive method based on cell radioactive labelling. The specific DNA precursor is labelled with a radioactive isotope. Then, the preparation is observed by autoradiography.

The most used is ^{3}H-TdR, thymidine labelled with tritium (Quastler and Sherman, 1959 *in* Nias and Fox, 1971).

But Baserga and Nemeroff in 1962 (*in* Nias and Fox, 1971) used a double labelling technique which involves two pulse labels. One with ^{3}H-TdR and the other one with ^{14}C-TdR. Three classes of labelled cells may be distinguished in the autoradiographs: those labelled only with ^{3}H, those with ^{14}C and those labelled with both of them.

Microscope method based on fixation, hydrolysis and staining of cell samples. According with the sample preparation and the staining, the chromosomes can be examined with light, fluorescent or electronic microscopy (F-Yu, 1999).

The main steps of this technique are the following:

- First, the samples are **fixed** to avoid structure tissue deterioration along time. The fixative destroys all cellular life eliminating all the water in the cell and substituting it with ethanol. A good fixative should have a fast action to block all the metabolic and cell cycle evolution, and it should coagulate and precipitate the chromatine, obtaining an increase of the differences between the chromosomes and the rest of cellular components refraction index. The basophil nature of the fixative let preserve the chromosomal structural integrity in the staining process.

  Fixation can be done using paraformaldehyde, neutral formalin, glutaraldehyde, etc., for optic and fluorescent microscopy, or osmium tetroxide for observation with electronic microscopy. The most employed fixatives are ethanol: acetic acid (3:1) and ethanol: chloroform: acetic acid (6:3:1) (Carnoy, 1884).

- After that, the samples are **hydrolysed** or **digested** to disperse cells and chromosomes. Digestion can be performed using carylase, macerozyme and lyticase (Cuq, 1995). The tissues are normally hydrolysed with HCl 1N, but at different time and temperature expositions (Nkongolo, 1994; Grob and Owens, 1994; Chiatante et al., 1997; Harrison et al., 1998; Müller et al., 1998).

- Then, cells can be **stained** with the fluorescent probe DAPI, Hoechst, Chromomycin, etc. (Cuq, 1985), with Wheigert´s haematoxylin (Bingham, 1999), with Toluidine blue (F-Yu, 1999), with aceto-orcein or with aceto-carmine (Feulgen's method; Feulgen and Rossenbeck, 1924; Feulgen and Imhauser, 1925).

- After staining, the samples are **squashed** applying pressure over the coverslip to get a single cell layer preparation.
- Finally, the stained chromosomes are scored at the **microscope**.

But there are also some variants of this technique:

Sometimes, dehydration with alcohol series (Bingham and Merritt, 1999) or with ethanol and propylene oxide (F-Yu et al., 1999) can be performed after the fixation. Then, the samples are embedded in Histooresin (Bingham and Merritt, 1999; Fernandez et al., 1999) or Araldite (F-Yu et al., 1999). After that, samples are cut in sections with a microtome. Finally, the staining is carried out without previous hydrolysis.

Another variant is to bleach the cytoplasm washing the samples with SO_2 water after the staining.

Finally, in the last variant, a dehydration with alcohol series and xylene (Chiatante et al., 1997; Harrison et al., 1998; Sundblad et al., 1998; Mazzuca et al., 1997) is performed after the squashing. Slides are mounted with immersion oil and xylene, DPX, Euparal or Canada Balsam in order to preserve the samples in the best conditions during so many years.

Another techniques for chromosome analysis, really different from the two previously mentioned, are:

- ⇒ The **Flow Cytometric Method** exposed in other chapter of this book. It allows physical sorting of fluorescent-labelled chromosomes for an ulterior study.
- ⇒ The **Binucleate Cell Method,** carried out by Clain and Brulfert (1980). This technique is based on a pre-treatment of the cells with a substance, which inhibits at telophase the cell-plate formation. So, we can get binucleate cells.

For any of these methods we can use total or partial synchronized samples. This can be made with the addition of chemical substances or using physical methods. For a detailed explanation of everyone of these methods, the review published by Nias and Fox (1971) can be consulted. In chemical process, tissues are embedded into a mitoclastic liquid:

- To block mitotic division at metaphase
- To contract the chromosomes

There are different agents for this pre-treatment: colchicine, 8-hydroxyquinoline, etc. The α-bromonaphthalene is the most used because of its simplicity, its activity and because it is very cheap. Concentrations and pre-treatment times are variables in function of the studied material.

OUR TECHNIQUE AND ITS IMPROVEMENTS

In our laboratory we have worked with seedling root meristems of *Lactuca sativa* L. var. Great Lakes, California.

First, the root meristems suffer a **synchronization** with hydroxyurea to obtain synchronized mitotic cells in a number enough to find out the differences between the control and the allelochemical-treated samples (see Fig. 4).

Then, at different times after washing of the inhibitor and cell cycle release, the roots are **fixed** for 24 h with acetic acid/chloroform/ethanol (6:3:1) and iron traces. Ethanol goes really fast into the tissue, precipitating the nucleic acids, and dehydrating the tissues without damages.

Acetic acid goes also really fast into the cells and keeps intact the chromosomic structure. And chloroform dissolves lipids, allowing the entry of the fixative into the tissues. Iron traces must be released slowly and continuously for 24 h. This metal acts as a "mordant", that is, it accelerates the staining process changing the isoelectric point of the tissue or allowing the stain binding to the chromosome.

Nevertheless, excess in the exposition time may produce a negative effect in the sample. If there are iron traces (acetic carmin) in the stain composition, the combination process of stain and mordant is denominated "quelation".

The samples are **stored** before analysis with fresh fixative and without iron traces, at -20 °C for at least 3 days (see Fig. 4).

After storage, samples are **hydrolysed** with HCl 1N for 25 min at 60 °C to allow the following dispersion of cells and chromosomes.

Samples are **stained** after hydrolysis with Schiff's reagent. With this step we will distinguish chromosomes stained wit a pink colour.

Finally, meristems are embedded in a drop of acetic and cut on a slide. Then, a drop of acetic carmin is put over the meristem and the slide is heated over a flame. After covering with the coverslip, the meristem is **squashed** applying pressure over it.

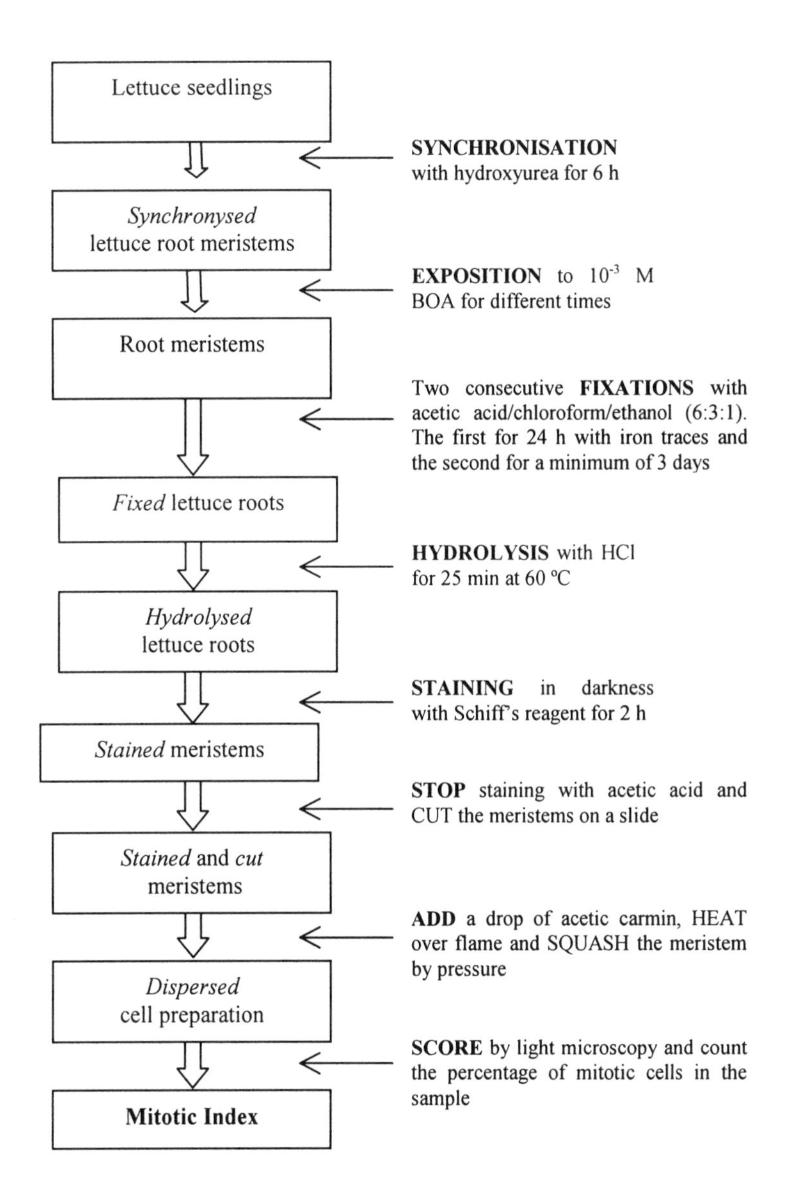

Figure 4. Representative scheme of mitotic index procedure

The squash separates cells into a monolayer sample, allowing distinguishing every cell and its chromosomes.

After a new heating, meristems are scored with light microscopy using the x40 objective and the **mitotic index** is estimated counting up a total number of 1000 cells in three slides of the same sample. The phase contrast can be used to facilitate the counting.

Acknowledgements

We are grateful to Dr. Marina Horjales (Faculty of Sciences, University of Vigo) for her help in the optimisation of mitotic index technique applied to lettuce.

REFERENCES

Alam S., Kabir G., Amin N.M., Islam M. Mitotic effect of leaf extracts of *Ipomoea carnea* on *Allium cepa*. Cytologia 1987; 52:721-724

Aliotta G., Cafiero G., Forentino A., Strumia S. Inhibition of radish germination and root growth by coumarin and phenylpropanoids. J Chem Ecol 1993; 19:175-183

Bajer A.S., Molé-Bajer J. *Spindle Dynamics and Chromosome Movements*. New York, USA: Academic Press, 1972

Baserga R., Nemeroff K. Two-emulsion radioautography. J Histochem Cytochem 1962; 10:628

Bhalla P.R., Kochhar T.S., Sabharwal P.S. Induction of mitotic abnormality in onion root tips by tobacco smoke condensate. Cytologia 1973; 38:707-712

Bingham I.J., Merritt G.J. Effects of seed ageing on early post-germination root extension in maize: a spatial and histological analysis of the growth-zone. Seed Sci Technol 1999; 27:151-162

Carnoy J.B. La biologie cellulaire; étude comparée de la cellule dans les deux regnes. I. Lierre. 1884

Chauhan L.K.S., Saxena P.N., Gupta S.K. Cytogenetic effects of cypermethrin and fenvalerate on the root meristem cells of *Allium cepa*. Environ Exp Bot 1999; 42:181-189

Chiatante D., Rocco M., Maiuro L., Scippa G.S., Di Martino C., Bryant J.A. Cell division and DNA topoisomerase I activity in root meristems of pea seedlings during water stress. Plant Biosyst 1997; 131:163-173

Clain E., Brulfert A. Hydroxyurea-induced mitotic synchronization in *Allium sativum* root meristems. Planta 1980; 150:26-31

Cobb A. "Further Targets for Herbicide Development" In *Herbicides and Plant Physiology*. A. Cobb, ed. London, England: Chapman and Hall, 1992

Colombo S.J., Glerum S.J., Webb D.P. Winter hardening in first-year black spruce (*Picea mariana*) seedlings. Physiol Plantarum 1988; 76:1-9

Cuq F., Brown S.C., Petitprez M., Alibert G. Effects of monocerin on cell cycle progression in maize root meristems synchronized with aphidicolin. Plant Cell Rep 1995; 15:138-142

Darzynkiewicz Z. "Metabolic and Kinetic Compartments of the Cell Cycle Distinguished by Multiparameter Flow Cytometry." In *Growth, Cancer and the Cell Cycle*. P. Skehan, S.J. Friedman, eds. Chifton: Humana Press, 1984

Einhellig F.A. "Allelopathy: Current Status and Future Goals." In *Allelopathy, Organisms, Processes and Applications*. Inderjit, K.M.M. Dakshini, F.A. Einhellig, eds., 1994

Evans H.J., Savage J.R.K. The effect of temperature on mitosis and on the action of colchicine in root meristem cells of *Vicia faba*. Exp Cell Res 1959; 18:51-61

Fernández S., Michaux-Ferriere N., Coumans M. The embryogenic response of immature embryon cultures of durum wheat (*Triticum durum* Desf.): histology and improvement by $AgNO_3$. Plant Growth Regul 1999; 28:147-155

Feulgen R., Rossenbeck H. Mikroscopisch-chemischer Nachweis einer Nucleinsäure vom Typus der Thymonucleinsäure. Hoppe-Seyl Z. 1924; 135:203-248

Feulgen R., Imhauser K. Über die für Nuklealreaktion und Nukleatkfärbung verantwortlich zu machenden Gruppen, etc. II. Hoppe-Seyl Z. 1925; 148:1-16

Fielder P., Owens J.N. A comparative study of shoot and root development of interior and coastal Douglas fir seedlings. Can J For Res 1989; 19:539-549

F-Yu D., Driss-Ecole D., Rembur J., Legué V., Perbal G. Effect of microgravity on the cell cycle in lentil root. Physiol Plantarum 1999; 105:171-178

Harrison J., Nicot C., Ougham H. The effect of low temperature on patterns of cell division in developing second leaves of wild-type and slender mutant barley (*Hordeum vulgare* L.). Plant Cell Environ 1998; 21:79-86

Hawkins B.J., Davradou M., Pier D., Shortt R. Frost hardiness and winter photosynthesis of *Thuja plicata* and *Pseudotsuga menziesii* seedlings grown at three rates of nitrogen and phosphorus supply. Can J For Res 1995; 25:18-28

Hemmerlin A., Bach T.J. Effects of mevinolin on cell cycle progression and viability of tobacco BY-2 cells. Plant J 1998; 14:65-74

Hess F.D. "Herbicide Interference with Cell Division in Plants." In *Target Sites of Herbicide Action*. P. Böger, G. Sandmann, eds. Boca Raton, Florida: CRC Press, 1989

Howard A., Pelc S.R. Nuclear incorporation of P32 as demonstrated by autoradiographs. Exp Cell Res 1951; 2:178-18

Lavania U.C. Duration of cell cycle, onset of S phase and induced mitotic synchronization in seeds of opium poppy, *Papaver somniferum* L. Indian J Exp Biol 1996; 34:773-775

Matthews B.F. Isolation of mitotic chromosomes from partially synchronized carrot (*D. carota*) cell suspension cultures. Plant Sci Letters 1983; 31:165-172

Mazzuca S., Bitonti M.B., Pranno S., Innocenti A.M. Nuclear metabolic changes in root meristems of *Lactuca sativa* induced by trigonelline treatment. Cytobios 1997; 89:39-50

Murín A. The effect of temperature on the mitotic cycle and its time parameters in root tips of *Vicia faba*. Naturwissenschaften 1966; 53:312-313

Nias A.H.W., Fox M. Synchronization of mammalian cells with respect to the mitotic cycle. Cell Tissue Kinet 1971; 4:375-398

Nkongolo K.K., Klimaszewska K. Karyotype analysis and optimization of mitotic index in *Picea mariana* (black spruce) preparation from seedling root tips and embryogenic cultures. Heredity 1994; 73:11-17

Olszewska M.J., Marciniak K., Kuran H. Analysis of mitotic synchrony induced by cold treatment in root meristems of *Vicia faba* L. Environ Exp Bot 1990; 30:373-382

O'Really C., Owens J.N. Long-shoot bud development, shoot growth, and foliage production in provenances of lodgepole pine. Can J For Res 1987; 17:1421-1433

Packa D. Cytogenetic effects of *Fusarium* mycotoxin on root tip cells of rye (*Secale cereale* L.), wheat (*Triticum aestivum* L.) and fields bean (*Vicia faba* L. Var. Minor). J Appl Genet 1997; 38:259-272

Packa D. Potential genotoxicity of *Fusarium* mycotoxins in *Vicia* and *Pisum* cytogenetic tests. J Appl Genet 1998; 39:171-192

Quastler H., Sherman J. Cell population kinetics in the intestinal epithelium of the mouse. Exp Cell Res 1959; 17:420

Shehab A.S. Cytological effects of medicinal plants in Qatar I. Mitotic effects of water extracts of *Pulicaria crispa* on *Allium cepa*. Cytologia 1979; 44:607-613

Shehab A.S. Cytological effects of medicinal plants in Qatar II. Mitotic effects of water extracts of *Teucrium pilosum* on *Allium cepa*. Cytologia 1980; 45:57-64

Sharp L.W. *Introduction to Cytology*. New York: Mc Graw-Hill, 1934

Smith J.A., Martin L. Do cells cycle? Proc Natl Acad Sci USA 1973; 70:1263-1267

Sundblad L.-G., Geladi P., Dunberg A., Sundberg B. The use of image analysis and automation for measuring mitotic index in apical conifer meristems. J Exp Bot 1998; 49:1749-1756

Teutonico R.A., Dudley M.W., Orr J.D., Lynn D.G., Binns A.N. Activity and accumulation of cell division-promoting phenolics in tobacco tissue cultures. Plant Physiol 1991; 97:288-297

Westin J., Sundblad L., Strand M., Hällgren J. Apical mitotic activity and growth in clones of Norway spruce in relation to cold hardiness. Can J For 1999; 29:40-46

Wimber D.E., Quastler H. A ^{14}C and ^{3}H-thymidine double labelling technique in the study of cell proliferation in tradescantia root tips. Exp Cell Res 1963; 30:8

Wisniewska H., Chelkowski J. Influence of deoxynivalenol on mitosis of root tip cells of wheat seedlings. Acta Physiol 1994; 16:159-162

Wonisch A., Tausz M., Müller M., Weidner W., De Kok L.J., Grill D. Treatment of young spruce shoots with SO_2 and H_2S: effects on fine root chromosomes in relation to changes in the thiol content and redox state. Water Air Soil Poll 1999; 116:423-428

Zetterberg A., Larsson O. Kinetic analysis of regulatory events in G_1 leading to proliferation or quiescence of Swiss 3T3 cells. Proc Natl Acad Sci USA 1985; 82:5365-5369

CHAPTER 7

DETERMINATION OF PHOTOSYNTHETIC PIGMENTS

Luís González
Dpto Bioloxía Vexetal e Ciencia do Solo. Universidade de Vigo. Spain

INTRODUCTION

Photosynthesis is the main process that can harvest energy derived from the sun. Plants and cyanobacteria capture the light of the sun and utilise its energy to synthesise organic compounds from inorganic substances, such as CO_2, nitrate and sulphate, to make their cellular material. Plant use solar energy to split water into oxygen and hydrogen. ATP and NADPH are the end products of these light reactions that take place in the photosynthetic reaction centres embedded in membranes, the thylakoids inside the chloroplast. ATP and NADPH are consumed for the synthesis of carbohydrates from CO_2 in the carbon fixation reactions (Buchanan, 2000).

In the chloroplast, two different functional units, called photosystems, harvest photon energy. The absorbed light energy is used to command the transfer of electrons through a series of compounds that act as electron donors and electron acceptors that reduce $NADP^+$ to NADPH. At the same time, photon energy is also used to produce a proton motive force across the thylacoid membrane, which is used to synthesise ATP (Heldt, 1997).

Photosynthetic pigments are responsible for absorbing and trapping light energy in the early steps of photosynthesis. They are chemical compounds that reflect only certain wavelengths of visible light but more important than their reflection of light is the ability of pigments to absorb certain wavelengths.

M.J. Reigosa Roger, Handbook of Plant Ecophysiology Techniques, 97–111.

Photosynthetic pigments are useful to plants because of they interact with light to absorb only certain wavelengths and the energy of this photons is captured for photosynthesis. Energy of the photon is proportional to its frequency:

$$E = h\nu = h\frac{c}{\lambda}$$

where:

h is the Planck constant ($6.6 \cdot 10^{-34}$ J·s)
ν is frequency
c is the velocity of the light ($3 \cdot 10^{8}$ m·s^{-1})
λ is the wavelength of light

According to this equation the energy of irradiated light is inversely proportional to the wavelength (Table 1) but plants use the range between 400 and 700 nm of the broad spectrum of electromagnetic waves (Fig. 1).

Table 1. Energy content of photons for light of different colours

Light colour	**Wavelength nm**	**Energy content Kj (mol photons)$^{-1}$**
Violet	400	298
Blue	440	271
Blue-green	500	238
Yellow	600	199
Bright red	650	183
Red	700	170

Consequently dark-red light, that is the highest wavelength that can still be utilised by plant photosynthesis, has about half the energy content of violet light at the beginning of the visible spectrum (Heldt, 1997).

Although almost all biological pigments can be reduced to no more than five or six major structural classes, there are three basic groups of pigments able to capture energy to carry out photosynthesis:

Chlorophylls: The chlorophylls are the pigments that give plants their characteristic green colour. They constitute about 4% of chloroplast dry mass. They have a similar molecular structure. The basic structure is a porphyrin ring (made of four pyrroles). Covalently bound with two atoms of N and co-ordinately bound to the other two atoms of the tetrapyrrole ring there is Mg^{2+} present in the centre of the ring. At ring d a propionic acid group forms an ester with a long chain named phytol (Figure 3). Phytol is formed from four isoprene units. This hydrophobic hydrocarbon tail renders the chlorophyll highly soluble in lipids and therefore promotes its presence in the membrane phase. In ring b, chlorophyll-*b* contains a formyl residue instead of the methyl residue in chlorophyll-*a*.

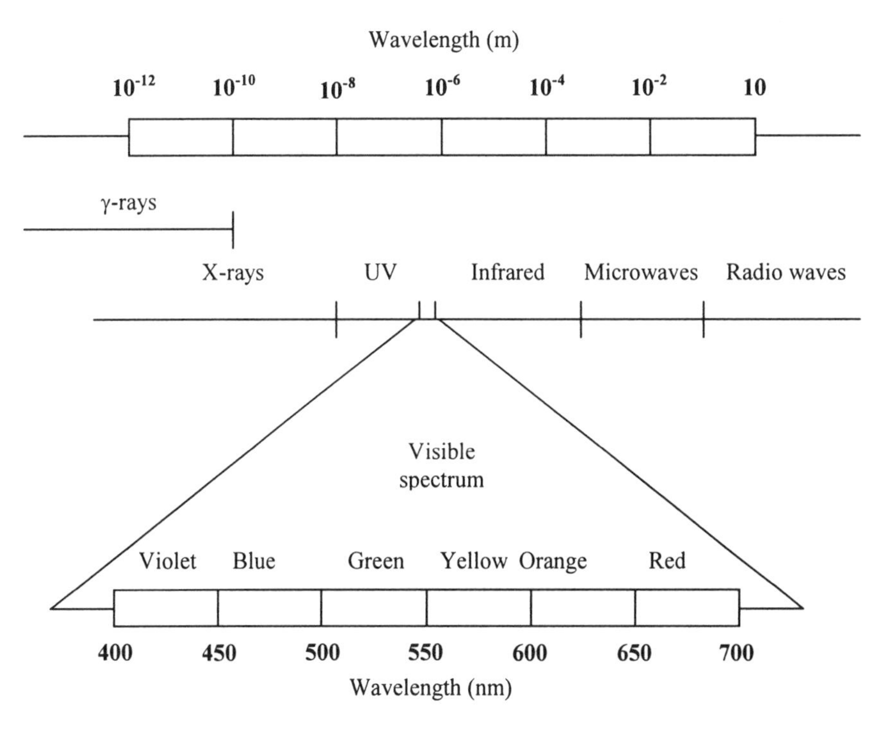

Figure 1. Spectrum of the electromagnetic radiation

The alternating double and single bonds of the porphyrin ring make chlorophyll an efficient light-absorbing molecule and determine the general shape of the absorption spectrum. The hydrophobic phytol provides the chlorophyll highly soluble in lipids and therefore support its presence in the membrane phase.

In plants (except in red algae and cyanobacteria), chlorophyll-*b* is presented about one-third of the content of chlorophyll-*a*. Only chlorophyll-*a* is a constituent of the photosynthetic reaction centres and thus we can consider it as the essential photosynthetic pigment.

The chlorophylls have a characteristic light absorption pattern. That for chlorophyll-*a* and chlorophyll-*b* is shown in Figure 2.

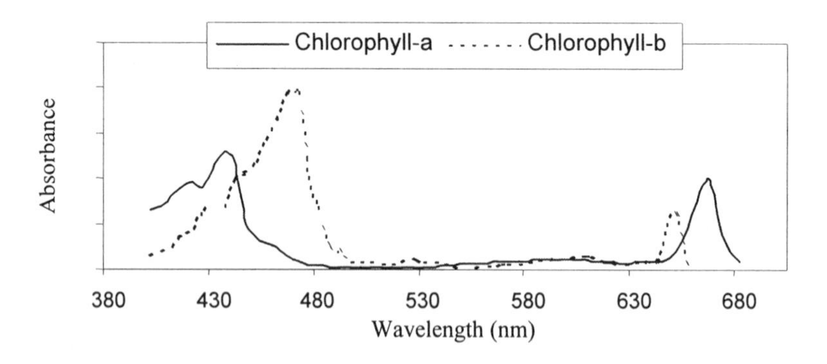

Figure 2. Absorption spectra of chlorophyll-*a* and chlorophyll-*b* dissolved in nonpolar solvent

Chlorophyll-*a* shows two major absorption bands: one in the red, with a maximum around 660 nm (in many cases, the spectra of photosynthetic pigments are substantially affected by the environment of the pigments in the photosynthetic membrane or, *in vitro,* by the solvent) and a major absorption peak in the blue to violet end of the visible spectrum (maximum around 435 nm). Another three minor peaks are shown in the yellow part of the spectrum. The main peak in the violet-blue range is characteristic of all cyclic tetrapyrroles and is known as the Soret band (Hendry, 1993).

Chlorophyll-*b* shows also two major absorption bands: one in the red-orange of the visible spectrum, with a maximum at 650 nm and the other one in the blue range.

The distribution of the various types of photosynthetic pigments is showed in Table 2. Chlorophyll-*a* is found in all photosynthetic eukaryotes and the other secondary chlorophyll (b, c, d or e) may have an evolutionary significance with a restricted presence in some groups of organisms.

Chlorophyll molecules are bound to chlorophyll-binding proteins. In this case the absorption spectrum of the bound chlorophyll may differ considerably from the absorption spectrum of the free chlorophyll and the attached proteins also confer both an orientation for light harvesting as well as for electron flow (Hendry, 1993). Thus, the function of the chlorophylls is the capture of light energy as the first step in its transformation into chemical energy (ATP) and reducing power (NADPH) utilised in the reduction of carbon dioxide to carbohydrates (Buchanan, 2000).

Table 2. Distribution of photosynthetic pigments. After Taiz and Zieger, 1998

	Chlorophylls						
Organism	*a*	*b*	*c*	*d*	*e*	**Carotenoids**	**Phycobiliproteins**
Eukaryotes							
Mosses, ferns, higher plants	+	+	-	-	-	+	-
Chlorophytes	+	+	-	-	-	+	-
Euglenophytes	+	+	-	-	-	+	-
Bacillaryophytes	+	-	+	-	-	+	-
Pyrrophytes	+	-	+	-	-	+	-
Pheophytes	+	-	+	-	-	+	-
Rodophytes	+	-	-	+	+	+	+
Prokaryotes							
Cyanobacteria	+	-	-	+	-	+	+
Prochlorophytes	+	+	-	-	-	+	-
Sulfur purple bacteria						+	-
Green bacteria						+	-
Heliobacteria						+	-

Carotenoids: They are the second most abundant group of pigments in plants, present in photosynthetic and non-photosynthetic tissue. The carotenoids are yellow or orange pigments and occasionally red in colour, found in most photosynthesising cells. Their colour in the cells is normally masked by chlorophyll that give plants their characteristic green colour.

Carotenoids are tetraterpenoids and are derived from a skeleton of 40 carbon atoms linked symmetrically with alternating unsaturated bonds (Figure 4). They are usually either hydrocarbons (carotenes) or oxygenated hydrocarbons (carotenols or xanthophylls).

All green tissues of higher plants contain broadly similar carotenoids located in the chloroplast membranes. In higher plants the most common carotenoids are β-carotene, and the xanthophylls lutein, violaxanthin and neoxanthin. Among the algae the carotenoids are more varied. Chlorophytes (green algae) contain carotenoids similar to those in higher plants but although almost all contain β-carotene,

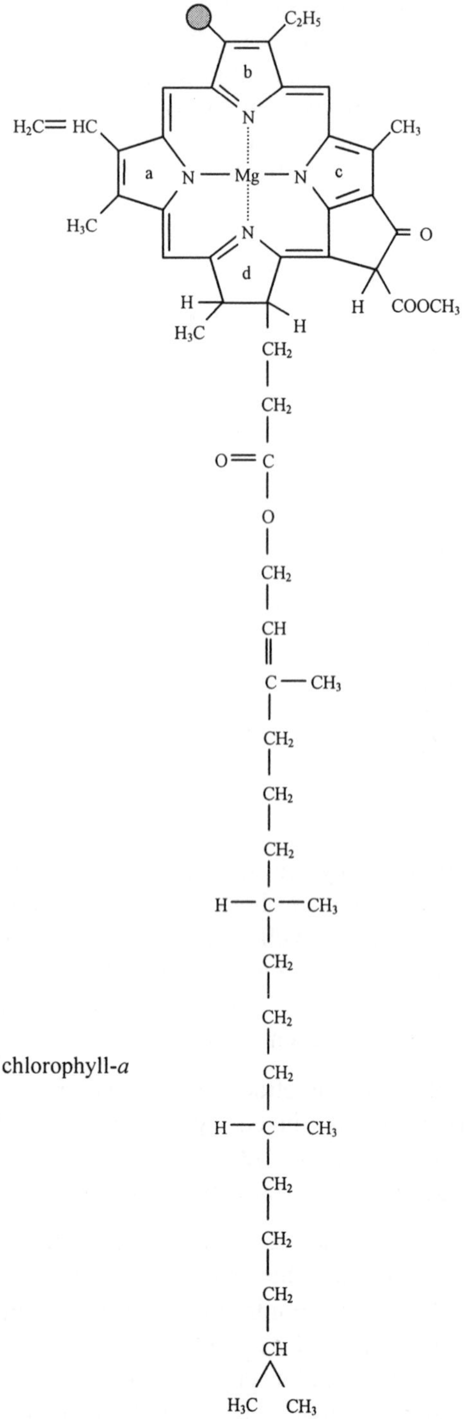

C_2H_5
b
$H_2C{=}HC$
N
CH_3
a N — Mg — N c
H_3C
N
O
d
H
H COOCH$_3$
H_3C
H
CH_2
CH_2
O=C
O
CH_2
CH
C — CH_3
CH_2
CH_2
CH_2
H — C — CH_3
CH_2
CH_2
CH_2
H — C — CH_3
CH_2
CH_2
CH_2
CH
H_3C CH_3

Chlorophyll-a: $-CH_3$
Chlorophyll-b: $-CHO$

Figure 3. Molecular structure of chlorophyll-*a* and chlorophyll-*b*

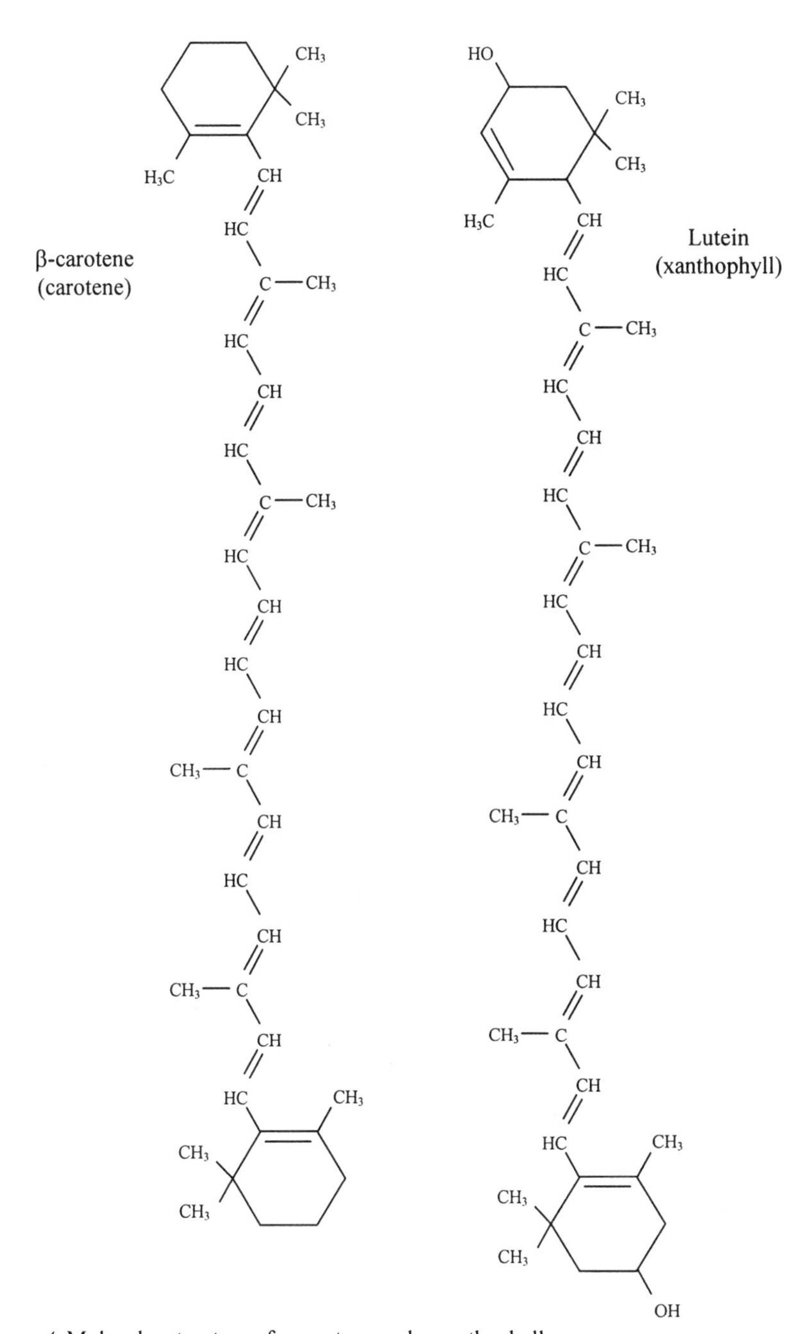

Figure 4. Molecular structure of a carotene and a xanthophyll

many algae have uncommon carotenoids or, such as the pyrophytes (dinoflagellates) and cyanophytes (blue-green algae), have carotenoids unknown in other organisms (Krinsky, 1990).

In photosynthesis the main function of these carotenoids is absorbing the photons whose wavelength corresponds to the green window between the absorption maxima of the chlorophylls (Figure 5). However, other important function of these carotenoids in the outer membrane of the chloroplast is to act as a protection against the formation of the harmful triplet state of the chlorophyll (destruction by light or photobleaching) (Hendry, 1993).

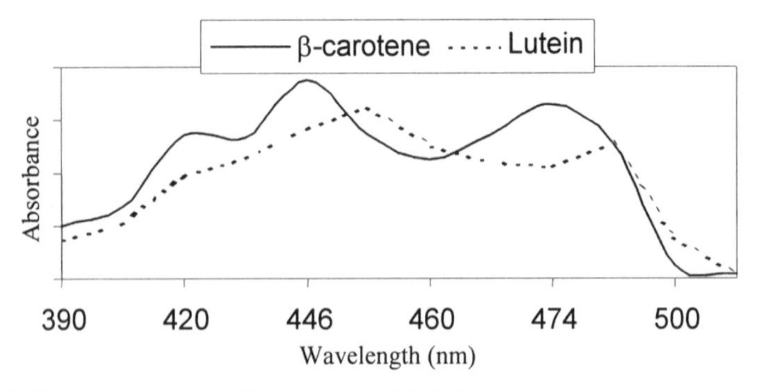

Figure 5. Absorption spectra of β-carotene and Lutein.

Phycobilins: These molecules are open-chained tetrapyrroles. Structurally are related to chlorophyll-*a*, but they do not have the phytol side chain and do not contain magnesium. The chromophores of phycobilins are covalently linked to polypeptides to form water-soluble phycobiliproteins by a thioether bond between an SH-group of the protein and the vinyl side chain of the phycobilin (Figure 6).

Phycobilineproteins are assembled into multimetric particles named phycobilisomes. They are constituents of the light-harvesting (LHCII) antenna of the cyanobacteria and red algae.

Phycobilisome morphology and composition vary with different organism (Heldt, 1997) but the basic structure of the phycobiliproteins consists of a heterodimer (α,β). Each of these subunits contains 1-4 phycobilins as a chromophore. Three of these heterodimers combined to form a trimer $(\alpha,\beta)_3$ and thus form the building block of a phycobilisome (Figure 7).

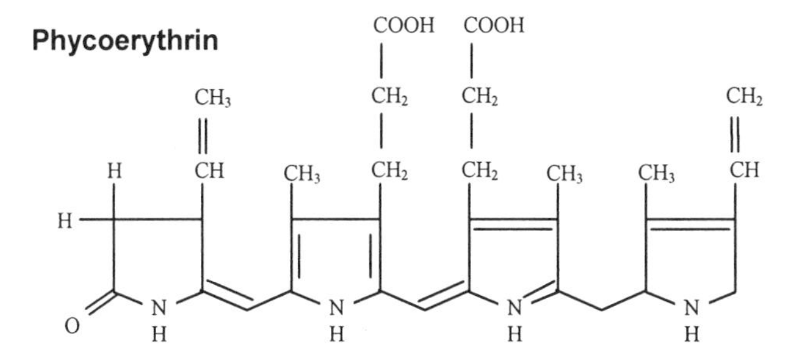

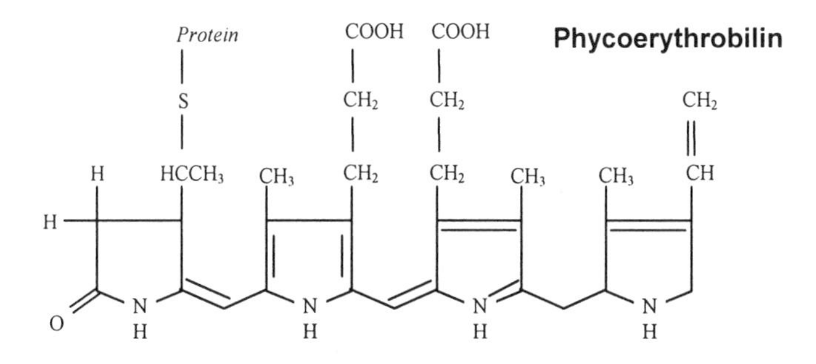

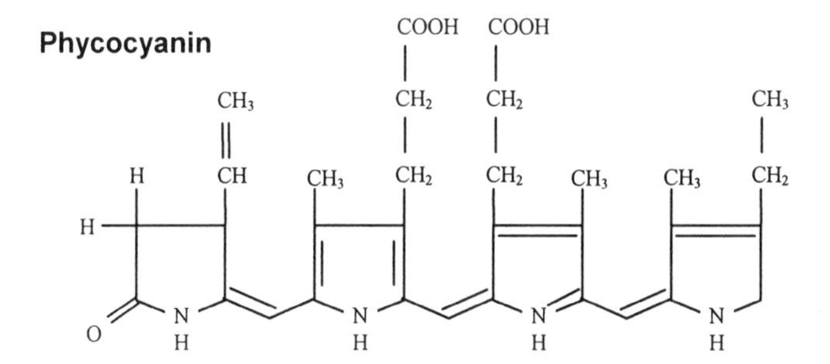

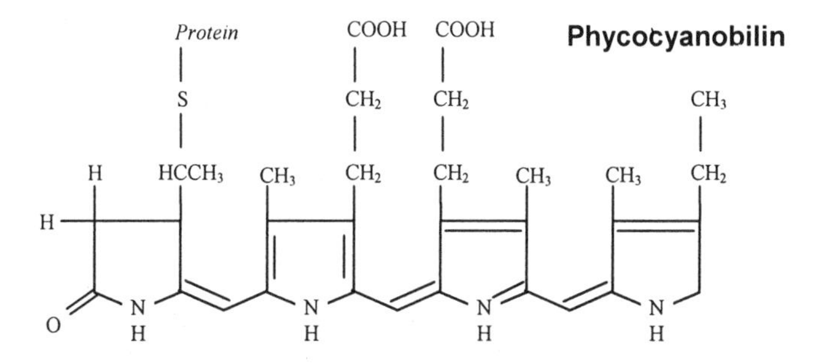

Figure 6. Molecular structure of two bilin pigments (chromophores) and biliproteins present in the phycobilisomes.

The function of this structural organisation is illustrated in the Figure 7 by the absorption of the various biliproteins. The light of shorter wavelength is absorbed in the periphery of the rods by phycoerythrin (PE) and the light of longer wavelength in the inner regions of the rods by phycocianin (PC). The core transfer the excitons to the reaction centre (Glazer, 1981).

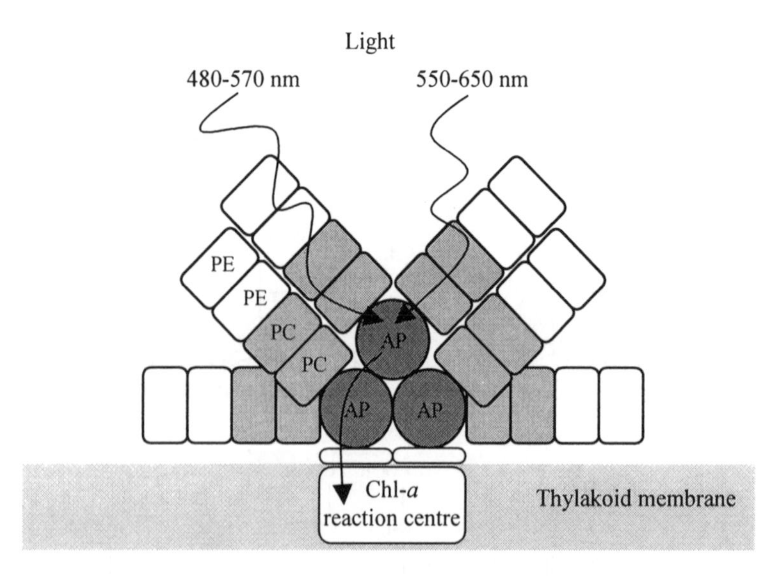

Figure 7. Scheme of a side-view of the structure of a phycobilisome. Each unit shown consists of three α- and three β-subunits. AP: Allophycocyanin; PC: Phycocyanin; PE: Phycoerythrin (after Glazer, 1981 and Heldt, 1997).

Phycobilins are presented in Cyanobacteria and red marine algae and these organisms can collect light of very low intensity (Figure 8). Due to this they are able to absorb green light very efficiently and they can survive in deep water. At these depths, only green light is available, as the light of the other wavelengths is absorbed by green algae living in the upper regions of the water (Larcher, 1995).

The study of these pigments is really interesting from an ecophysiological point of view and it gives us information about the diversity, productivity, distribution, nutrient limitation, degradation and even many taxonomic considerations that are based on the distribution of certain photosynthetic pigments.

The rate of photosynthesis depends on the available irradiance or in the end the absorbed irradiance. Changes in the abundance and composition of the photosynthetic pigments are often related with photoacclimation that is a long-term acclimation to irradiance. Chlorophyll per cell or per unit surface area can increase five- to tenfold as irradiance decreases (Ramus, 1990). The response to this phenomenon is not a linear function of irradiance. At really low light levels, cells often become a bit chlorotic but chlorophyll reach a maximum with slightly higher irradiance. A continuous increases in irradiance lead to a decrease in the cellular content of chlorophyll, until a minimum value is reached (Falkowski and Owens, 1980). The irradiance values that induce these effects are specific for different plant species, and the chlorotic response is not universal (Geider et al. 1986).

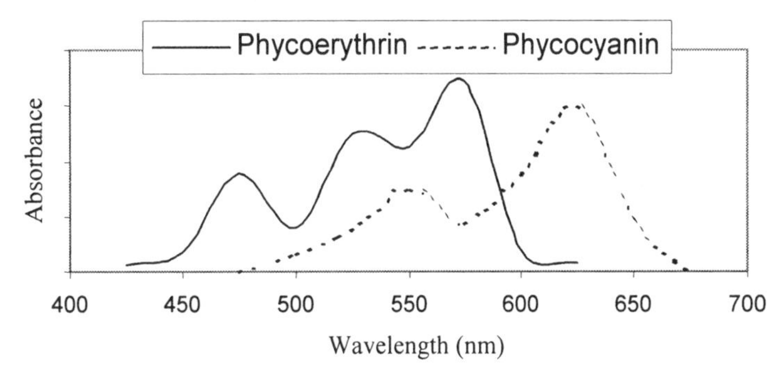

Figure 8. Absorption spectra of phycoerythrin and Phycocianin.

Changes in pigment composition, resulting from photoacclimation, modify the capacity of absorption of cells:

First, cells acclimated to high irradiance levels generally have high carotenoid concentrations relative to chlorophyll *a*. Carotenoids such as β-carotene and zeaxanthin do not transfer excitation energy to the reaction centre, therefore they protect the cell from excess of light.

Second, when cells acclimate to low irradiance levels, increase in pigmentation that is associated with a decrease in the optical absorption cross-section. This effect is due mainly to the self-shading of the chromophores between layers of thylakoid membranes: the more membranes, the lower the optical absorption cross-section. Despite cells accumulate chlorophyll, each chlorophyll molecule becomes less effective in light absorption (Falkowski and Raven, 1997).

PIGMENTS MEASUREMENTS

Spectrophotometry or fluorimetry are the classical methods to determine chlorophyll pigments and other secondary photosynthetic pigments but in recent years, there have been numerous efforts to improve the quality of HPLC separation. Highly complex pigment matrices of plant samples and similar absorbance maxima between related pigments makes the use of HPLC the method of choice for the accurate quantitative determination of chlorophylls and carotenoids (Jeffrey et al, 1997). There are a lot of protocols for a classic spectrophotometric pigment determination (Moss, 1967 and Jeffrey and Humphrey, 1975)

Method of extraction

Pigments were analysed in plants since the beginning of the last century and a wide range of solvents were used for extraction: 90 % aqueous acetone, acetone-methanol-water (80:15:5), dimethylsulphoxide (DMSO), DMSO-90% acetone, *N,N*-dimethylformamide (DMF), dichloromethane-methanol (9:1), cold methanol, chloroform and other organic solvents and other less used combinations. Instead of this, aqueous acetone (90%) and cold methanol are the most common solvents used for high plants and algae.

There are some organisms that present special difficulties to be extracted (e.g. chlorophytes and some cyanobacteria) and then will be necessary to check or made appropriate modifications in the extraction method selected to improve the process for every species or even for different experiments, adapting the new revision to the user requirement. For most plant samples extraction with 90 % of acetone appears to give acceptable results.

During the extraction process we need take care with some aspects of the management. In some extractants as methanol allomerisation of chlorophyll to the corresponding lactone compounds can occur. This transformation is very fast and this chemical process should be checked previously to extraction. The storage of the samples is other important factor in qualitative analysis of photosynthetic pigments. Thus, it is highly recommended to extract as soon as possible after collection of the samples; the extracts can then be stored deep-frozen for up to five months with only minor changes.

Because of the light influence on the pigments, extraction should be carried out in subdued light at low temperatures.

To improve the extraction process sometimes it is useful to disintegrate the sample by ultrasonication or grinding with liquid nitrogen. This process increases the number of particles that can interfere with the photometric method and it will be important to remove the solid elements from the liquid in which the photosynthetic pigments are suspended. Low temperature centrifugation (4 °C) for 15 minutes at 10000 rpm is sufficient to remove particulate debris in the solution of the extract. The supernatant may be analysed by HPLC without any further clean-up.

HPLC analysis

The general HPLC technique has been described previously in chapter 18 of this book. Reversed-phase columns have been employed in most published HPLC techniques for pigment analysis, usually monomeric or polymeric octadecyl silica (Wallerstein and Liebezeit, 1999).

Chromatography is performed using a linear mobile phase gradient from methanol-water mixtures. The specific composition is dependent on the particular separation problem, commonly using ammonium acetate, acetone, acetonitrile or ethyl acetate. Thus, it is not possible to give a general solvent composition or gradient programme for photosynthetic pigment analysis.

It was explained previously in the HPLC technique procedure the advantages of the recent development of the photodiode array for photometric detection which allow on-line recording of absorption spectra and the identification of unknown peaks becomes much easier. Thus, HPLC equipment for pigment analysis is connected to scanning detector module interfaced with a spectrofluorometer that allow the identification of the most of the photosynthetic pigments using some of its physical characteristics: fluorescence and light absorbance.

Identification and quantification is carried out by the use of standards that are commercially available.

REFERENCES

Blankenship R.E. Origin and early evolution of photosynthesis. Photosynth Res 1992; 33:91-111

Buchanan B.B., Gruissem W., Jones R.L. *Biochemistry and Molecular Biology of Plants*. Rockville, Maryland: American Society of Plant Physiologists, 2000

Falkowski P.G., Raven J.A. *Aquatic Photosynthesis*. Massachusetts: Blackwell Science, 1997

Falkowski P.G., Owens T.G. Light-shade adaptation: two strategies in marine phytoplankton. Plant Physiol 1980; 66:592-595

Frank A.H., Cogdell R.J. Carotenoids in photosynthesis. Photochem Photobiol 1996; 63:257-264

Garrido J.L., Zapata M. High performance liquid chromatographic separation of polar and non-polar chlorophyll pigments from algae using a wide pore polymeric octadecyl silica column. HRC-J High Res Chrom 1993; 16:229-233

Geider R.J., Osborne B.A., Raven J.A. Growth, photosynthesis and maintenance metabolic cost in the diatom *Phaeodactylum tricornutum* at very low light levels. J Phycol 1986; 22:39-48

Glazer A.N. "Photosynthetic Accessory Proteins with Bilin Prosthetic Groups." In *The Biochemistry of Plants*. Vol 8. M.D. Hatch, N.K. Boardman, eds. New York, USA: Academic Press, 1981

Goodwin T.W. *Plant Pigments*. London: Academic Press, 1988

Green B.R., Durnford D.G. The chlorophyll-carotenoid proteins of oxygenic photosynthesis. Annu Rev Plant Physiol Plant Mol Biol 1996; 47:685-714

Hall D.O., Rao K.K. *Photosynthesis*. Cambridge: Cambridge University Press, 1999

Heldt H-W. *Plant Biochemistry and Molecular Biology*. Oxford: Oxford University Press, 1997

Hendry G.A.F. "Plant Pigments." In *Plant Biochemistry and Molecular Biology*. P.J. Lea, R.C. Leegood, eds. Chichester: John Wiley and Sons, 1993

Jeffrey S.W., Mantoura R.C.F., Wright S.W. *Phytoplankton Pigments in Oceanography: Guidelines to Moderm Methods*. Paris: Unesco, 1997

Jeffrey S.W., Humphrey G.F. New spectrophotometric equations for determining chlorophyll *a, b, c1* and *c2* in higher plants, algae and natural phytoplankton. Biochem Physiol Pfl 1975; 167:191-194

Krinsky N.I., Mathews-Roth M.M., Taylor R.F. *Carotenoids: Chemistry and Biology*. New York: Plenum Press, 1990

Larcher W. *Physiological Plant Ecology*. Berlin: Springer-Verlag, 1995

Moss B. A spectrophotometric method for stimation of percentage degradation of chlorophyll to phaeopigments in extracts of algae. Limnol Oceanogr 1967; 39:335-340

Ramus J.A. Form-function Analysis of photon capture for seaweeds. Hydrobiologia 1990; 204/205:65-71

Schmid H., Stich H.B. HPLC-analysis of algal pigments: comparison of columns, column properties and eluents. J Appl Phycol 1995; 7:487-494

Taiz L., Zieger E. *Plant Physiology*. Sunderland: Sinauer Associates, 1998

Vogelmann T.C., Nishio J.N., Smith W.K. Leaves and light capture, light propagation and gradients of carbon fixation in leaves. Trends Plant Sci 1996; 1:65-70

Wright S.W., Shearer J.D. Improved HPLC method for the analysis of chlorophylls and carotenoids from marine phytoplankton. Mar Ecol-Prog Ser 1991; 77:183-196

Wallerstein P., Liebezeit G. "Determination of Photosynthetic Pigments." In *Methods of Seawater Analysis*. K. Grasshoff, K. Kremling, M. Ehrhardt, eds. Weinheim: Wiley-VCH, 1999

CHAPTER 8

GAS EXCHANGE TECHNIQUES IN PHOTOSYNTHESIS AND RESPIRATION INFRARED GAS ANALYSER

Pilar Ramos Tamayo[1], Oliver Weiss[1,2] and Adela M. Sánchez-Moreiras[1]
[1]*Dpto Bioloxía Vexetal e Ciencias do Solo. Universidade de Vigo. Spain.*
[2]*Institut für Pflanzenbau. Rheinische Friedrich-Wilhelms-Universität. Bonn. Germany*

INTRODUCTION

The term photosynthesis describes the metabolic process by which plants synthesise organic compounds from inorganic raw materials in the presence of sunlight. Photosynthesis can be regarded as a process of converting radiant energy of the sun into chemical energy of plant tissues.

The ultimate source of all metabolic energy in our planet is the sun and photosynthesis is essential for maintaining all forms of life on earth. All live organisms on Earth need energy for growth and maintenance. Thus, higher plants, algae and certain types of bacteria capture this energy directly from the sunlight and utilise it for the biosynthesis of essential food materials. In spite of the animals that cannot use directly solar radiation as a source of energy, they can obtain the energy by eating plants or by eating other animals that have eaten plants.

The plant photosynthetic apparatus contain the necessary pigments in leaf able to absorb light and channel the energy of the excited pigment molecules into a series of photochemical and enzymatic reactions. Light energy is absorbed by protein-bound chlorophylls and carotenoids that are located in light-harvesting complexes and the energy migration to photosynthetic reaction centres results in electron excitation and transfer to other components of the electron transfer chain (Hall and Rao, 1999).

M.J. Reigosa Roger, Handbook of Plant Ecophysiology Techniques, 113–139.

In higher plants the chlorophyll is contained in the chloroplast. Solar energy trapped by PSII and PSI is transformed to chemical energy through photosynthetic electron transport to provide the ATP and NADPH and carbon skeletons for all the other main cellular processes of nitrogen, sulphur and lipid metabolism.

Figure 1 shows the two essential phases of the photosynthetic process, both photochemical and biochemical phases involved in the fixation of CO_2 in starch and sucrose.

In 1957, Arnon, Allen and Whatley isolated chloroplasts from spinach leaves and were able physically to separate the photochemical and biochemical phases and to show the light-dependent formation of ATP and $NADPH_2$, which then acted as the energy sources for the subsequent dark fixation of CO_2.

The photochemical phase of overall CO_2 fixation to the level of carbohydrate has been shown to occur in the grana lamellae (or thylakoids) of the chloroplast, while the biochemical phase occurs in the stroma of the chloroplast.

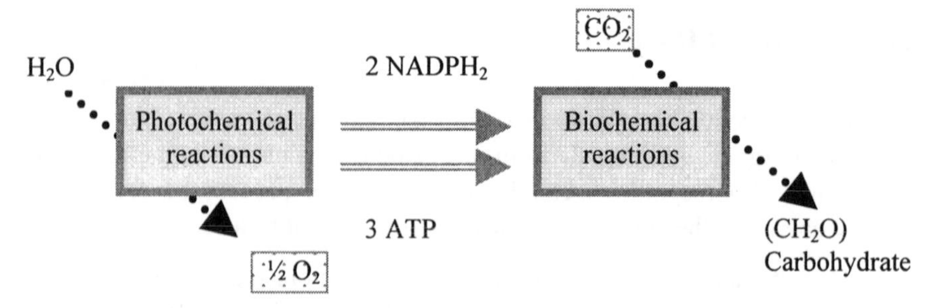

Figure 1. Scheme of the two photosynthetic reactions (photochemical and biochemical reactions). Principal products and gases interchange. Other molecules can also be photosynthetically reduced.

Stomata are the physical responsible foliar structures for the gases interchange between the outside atmosphere and the internal medium, (CO_2, O_2 and water transpiration) and stomatal aperture and closure result from turgor differences between guards cells and the surrounding subsidiary or epidermal cells. Every environment factor that can affect the guard cells turgor (modulated by ABA), will affect the photosynthesis too (Hsiao, 1973; Hussain et al., 1999).

Ecophysiological aspects with regards to photosynthesis

Water is the most important constituent of biological system, as life originated in this medium and it is only in the later part of the evolution that the cells of higher plants have evolved to survive and grow outside the continuous water stream. Thus, it is not surprising that prolonged water deficit virtually affects all metabolic processes.

- There are **constitutive adaptations** that include elaborate mechanisms involving thickened wax cuticle, C_4 and CAM (*Crassulacean acid metabolism*) metabolisms for carbon fixation, reduced surface area of leaves and shrunken stomata, so as to maximise water use efficiency and minimise water loss (Boyer, 1996).
- Secondly, there are **induced physiological responses** to periods of water stress unpredictively in the plant life cycle. This includes several metabolic changes, production and accumulation of osmolytes, alterations in membrane properties, denaturation of proteins, detoxifying enzymes and chaperones synthesis, etc. (Shinozaki and Yamaguchi-Shinozaki, 1997; Turner, 1997).

Photosynthesis is the driving force of plant productivity and the ability to maintain the rate of photosynthetic carbon dioxide and nitrate assimilation under environmental stresses is fundamental to the maintenance of plant growth and production (Tilman et al., 1996). The edaphic and atmospheric factors are seldom totally favourable to the plants. A range of different abiotic and biotic stresses are encountered by plants during their life cycle.

Many strategies of photosynthetic adaptation and acclimation are elicited in plants in response to ever-changing environmental stimuli. Dynamic modulations of the composition, function and structure of the photosynthetic apparatus occur in response to abiotic stresses, light, temperature and the availability of carbon dioxide, water and nutrients. These, highly regulated, dynamic responses to environmental stimuli are vital because plants are extremely limited in their ability to change their immediate environment. The light, for example, affects plant growth, development and stress tolerance in two ways, light is the primary source of energy for photosynthesis, but it is also perceived as signal for the regulation of expression of plastid and nuclear genes (Anderson, 2000).

Water availability in higher plants is the most important limiting factor of photosynthesis and in consequence of the plant productivity (Nilsen and Orcutt, 1996). When the external hydric potential becomes very negative, a sequence of processes starts the cellular turgor decrease, the cellular expansion is retarded and the growth declines. The stomata close when the hydric deficit increases and the CO_2 capture is restricted. This succession of processes are modulated by a hormonal signal, the abscisic acid (ABA), that causes these responses to protect the plant from adverse conditions (Thompson et al., 1997). So that under water deficit the plant closes their stomata avoiding water losses by transpiration in spite of the decrease of photosynthetic yield (Hsiao, 1973).

Nowadays, it doesn't exist a unique concept about the direct cause of the photosynthetic efficiency reduction under water stress (Lawlor, 1995). It seems to be clear that the photosynthesis limitation can be due to stomatal and/or nonstomatal factors (Ort et al., 1994; Pankovic et al., 1999).

- Much of the reduction in CO_2 assimilation in light during water stress is due to **stomatal closure**. The sequence is as follows: ABA increases, the ionic fluxes rapidly are altered with ion losses in the guard cells (K^+, Cl^- and malate), the turgor of guard cells decreases and it follows a reduction in the stomatal aperture (Kaiser, 1987).
- Water limitation can result in **feedback inhibition of photosynthesis** by carbohydrate accumulation in leaves. This can be related with observations in plant leaves under drought like: a decrease in carboxylation capacity of Rubisco (ribulose-1,5-biphosphate carboxylase oxidase), or a reduction in the amount of functional Rubisco or a decrease in regeneration capacity of RuBP (ribulose biphosphate) (Wise et al., 1991; Kanechi et al., 1995; Tezara and Lawlor, 1995). Potential effect of water stress on photosystem II (photoinhibition) seems to be not necessarily entirely adverse; the photosynthetic apparatus responds to the particular changes in energy balance by redox molecular sensing/signalling mechanisms sensed at specific points in the photosynthetic electron transport chain, so that energy supply and consumption turns again in balance (Jefferies, 1994; Heber et al., 1996; Anderson, 2000).

In relation to the CO_2 fixation, three types of **CO_2 reduction pathways** can be distinguished, C_3, C_4 and CAM (*Crassulacean Acid Metabolism*). The existence of these metabolic pathways has important ecophysiological implications. The alternative CO_2 reduction pathways (C_4 and CAM) are physiological adaptations to stressful environments. The C_4 physiology is associated with species in warmer and drier sites, while the CAM physiology is commonly found in taxa of extremely dry environments or sites with extreme seasonal drought. However, there are many facultative CAM species that are located in other environments, including aquatic environments (Nilsen and Orcutt, 1996). Figure 2 shows some of the distinguishing characteristics between these three CO_2 reduction pathways.

C_3 plants have a unique metabolic pathway for CO_2 fixation. It has the carbohydrate ribulose biphosphate (RuBP) as an exclusive initial acceptor and phosphoglyceric acid (three carbons) as first molecule for the fixation of CO_2. C_3 plants are typically from temperate climates and they have an intermediate productivity.

Many tropical grasses and plants such sugar cane and maize are able to fix CO_2 initially into four-carbon compounds like oxaloacetate, malate and aspartate, in addition to the CO_2 fixation that occurs via the Calvin cycle. In the leaves of C_4 plants the primary carboxylation to C_4 acids occurs in the mesophyll and the decarboxylation and secondary CO_2 fixation to PGA (3-carbon phosphoglyceric acid) reactions in the bundle-sheath (See Figure 2).

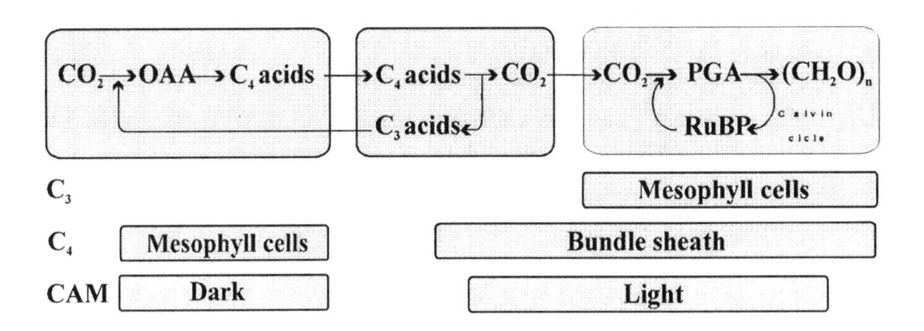

Figure 2. Illustration of some differences in photosynthetic CO_2 assimilation between C_3, C_4 and CAM species; specially, the spatial segregation of the two carboxylations in C_4 and the temporal segregation in the CAM plants. OAA, oxaloacetate; PGA, 3-carbon phosphoglyceric acid; RuBP, ribulose biphosphate (From Hall and Rao, 1999).

C_4 plants possess a carbon dioxide pump, which generates an elevated concentration of carbon dioxide in the vicinity of Rubisco. The carbon dioxide pump is based upon a cycle of carboxylation and decarboxylation involving the generation of C_4 acids, and split between two compartments: the mesophyll cells and the bundle-sheath cells. The bundle-sheath cells form a relatively gas-tight compartment in which carbon dioxide is concentrated for assimilation via Rubisco and the Calvin cycle. The biochemistry of the C_4 pathways is, therefore, tightly integrated with anatomical adaptations (Leegood, 1993).

It is important to point out that structural and physiological changes have been observed in some C_3 plants when they are exposed to arid and warm environments that approach them to C_4 plants (i.e. development of *Kranz anatomy* and/or dicarboxylic acid metabolism expression). In this way C_3-C_4 species with intermediate properties are related (Sharkey, 1993).

CAM occurs in many succulents and tropical epiphytes and the nocturnal stomatal opening associated with dark CO_2 fixation increases the water use efficiency of photosynthesis. With an adequate water status, some CAM plants may behave like C_3 species (Ting, 1985).

The exchange of carbon dioxide (CO_2) and water vapour between plants and the atmosphere is regulated in the long term (days to weeks) by changes in leaf area and by the development of a photosynthetic apparatus in the leaf mesophyll, and in the short term (hours to days) by adjustment of photosynthetic capacity and changes in stomatal aperture.

CO_2 uptake for photosynthesis and water vapour loss of transpiration take place through the same pores in the epidermis, the stomata, which inevitably will result in a proportionally greater water loss than CO_2 uptake. Therefore the regulation of gas exchange is important for plant performance, to maintain growth without desiccation (Schulze, 1986). And, of course, this regulation becomes vitally relevant under adverse conditions.

Plants respond to resource limitation with growth and allocation responses that tend to increase the efficiency with which they use the limiting resources. Thus, for example, in response to drought, plants typically reduce transpiration rates by partial stomatal closure; this reduces the water loss more than the photosynthetic carbon gain, increasing water use efficiency (i.e. carbon gain per unit water loss). Similarly, shade plants have low rates of dark respiration and photorespiration, characteristics that reduce carbon loss under carbon stress so that they use photosynthetically gained carbon more efficiently (Chapin et al., 1987).

The efficiency of **photosynthesis of the whole plant** is crucial to agriculture, forestry, ecology, etc. when it comes to analysing productivity for food and fuels and many other product users. The quality and quantity of photosynthetic incident light (PAR), temperature and water stress, availability and utilisation of mineral nutrients, photorespiratory losses, presence of pollutants in the atmosphere (NO_2, SO_2, O_3) and in the soil (heavy metals), etc., are some of the factors that affect plant productivity. How these factors interact with the changing environment is now the subject of much practical and basic research.

An important ecophysiological factor, which can produce plant stress and affect photosynthesis, is the presence of **allelochemicals**. Specific allelochemicals have shown interferences with stomatal function, carbon fixation and distribution, and respiration (Einhelling, 1986).

An interesting example is the *sorgoleone* Numerous studies have shown that grain sorghum (*Sorgum bicolor* L., Moench) residues can suppress weeds for at least 8 weeks after cover crop kill, and when turned under, have inhibited weed growth throughout the following season. Several *Sorghum* species have shown strong allelopathic interference. Sorgoleone is an oxidised form of a hydrophobic p-benzoquinone isolated from grain sorghum root exudates (Netzly and Butler, 1986). Moreover, this compound has been isolated from soil and can persist for a substantial period in soil, so that it demonstrates that sorgoleone has a role in *Sorghum* allelopathy (Rasmussen et al., 1992).

Many herbicides show a common mechanism of photosynthesis inhibition. This is based on the blockage of photosystem II (PS II) electron transport by binding to second stable electron acceptor site (Q_B, oxidised form of plastoquinone) at the D1 protein. This type of inhibition is characteristic of DCMU (diuron) and others such as s-triazines, phenylureas, triazinones, ureas, uracils, biscarbamates and pyridazinones, often termed as classical diuron-type herbicides. Sorgoleone acts in a similar way, as a potent inhibitor of photosynthetic electron transport by binding to the same Q_B niche of the D1 protein (Nimbal et al., 1996; González et al., 1998).

THE TECHNIQUE

'What happens in photosynthesis is that light energy is used by the plant to break the water molecule into hydrogen atoms and oxygen atoms. Then, if the chloroplasts are intact and contain all their enzymes, the hydrogen combines with carbon dioxide (CO_2) to form glucose and starch, while the oxygen passes into the air' (Asimov, 1989). In 1992, Salisbury and Ross introduced the chapter about photosynthesis as follows: 'Photosynthesis is essentially the only mechanism of energy input into the living world'. How this energy input happens was seen above. Now, we are going to see how photosynthesis can be measured. But in this way, there is a great problem: A discrete sensor, which is able to measure photosynthesis, is not available (Field et al, 1989).

We should not forget that energy derived from light can be used to reduce other molecules than CO_2, but here we will only discuss carbon net photosynthesis. Despite of the fact that carbon photosynthesis is defined as the carbon dioxide assimilation and is expressed as μmol CO_2 m^{-2} s^{-1}, it is always a calculated parameter. The driving force is the amount of photosynthetically active radiation (PAR) per surface and per unit time. So, photosynthesis measurement is determined by measurements of CO_2 concentrations, gas flow, and we will simultaneously measure others parameters like temperature or water vapour pressure.

Radiation – The driving force

Photosynthesis is a photochemical process driven by the absorption of photons by chlorophyll in the wavelength from 400 to 700 nm (Pearcy, 1989). In this range the major part of plant leaves have high absorption. But photons of different wavelength have different energy content, according to the Planck's law. The energy content of a 700 nm photon (171 kJ mol^{-1}) is only nearly the half of the energy content of a 400 nm photon (299 kJ mol^{-1}).

Despite of the big difference in the energy content of these photons, the photosynthetic rate should be independent of this fact, because any energy excess of the photons in the chlorophyll will be dissipated as heat or fluorescence (see Chapter 10 and 11). It seems that the photosynthetic rate depends more on the number or quanta of photons absorbed than the amount of energy (Pearcy, 1989). So, a red photon will have the same effect as a blue photon, which contains much more energy. In table 1 some definitions and the most relevant SI units are presented.

Table 1. Definitions and SI units for radiation and related quantities (Salisbury and Ross, 1992; Larcher, 1994; Pearcy, 1989)

Quantity	SI unit	Description
Radiant Energy Flux	J s^{-1}	The radiant energy falling upon a surface per unit time
Irradiance	W m^{-2}	The radiant energy flux incident on a unit surface
Photosynthetically Active Radiation (PAR)	W m^{-2}	The radiation in the 400 – 700 nm waveband
Photosynthetic Photon Flux Density (PPFD)	mol m^{-2} s^{-1} (Usually the unit μ mol m^{-2} s^{-1} is used, but it is not a SI unit)	The incident photon flux density of PAR: the number of photons (400 – 700 nm) incident per unit time on a unit surface.

The photosynthesis measurement

There are many different ways to measure photosynthesis. Techniques like growth analysis or sophisticated *in vitro* biochemical and biophysical techniques have been applied. But especially the gas exchange method provides the information to study the macroscopic plant response to the environment (McDermitt et al., 2000).

Therefore, agronomists, crop physiologists, plant ecophysiologists and horticulturists have applied gas exchange techniques to study the plant response to experimental treatments or environments. Plant physiologists and biochemists also used this technique to get knowledge about the exact chemical and physical mechanisms, which are causal for the plant response. But, how does it work?

As mentioned above, there is no discrete photosynthesis sensor. And photosynthesis is a calculated parameter based on CO_2 concentration measurements, gas flow, and others parameters. Therefore, it seems understandable that usually a system is used to determine photosynthesis – a gas-exchange system (Field et al, 1989).

In the nineties the calculation of photosynthesis was based on the measurement of the CO_2 rate changes of a leaf enclosed in a largest or shortest chamber. This system is called 'closed system' (see Fig. 3). Air circulates a closed circuit through cuvettes. When a leaf is enclosed in the chamber, the CO_2 concentration in this chamber will decline until the experiment is finished or until the CO_2 compensation point is reached (Davis et al., 1987; McDermitt et al., 1989).

If the enclosed plant leaf makes photosynthesis, the CO_2 concentration will decrease. The decrease of CO_2 concentration per unit time is the base of calculation of net photosynthesis:

$$A = \frac{(c_b - c_s) \times V}{L \times \Delta t}$$

where: A is the net CO_2 assimilation rate, expressed in mol CO_2 m^{-2} s^{-1}

c_b is the CO_2 concentration at the beginning of the measurement, expressed in mol mol^{-1}

c_s is the CO_2 concentration at the end of the measurement, expressed in mol mol^{-1}

L is the leaf area, expressed in m^{-2}

V is air volume inside of the system, expressed in mol

Δt is the duration of the measurement, expressed in s

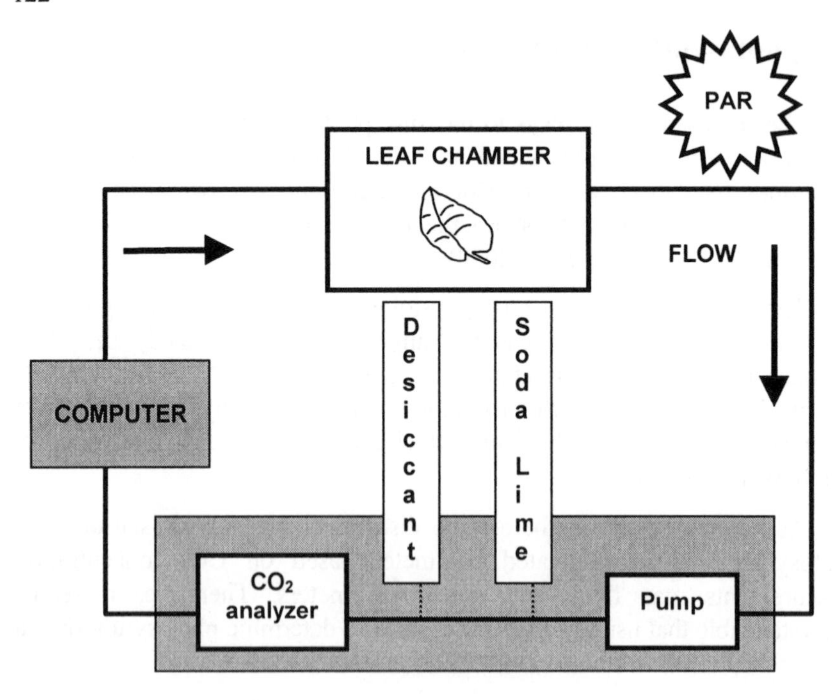

Figure 3. Scheme of closed system to measure photosynthesis, for example used by the Li-Cor 6200

An analyser (closed system) is represented schematically in Fig.3. It consists of four parts:

1. **Leaf Chamber:** Photosynthesis depends on many factors. The most important factors are PAR (Photosynthetically Active Radiation), the concentration of CO_2 and the water vapour pressure in the air. That's the reason by what the leaf, whose photosynthesis will be measured, must be normally enclosed in a chamber and as many as possible external factors must be controlled. The leaf chamber can be made by several materials, the chamber of a LI-6200, for example, is made of acrylic glass. Inside the chamber are a lot of more or less utilitarian sensors and ventilators. Generally, the ventilators are to homogenise the CO_2-concentration. The sensor heads check for several ambient parameters, for example leaf and air temperatures, relative air humidity and also the PAR.

2. **Pump:** Its function is to provide the system with a continue airflow along the closed circuit between the leaf chamber and the gas analyser. The air stream could be detoured to pass through a desiccant to avoid excess of

humidity and/or through a soda lime to catch any CO_2, before its entry into the gas analyser. This is only required for calibration.

3. **Infrared Gas Analyser (IRGA):** It makes exact measurements of the whole CO_2 and performs every second one measurement of the air stream, which comes from the leaf chamber. In general accuracy and resolution of 1 μmol mol^{-1} (1 ppm) is required, especially in a differential system. The heart of any gas exchange analyser is the infrared gas analyser (IRGA). The basic principle is the following: CO_2 is a strong absorbent of infrared radiation. So, an infrared source shines through a sample chamber and on to a detector. The energy at the detector is the total energy entering in the system minus that absorbed in the sample chamber (Field et al., 1989).

   The main problem is the discrimination between CO_2 and other gases that absorb in the IR. Therefore, most modern IR uses a broadband (non-disperse) source and a detector sensitive only to the CO_2 absorption bands. Recently, practically all CO_2 IRGA incorporated either a parallel or series Luft detector, which consists of two CO_2 filled chambers, separated by a flexible membrane (Luft et al., 1967). Different IR absorption in the chambers can be measured by connecting the membrane to a plate. So, one of the two detector chambers is exposed to IR passing through the sample cell, while the other one is exposed to IR coming through a reference cell.

   Another problem must be taken under consideration: Both CO_2 and H_2O vapour absorb IR, even at different wavelength. So, different strategies have been applied. Or the incoming air stream have been completely dried before reaching the IRGA, or the IRGA has been provided with filters to eliminate the wavelength band for H_2O, or the effect of water vapour has been compensated by calculation. So, the best way seems to be to dry the incoming air stream (Field et al, 1989).

4. **Computer:** It is provided with a program to store and handle the data. It permits the modification of inner parameters of the system in function to the use of the equipment. It disposes of storing data memory necessary for the measurement, and permits the introduction of codes in any measurement for a later and correct data treatment.

Another kind of system for photosynthesis measurement was also developed. It is called 'open system' (see Fig. 4).

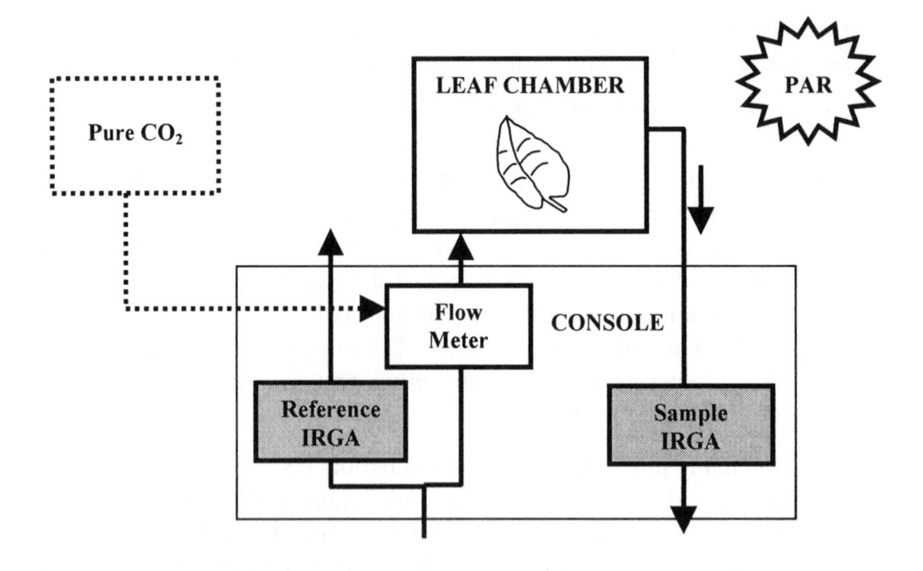

Figure 4. Scheme of a traditional, differential open system. The broken line presents a compensated open system (Li-Cor, 1998)

And what is an open system? The measurement of photosynthesis in this system is based on the differences of CO_2 and H_2O in an air stream, which flows through a chamber where a leaf is placed in.

In an open system, especially a differential system, fresh air from outside of the console will be aspirated and introduced into the system during the whole measurement. Photosynthesis will be calculated from the CO_2 depletion, which occurs when a known airflow rate passes the leaf chamber where a photosynthesising leaf is enclosed (Field et al, 1989; Steubing and Fangmeier, 1992).

$$A = \frac{u_e \times c_e - u_a \times c_a}{L}$$

where: A is the net CO_2 assimilation rate, expressed in mol CO_2 m^{-2} s^{-1}

u_e is the airflow rate in the entry of the system, expressed in mol s^{-1}

c_e is the CO_2 concentration at the entry of the system, expressed in mol mol^{-1}

u_a is the airflow rate at the exit of the system, expressed in mol s^{-1}

c_a is the CO_2 concentration at the exit of the system, expressed in mol mol^{-1}

L is the leaf area, expressed in m^{-2}

Some gas exchange analysers can control microclimatic conditions in the leaf chamber (humidity, temperature, PAR, CO_2 concentration). In the case of compensating systems CO_2 depletion by photosynthesis is compensated by CO_2 injection, and so the CO_2 concentration in the air exiting from the chamber is the same as that in the air stream that enter in the chamber. Therefore, the CO_2 addition means exactly the net photosynthesis rate (Field et al, 1989; Steubing and Fangmeier, 1992)

In an open system measurement, even more as in a closed system, all depends on an accurate CO_2 measurement. Under normal outdoor conditions 78 $\%_{vol}$ of the air is nitrogen, 21 $\%_{vol}$ is oxygen, 0.95 $\%_{vol}$ are some inert gases and only about 0.035 $\%_{vol}$ or 350 $\mu mol\ mol^{-1}$ is CO_2. But the potential for fluctuations is high. So your breath could reach a CO_2 concentration up to 50,000 $\mu mol\ mol^{-1}$ (Li-Cor, 1998).

Light measurement

Radiation seems to be the most important environmental factor in photosynthesis, but also the most difficult to measure (Li-Cor, 1998).

To measure the photosynthetically active radiation normally two types of sensors are used (Pearcy, 1989):

a) **Quantum sensor**: Its relatively low cost has made it standard for measurement of PPFD. It is a filtered silicon-blue photocell covered by a 400-700 nm broadband interference filter, some piece of heat absorbing glass and a coloured gelatine filter. On the top of the housing an opaque white plastic disk is mounted, which has the function of a diffuser. The heat absorbing glass provides further blockage beyond 800 nm. The coloured gelatine filter corrects the spectral sensitivity so the sensor will respond more closely to an ideal photon response. Different firms offer models based on these characteristics.

b) **GaAsP photocells**: They are very cheap, and so where a large amount of sensors is required the gallium arsenide phosphide (GaAsP) photocells will be very attractive. This sensor uses the photovoltaic effect to generate a voltage when it is exposed to light. It consists in a bare plate coated with a layer of GaAsP covered by a thin layer. The absorption of a photon with sufficient energy excites an electron, creating a positively charged hole. It begins a drift to wards the junction of the layers, and the result is a negatively charged GaAsP layer and another, positively charged layer. This kind of sensor has the peak sensitivity at 610 nm, with a spectral response from 300 to 680 nm.

Although the errors can be within a tolerable range, the spectral error exists.

Sunlight is an excellent light source, but only on sunny settings. In understories or at cloudy days frequent measurements must be taken to minimise light differences. The slow response of the gas exchange equipments in relation to the changes in light and the responses of the light sensor present a special problem if PAR measurements are made in the conjunction of photosynthesis measurements (Pearcy, 1989).

So, sunlight is also unpredictable. It varies during the day and can be obscured by clouds. Therefore, artificial light sources are typically used for controlled environmental experiments (Field et al., 1989).

Common errors

Some errors affect the accuracy of gas exchange measurements. These errors are normally related with a wrong, careless or negligent calibration, with the wrong use of a leaf chamber and with the consequences of CO_2 leaks and diffusion through chamber materials and tubing (McDermitt et al, 2000).

When we place a leaf into a chamber to measure photosynthesis in it, we must take in account that the microenvironmental conditions in the leaf surface will change at this moment. So, the light intensity drops 10%, the temperature varies and the wind speed increases considerably (Nobel, 1983). We can make it fast and with care, but we can not avoid that in the moment that we place the leaf within the chamber the plant physiology changes. Therefore, when we use a chamber we must to be able to correct our values with some equation to reach the starting environmental conditions outside the chamber. Adjustments to the parameter values and to the response curves are necessary.

A main error in the calibration is to believe that the products (desiccant, soda lime) are OK while they are saturated of water or CO_2. It can alter the measure as the apparatus function. So, we must sure that these products are changed before starting the measurements and in the right order. The hurry for starting the measurements is usually another common error. The gas exchange system needs a time for calibration to obtain stable conditions and all the parameters must be checked.

The negative effects of CO_2 leaks are another big problem. McDermitt et al. (2000) evaluated CO_2 diffusion through gasket material and tubing used in LI-6400 Portable Photosynthesis System (see below) and could see that the permeability of the CO_2 in these organic polymers was higher than that of other atmospheric gases. In the same way, experiments made by them demonstrated that effects of leaks of CO_2 are increased when assimilation rates are evaluated in small leaf areas. So, we must to take in account the material, the pressure, the temperature and the chamber size to adjust the photosynthesis value to the real CO_2 concentration.

TWO DIFFERENT GAS EXCHANGE INSTRUMENTS

The two next commented apparatus are able to measure net carbon photosynthesis in plant leaves as well in the laboratory as in the field. Li-Cor 6200 and Li-Cor 6400 are two different systems to obtain the 'same' measurement in two different ways. Both can be applied in a broad range of different fields, as in:

- Ecology, with the possibility to know the photosynthetic behaviour of a same species in different ecosystems or of different species in different ecosystems
- Agronomy, environmental stress, productivity, the influence of the soil conditions in the crop status, the study of the herbicide effects
- Entomology, the effect of the pest as of the insecticides on the crop evolution
- Biotechnology, comparisons between different genotypes, etc.

LI-6200. A closed system

LI-6200 (Li-Cor model 6200, Lincoln, NE) was one of the first CO_2 analyser, based in a closed system for CO_2, which can measure in a short time interval (10-20 sec) the net interchange between a leaf enclosed in a leaf chamber and the atmosphere inside it.

Other measurements that can be determining the final value of photosynthesis, like the photosynthetically active radiation (PAR), relative humidity, leaf temperature, etc., can be also obtained here. An improvement of this apparatus according to previous gas exchange systems is the possibility to get transpiration measurements. A decrease in the water potential of leaves can be related with a photosynthetic reduction (Boyer, 1989). The water loss by the leaf enclosed in the chamber is determined measuring the necessary dry airflow to maintain a constant relative humidity inside the chamber. So, it is possible to obtain an idea about the stomatal aperture/closure seeing the transpiration values (mol m^{-2} s^{-1}).

LI-6200 is a compact, small size and portable instrument, which allows measuring directly in the field (Davis et al., 1987; McDermitt et al., 1989). The leaf chambers are interchangeable and available in different sizes, which permits measuring whole leaves from very different plants. These chambers hold a light sensor in the outside and a fan (for circulating air) in the inside, and they are made from acrylic glass and Teflon to minimise CO_2 and H_2O exchange by chamber materials (Li-Cor, 1990).

LI-6200 is prepared for a wide range of photosynthetic measurements conditioned by the so diverse photosynthetic efficiency of the different plant species. By other hand, the rapid and easy way of measurement permits to maintain almost intact the environmental conditions of the leaf before its entry in the leaf chamber.

Measuring

So, for measuring (see Fig. 5), and after a heating time where calibration and checking must be done, we have only to clean the leaf, to put it in the leaf chamber and close it, to enter in the software the data of the leaf area enclosed in the chamber, to take in account the starting CO_2 value, and when the CO_2 concentration is lowing, to start the measuring. All data are stored in the memory and later can be analysed in the computer (see Table 2).

Before starting the measurement it is also necessary to select the desired program in the software of the apparatus. That is so because this gas exchange system allows also the soil respiration measurement with a soil chamber (LI-6000-09).

But some things are necessary to take in account when we work with this apparatus (LI-6200). For example, there is an important influence of handling, because the person who takes the measure is who decides the appropriate moment to start this measurement. By other hand, there is no control on the light, the temperature, the CO_2 concentration or the PAR into the chamber used by the system to calculate the photosynthetic value. So, if the measure is made in the field along several days, the user must be careful to select those days with similar environmental conditions or to keep in mind in the data interpretation the environmental differences.

In the same way, altered conditions into the chamber (an increase of the temperature or a shadow by the complicate angle necessary to measure the leaf) make that the microenvironmental conditions around the leaf surface could change along the measurement, and so the finishing conditions will be different from the starting conditions into the chamber.

Table 2. A typical data sheet for LI-6200 Portable Photosynthesis System after photosynthesis measuring of field grown maize, where: (P) is pressure; (Vt) is total system volume; (Vg) is IRGA volume; (A) is leaf area within the chamber; (BC) is boundary layer conductance of a leaf side; (STMRAT) is and estimation of the ratio of stomatal conductance of one leaf side to the other side; (Fx) is the maximum flow rate that can be reached through the desiccant; (Kabs) is a water absorption factor; (TIME) is the measurement time in sec; (QNTM) is the quantum sensor value; (TAIR) is the air temperature in the chamber; (TLEAF) is the leaf temperature; (CO2) is the CO_2 concentration in ppm; (FLOW) is the flow rate through the desiccant; (RH) is the relative humidity; (EAIR) is the vapor pressure of the air in the leaf chamber; (PHOTO) is the net photosynthesis; (COND) is stomatal conductance; (CINT) is intercellular CO_2 concentration; (RS) is stomatal resistance; (CS) is stomatal conductance.

PAGE	TIME							
35	Jul 10:39							
P (mb)	Vt (cc)	Vg (cc)	A (cm²)	BC(mol)	STMRAT	Fx (μmol)	Kabs	
979.9	1174	154.0	20.0	1.300	0.5	1300	1.100	
Crop Maize	Anthesis	Leaf Nº 4						
OB	TIME	QNTM	TAIR	TLEAF	CO2	FLOW	RH	EAIR
1M	4.825	1612	25.57	25.75	371.9	1076	30.39	9.960
1R	5.500	6.986	0.1192	0.7209	7.560	3.620	2.282	
OB	PHOTO	COND	CINT	RS	CS			
1	31.79	0.1125	103.7	3.507	0.2851			

The question is: Is there any way to measure photosynthesis more accurately?

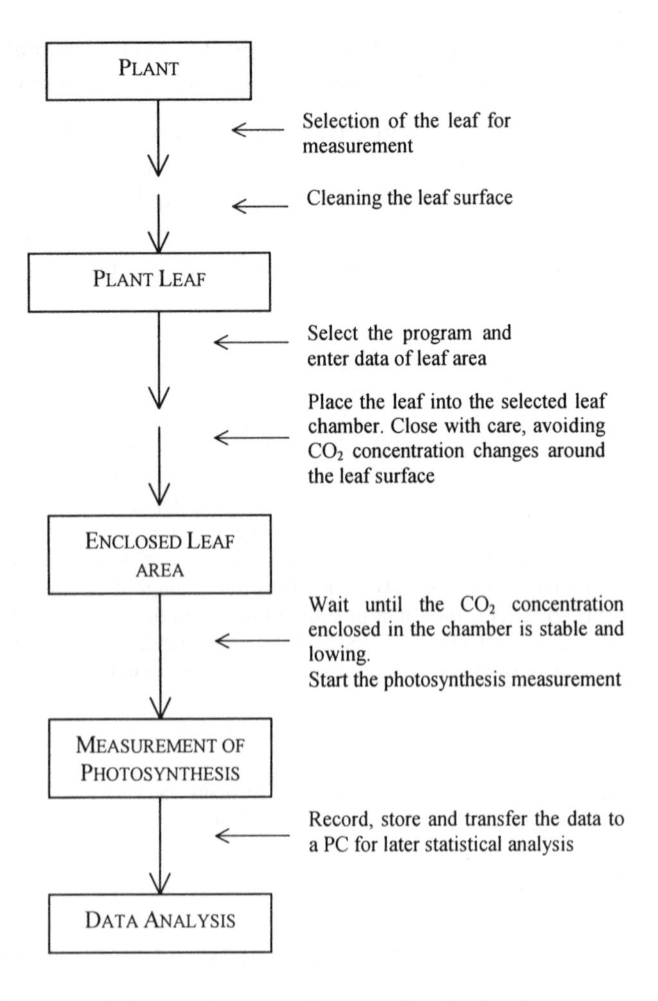

Figure 5. Scheme of the photosynthetic measurement procedure in plant leaves

LI-6400. An open system

As we have previously said, the photosynthetic measures can be influenced by several external and internal factors (Norman and Hesketh, 1980; Kirschbaum and Pearcy, 1988; Al-Khatib and Pausen, 1990; Allard et al., 1991). Controlling total conditions required for a true and stable photosynthesis measurement is very difficult. This inconvenience will be even worst if the measure is done in the field, where it is impossible to have control over environmental parameters as temperature, high light, sun fleck, humidity, etc.Therefore, it is important that the selected apparatus be able to measure, monitor and control all these conditions. That is one of the most prominent advantages of the LI-6400 on the LI-6200, the capacity to fix a desired value of light, CO_2, humidity, temperature, or other parameter.

Carbon dioxide response curves give us important information about the physiology and biochemistry of the photosynthesis system (Davis et al., 1987; McDermitt et al., 1989). Thanks to this control, response curves can be made and displayed by the LI-6400 software, and their knowledge can help to the person that is recording the measurements to correct default preset measuring conditions.

LI-6400 (Li-Cor model 6400) is an open system for laboratory or field photosynthesis measurement of whole leaves (see Fig. 6). It has the step up of having the gas analysers in the sensor head, eliminating the connection tube between the chamber where the leaf is placed and the console where the value is calculated.

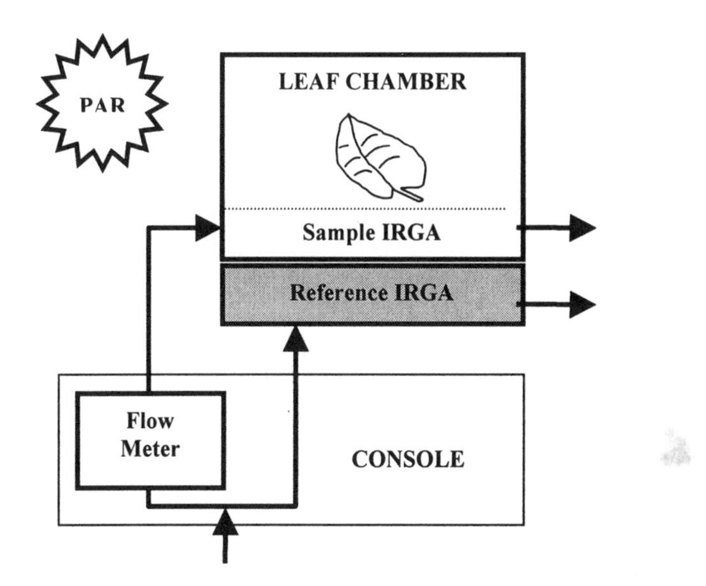

Figure 6. Scheme of the developed open system used by the Li-Cor 6400

This instrument has two gas analysers in the sensor head, which allow obtaining absolute CO_2 and H_2O values as well from the reference as from the sample. With this progress, gas losses, water condensation or contamination in the return tube are eliminated and the measurement gives a dynamic knowledge about what is happening in the leaf in every moment, in real time and in real microenvironmental conditions (Li-Cor, 1995; 1998).

Measurements in the sensor head come from 'the difference between intensity of absorbing wavebands and nearby non-absorbent wavebands in the same optical path' (Li-Cor, 1995).

Control on the measurement

Likewise, this system has the possibility to fix all desired parameters, and so the user can control the environmental conditions in the leaf chamber.

Until twelve parameters can be recorded and displayed by LI-6400 at the same time while the measurement of photosynthesis is obtained. It allows possible corrections in the measuring conditions giving more real and exact knowledge about the effects of the treatment or the ecological conditions on the plant. The parameters subject to be controlled with the LI-6400 are (Li-Cor, 1998):

- **Light**. The apparatus gives the option of selecting the type of used light source (halogen, sun, fluorescent, etc.) thanks to two quantum sensors that provide measurements about photosynthetically active radiation and that allows corrections with calibrating coefficients, ensuring a more true light measurements.

  It is possible to use a Red/Blue light lamp (see Fig. 7), with which we can establish a fixed PPFD for the leaf or a range of PPFD series in the chamber (0-2000 μmol m^{-2} s^{-1}). It has been demonstrated that the use of an additional blue light is essential for studying stomatal kinetics (Zeiger et al., 1987), because the blue light is involved in the control of the stomatal aperture.

  It has a very little size, it is perfectly coupled to the leaf chamber and eliminates possible falls in the measurement caused by the shade, the weather of this concrete day, the season, the period of the day (the radiation differs along the day), or the handling. This optional LI-6400-02B LED source is controlled by a third sensor installed in the sensor head.

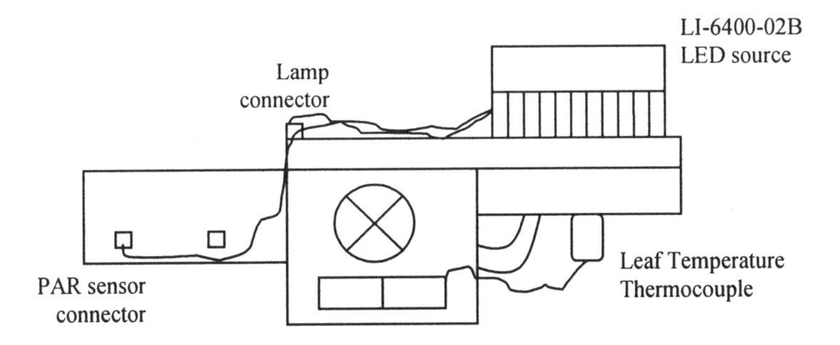

Figure 7. Schematic representation of the LED source connected to the leaf chamber and theirs connectors

- **CO_2 Concentration**. LI-6400 gives to the user two options to control the CO_2 concentration. One of them is the manual control with the help of a tube of soda lime (that catches the CO_2), which is limited to the capacity of this substance to catch the CO_2 and to the incoming CO_2 concentration. The second one is the use of the optional CO_2 mixer (LI-6400-01 CO_2 mixer). Mixing scrubbed air and pure CO_2 (in 12 g cylinders) can control the CO_2. This option allows constant CO_2 concentration in the leaf chamber (known as 'sample CO_2') or maintains constant incoming CO_2 (known as 'reference CO_2') close to the environmental value. With this injector, CO_2 and H_2O absolute values can be separately controlled.

- **Temperature**. Block or leaf temperatures (by way an indirect measurement of the air temperature) can be maintained constant by means of internal peltier coolers. So, leaf temperature increases can be eliminated without uses of additional instruments that complicate the management.

- **Humidity**. The humidity can be also maintained constant along the measurements. Two mechanisms are needed for it: the control of the incoming humidity (known as 'reference H_2O') drying the incoming air with the desiccant or moistening it, and the air flow that goes through the chamber (known as 'sample H_2O'). When the humidity value is fixed, the flow value varies and must be manually adjusted.

- **Measuring Time**. A predetermined time to provide measurements can be fixed before starting measurement. So, data will be recorded and curves will be made in function to this preset time period.

Measuring

For starting the work with the LI-6400 a previous warming (10 min), as with LI-6200 (30 min), is necessary. At this time, checking of leaf temperature, CO_2 flow, light sensors, leaf fan, etc. should be done (see Fig. 8).

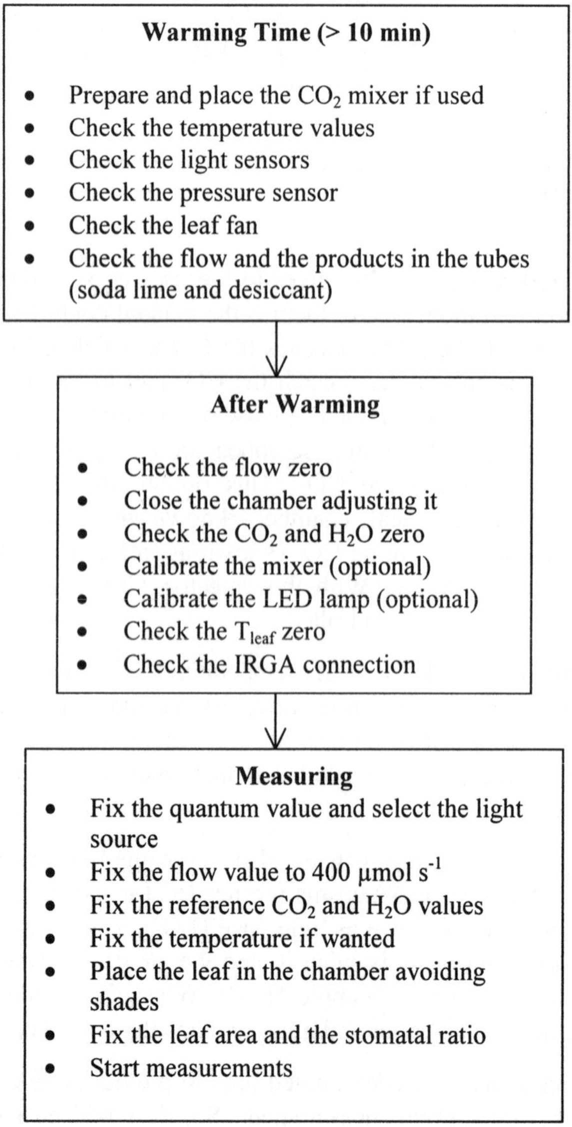

Warming Time (> 10 min)

- Prepare and place the CO_2 mixer if used
- Check the temperature values
- Check the light sensors
- Check the pressure sensor
- Check the leaf fan
- Check the flow and the products in the tubes (soda lime and desiccant)

↓

After Warming

- Check the flow zero
- Close the chamber adjusting it
- Check the CO_2 and H_2O zero
- Calibrate the mixer (optional)
- Calibrate the LED lamp (optional)
- Check the T_{leaf} zero
- Check the IRGA connection

↓

Measuring

- Fix the quantum value and select the light source
- Fix the flow value to 400 μmol s^{-1}
- Fix the reference CO_2 and H_2O values
- Fix the temperature if wanted
- Place the leaf in the chamber avoiding shades
- Fix the leaf area and the stomatal ratio
- Start measurements

Figure 8. Scheme of the photosynthetic measurement procedure in plant leaves with a LI-6400 Portable Photosynthesis System

Once finished the warming and after calibrating the LED lamp (optional), the CO_2 mixer (optional), and the zero of sample and reference H_2O and CO_2 flow, we can start to establish the fixed parameter values for our experiment.

So, it is possible to fix the block and leaf temperature value. As the leaf temperature is an indirect measure it is better to fix the block temperature in a value near to the starting temperature. If we have an LED light source we can fix also the desired PPFD. In any case the light source type must be selected at this moment (sun+sky, quartz halogen, LED lamp, fluorescent, etc). It is necessary to enter also the value of the leaf area enclosed in the chamber and the stomatal ratio. The stomatal ratio is the stomatal ratio in one leaf side with respect to the other side (0-1).

Before fixing the flow it is important to decide if we prefer to maintain constant the humidity or the CO_2 flow. If we want to have a stable humidity for the measurement the apparatus will give us the humidity value at that moment and then we can enter the value (mmol mol^{-1}) that we want to have along the experiment. When the humidity is controlled we must to manually adjust the flow, because it changes.

For fixing the CO_2 reference value the apparatus gives us two options. If a CO_2 mixer is installed and we are going to use it, the reference value must be close to the ambient conditions (± 400 μmol mol^{-1}). But if no CO_2 mixer is connected, then we must to use the Soda lime until obtaining the desired CO_2 concentration.

Finally, we can fix the time interval to obtain the photosynthesis measurements and the response curves that we want to record.

Once the leaf is placed into the selected leaf chamber and the CO_2 flow is stable the data recording can start. During this time we can display the parameter values of our interest and correct them if necessary.

Data can be stored in a data memory and transferred to a PC with a data transfer program for later analysis.

REFERENCES

Al-Khatib K., Paulsen G.M. Photosynthesis and productivity during high-temperature stress of wheat genotypes from major world regions. Crop Sci 1990; 30:1127-1132

Allard G., Nelson C.J., Pallardy S.G. Shade effects on growth of tall fescue: II. Leaf gas exchange characteristics. Crop Sci 1991; 31:167-172

Anderson J.M. "Strategies of Photosynthetic Adaptations and Acclimation." In *Probing Photosynthesis: Mechanisms, Regulation and Adaptation.* M. Yunus, U. Pathre, P. Mohanty, eds. London: Taylor and Francis, 2000

Asimov I. *How we did Find out about Photosynthesis?* New York: Walker and Company, 1989

Boyer J.S. Advances in drought tolerance in plants. Adv Agronomy 1996; 56:187-217

Chapin III F.S., Bloom A.J., Field C.B., Waring R.H. Plant responses to multiple environmental factors. Bioscience 1987; 37:49-57

Davis J.E., Arkebauer T.J., Norman J.M., Brandle J.R. Rapid measurement of the assimilation rate *versus* internal CO_2 concentration relationship in green ash (*Fraxinus pennsylvanica* Marsh): the influence of light intensity. Tree Physiol 1987; 3:387-392

Einhellig F.A. "Mechanisms and Modes of Action of Allelochemicals." In *The Science of Allelopathy,* A.R. Putnam, C.S. Tang, eds. New York: John Wiley and Sons, 1986

Field C.B., Ball J.T., Berry J.A. "Photosynthesis: Principles and Field Techniques" In *Plant Physiological Ecology.* R.W. Pearcy, J.R. Ehleringer, H.A. Mooney, P.W. Rundel, eds. London: Chapman and Hall, 1989

Garcia R.L. "Canopy Photosyntesis of Soybeans" In *Assimilation and Allocation of Carbon in Determinate and Indeterminate Soybeans.* Nebraska, 1991

González V.M., Kazimir J., Nimbal C., Weston L.A., Cheniae G.M. Inhibition of a photosystem II electron transfer reaction by the natural product sorgoleone. J Agr Food Chem 1998; 45:1415-1421

Hall D.O., Rao K.K. *Photosynthesis.* Cambridge: Cambridge University Press, 1999

Heber U., Bligny R., Streb P., Douce R. Photorespiration is essential for the protection of the photosynthetic apparatus of C_3 plants against photoinactivation under sunlight. Bot Acta 1996; 109:307-315

Hsiao T.C. Plant responses to water stress. Annu Rev Plant Physiol 1973; 24:519-570

Hussain A., Black C.R., Taylor I.B., Mullholland B.J., Roberts J.A. Novel approaches for examining the effects of differential soil compaction on xylem sap abscisic acid concentration, stomatal conductance and growth in barley (*Hordeum vulgare* L.). Plant Cell Environ 1999; 22:1377-1388

Jefferies R.A. Drought and chlorophyll fluorescence in field-grown potato (*Solanum tuberosum*). Physiol Plantarum 1994; 90:93-97

Kaiser W.M. Effects of water deficit on photosynthetic capacity. Physiol Plantarum 1987; 72:142-149

Kanechi M., Kunitomo E., Inagaki N., Maekawa S. "Water Stress Effects on Ribulose-1,5-bisphosphate carboxylase and its Relationship to Photosynthesis in Sunflower Leaves." In *Photosynthesis: from Light to Biosphere,* P. Mathis, ed. The Hague: Kluwer Academic Publishers, 1995

Kirschbaum M.U.F., Pearcy R.W. Concurrence measurements of oxygen and CO_2 exchange during light flecks in *Allocacia macrorrhiza.* Planta 1988; 174:527-533

Larcher W. *Ökophysiologie der Pflanzen. 5. Auflage*. Stuttgart, Germany: Verlag Eugen Ulmer, 1994

Lawlor D.W. "The Effects of Water Deficit on Photosynthesis." In *Environment and Plant Metabolism: Flexibility and Acclimation*. N. Smirnoff, ed. Oxford: Bios Scientific, 1995

Leegood R.C. "Carbon Dioxide-concentrating Mechanisms." In *Plant Biochemistry and Molecular Biology*. P.J. Lea, R.C. Leegood, eds. West Sussex: John Wiley and Sons, 1993

Li-Cor, Inc. *Using the Li-6200. Portable Photosynthesis System*. Lincoln: Li-Cor, 1990

Li-Cor, Inc. *Using the Li-6400. Portable Photosynthesis System. Feature and Spec's*. Lincoln: Li-Cor, 1995

Li-Cor, Inc. *Using the Li-6400. Portable Photosynthesis System*. Lincoln: Li-Cor, 1998

Luft K.F., Kesseler G., Zorner K.H. Nicht dispersive Ultrarot-Gasanalyse mit dem UNOR. Chem Ing Tech 1967; 39:937-945

McDermitt D.K., Garcia R.L., Welles J.M., Demetriades-Shah T.H. "Common Errors in Gas Exchange Measurements." In *Probing Photosynthesis*. M. Yunus, U. Pathre, P. Mohanty, eds. London and New York: Taylor and Francis, 2000

McDermitt D.K., Norman J.M., Davis J.E., Arkebauer T.J., Ball J.T., Welles J.M., Roemer S.R. "CO_2 Response Curves Can Be Measured with a Field Portable Closed Loop Photosynthesis System." In *International Symposium on Forest Tree Physiology*. Nancy, 1988

Netzly D.H., Butler L.G. Roots of sorghum exudate hydrophobic droplets contains biologically active components. Crop Sci 1986; 26:775-778

Nilsen E.T., Orcutt D.M. *Physiology of Plants under Stress. Abiotic Factors*. New York: J Wiley and Sons, 1996

Nimbal C.I., Yerkes C.N., Weston L.A., Weller S.C. Herbicidal activity and site of action of natural product sorgoleone. Pestic Biochem Phys 1996; 54:73-83

Nobel P.S. *Biophysical Plant Physiology and Ecology*. W.H. Freeman, ed. New York, 1983

Norman J.M., Garcia R., Verma S.B. Soil surface carbon dioxide fluxes and the carbon budget of a grassland. J Geophys Res 1992; 97:18845-1885.

Norman J.M., Hesketh J.D. "Micrometeorological Methods for Predicting Environmental Effects on Photosynthesis" In *Predicting Photosynthesis for Ecosystems Models. Vol. I*. J.D. Hesketh, J.W. Jones, eds. Boca Raton-Florida: CRC Press, 1980

Ort D.R., Oxborough K., Wise R.R. "Depressions of Photosynthesis in Crops with Water Deficits." In *Photoinhibition of Photosynthesis from Molecular Mechanisms to the Field*, N.R. Baker, J.R. Bowyer, eds. Oxford: Bios Scientific, 1994

Pankovic D., Sakac Z., Kevresan S., Plesnicar M. Acclimation to long-term water deficit in the leaves of two sunflower hybrids: photosynthesis, electron transport and carbon metabolism. J Exp Bot 1999; 50:127-138

Pearcy R.W. "Radiation and Light Measurements" In *Plant Physiological Ecology*. R.W. Pearcy, J.R. Ehleringer, H.A. Mooney, P.W. Rundel, eds. London: Chapman and Hall, 1989

Rasmussen J.A., Hejl A.M., Einhellig F.A., Thomas J.A. Sorgoleone from root exudate inhibits mitochondrial functions. J Chem Ecol 1992; 18:197-207

Salisbury F.B., Ross C.W. *Plant Physiology*. F.B. Salisbury, ed. Belmont, California: Wadsworth, 1992

Schulze E.-D. Carbon dioxide and water vapor exchange in response to drought in the atmosphere and in the soil. Annu Rev Plant Physiol 1986; 37:247-274

Sharkey T.D. "Fotosíntesis. Absorción y Utilización del Dióxido de Carbono en un Contexto Ecológico." In *Fisiología y Bioquímica Vegetal*. J. Azcón-Bieto, M. Talón, eds. Madrid: McGraw-Hill-Interamericana of Spain, 1993

Shinozaki K., Yamaguchi-Shinozaki K. Gene expression and signal transduction in water-stress response. Plant Physiol 1997; 115:327-334

Steubing L., Fangmeier A. *Pflanzenökologischen Praktikum: Gelände– und Laborpraktikum der terrestrischen Pflanzenökologie*. L. Steubing, ed. Stuttgart: Ulmer, 1992

Tezara W., Lawlor D.W. "Effects of Water Stress on the Biochemistry and Physiology of Photosynthesis in Sunflower." In *Photosynthesis: from Light to Biosphere*. P. Mathis, ed. The Hague: Kluwer Academic Publishers, 1995

Thompson S., Wilkinson S., Bacon M.A., Davies W.J. Multiple signals and mechanisms that regulate leaf growth and stomatal behaviour during water deficit. Physiol Plantarum 1997; 100:303-313

Tilman D., Wedin D., Knops J. Productivity and sustainability influenced by biodiversity in grassland ecosystems. Nature 1996; 379:718-720

Ting I.P. Crassulacean acid metabolism. Annu Rev Plant Physiol 1985; 36:595-622

Turner N.C. Further progress in crop water relations. Adv Agron 1997; 58:293-338

Wise R.R., Sparrow D.H., Ortiz-López A., Ort D.R. Biochemical regulation during the mid-day decline of photosynthesis in field-grown sunflower. Plant Sci 1991; 74:45-52

Zeiger E., Iino M., Shimazaki K.-I., Ogawa T. "The Blue-Light Response of Stomata: Mechanism and Function." In *Stomatal Function*. E. Zeiger, D.D. Faquhar, I.R. Cowan, eds. Stanford, California: Stanford University Press, 1987

General references

Hall D.O., Rao K.K. *Photosynthesis*. Cambridge: Cambridge University Press, 1999

Lawlor D.W. "The Effects of Water Deficit on Photosynthesis." In *Environment and Plant Metabolism: Flexibility and Acclimation*, N. Smirnoff, ed. Oxford: Bios Scientific, 1995

Li-Cor, Inc. *Using the Li-6200. Portable Photosynthesis System*. Lincoln: Li-Cor, 1990

Li-Cor, Inc. *Using the Li-6400. Portable Photosynthesis System. Feature and Spec's*. Lincoln: Li-Cor, 1995

Li-Cor, Inc. *Using the Li-6400. Portable Photosynthesis System*. Lincoln: Li-Cor, 1998

Nilsen E.T., Orcutt D.M. *Physiology of Plants under Stress. Abiotic Factors*. New York: J. Wiley and Sons, 1996

Yunus M., Pathre U., Mohanty P. *Probing Photosynthesis. Mechanisms, Regulation and Adaptation.* London: Taylor and Francis, 2000

CHAPTER 9

USE OF OXYGEN ELECTRODE IN MEASUREMENTS OF PHOTOSYNTHESIS AND RESPIRATION

Luís González[1], Carlos Bolaño[1] and François Pellissier[2]
[1]*Depto Bioloxía Vexetal e Ciencias do Solo. Universidade de Vigo. Spain*
[2]*Laboratoire de Dynamique des Ecosystemes d'altitude. LDEA. Université de Savoie. France*

INTRODUCTION

In order to grow, plants must be able to convert the energy from the sun into a useful form (photosynthetic process). In higher plants the final products of photosynthesis are sucrose and starch or fructans. Sucrose is the major form into which carbon from carbon dioxide is assimilated for transport throughout the plant, while starch and fructans are the major forms in which carbon is stored. Oxygen is produced during photochemistry process in photosynthesis. The oxidation of photosynthates is converted to pyruvate or malate to produce energy or carbon skeletons for biosynthesis coupled with oxygen consumption (Bryce and Hill, 1993).

Quantification of oxygen release in presence of light and oxygen consumption in absence of light by a plant is a good approach to measure photosynthesis and respiration (Hall and Rao, 1999).

Photosynthesis and respiration constitute two coupled physiological processes (Figure 1). Photosynthesis is essential to maintain all the life forms in the Earth and respiration is an anabolic process that consumes the products of the photosynthesis. So that, it is very important in plant ecophysiology to know the relationship between photosynthesis and

M.J. Reigosa Roger, Handbook of Plant Ecophysiology Techniques, 141–153.

respiration, because these parameters are frequently affected by changes in the plant environment (Bazzaz, 1996).

The measurement of oxygen evolution, uptake and liberation, during a reaction in a closed system is one of the easiest (and cheapest!) ways of following the processes of photosynthesis and respiration in plants. The oxygen electrode works based on the principle of Clark; that is, using a polarographic measurement of the electricity that flows between an anode and a cathode and it is sensitive enough to detect oxygen concentrations in the order of 0.01 mmol.

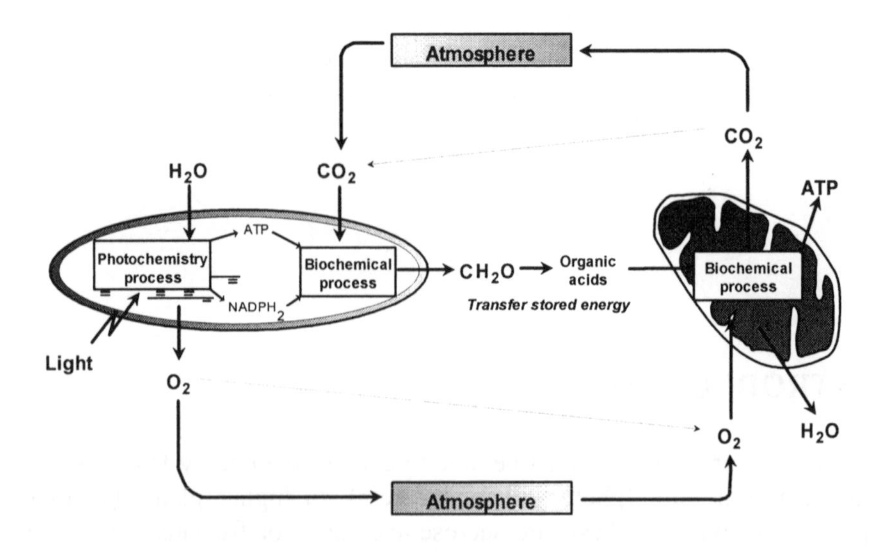

Figure 1. Coupling of the physiological processes photosynthesis and respiration.

This electrode consists on a cell which contains a circular silver wire anode and a platinum wire cathode, bathed in a saturated potassium chloride (KCl) solution. The electrodes are separated from the reaction mixture by an O_2 gas-permeable teflon membrane. The reaction mixture in the appropriate container is stirred constantly with a small magnetic stirring rod (little flea). A voltage applied across the two electrodes undergoes oxygen electrolytic reduction in the solution (Figure 2).

The sample inside the reaction chamber maintains a balance between photosynthesis and respiration releasing and taking oxygen. Any variation in concentration is reflected in the electrolyte, as a consequence of this and related directly, the electric current between cathode and anode also changes. Variation in the oxygen concentration inside the chamber is measured by connecting the electrode to a recorder throughout a control box. Taking into account that oxygen solubility is temperature dependent, the whole electrode and reaction chamber is kept to a constant temperature by mean of a water jacket attached to a temperature controlled water flow (Figure 3).

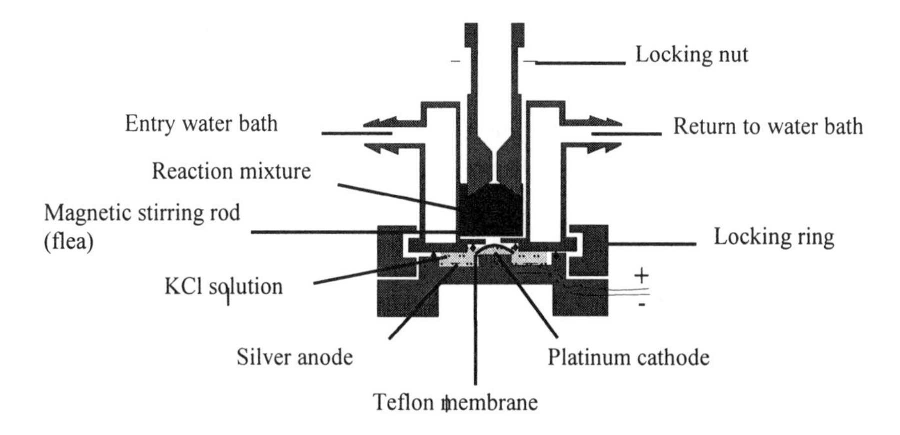

Figure 2. Oxygen electrode components

The basic Clark-type electrode is generally used to measure oxygen exchanges activities of isolated chloroplasts and of algal and cyanobacterial cells in the laboratory. Modified versions of the electrode can be used to measure oxygen evolution from seedlings and leaves and linked with other apparatus like a light controlled source or a pulse modulated chlorophyll fluorescence instrument, it will allow us more detailed study about photosynthesis.

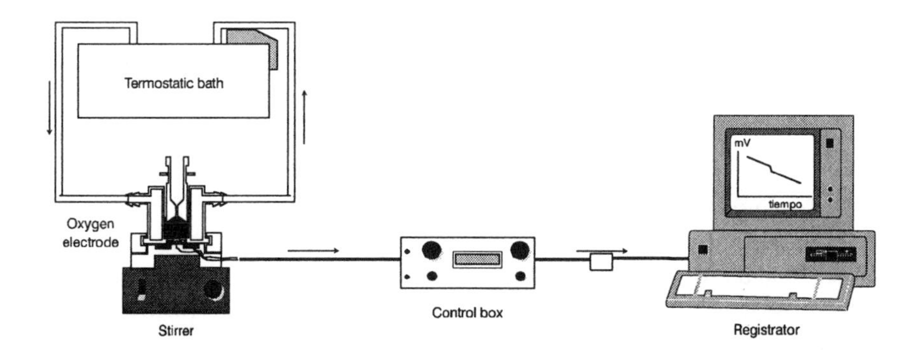

Figure 3. Measurement equipment associated to the oxygen electrode.

PHYSIOLOGICAL ASPECTS

Photosynthesis and respiration are two physiological processes that we can measure by oxygen quantification. Biological oxidation can be seen as a reversal of the photosynthesis process. (Figure 4).

The photosynthesis pigment (P680) of photosystem II (PSII) is a pair of two chlorophyll molecules ([Chl a]$_2$) Their excitation by excitons results in a charge separation. An electron is transferred from the chlorophyll pair to plastoquinone (PQ). This plastoquinone accepts two electrons one after the other and is thus reduced to hydroquinone (PQH_2) (Heldt, 1997).

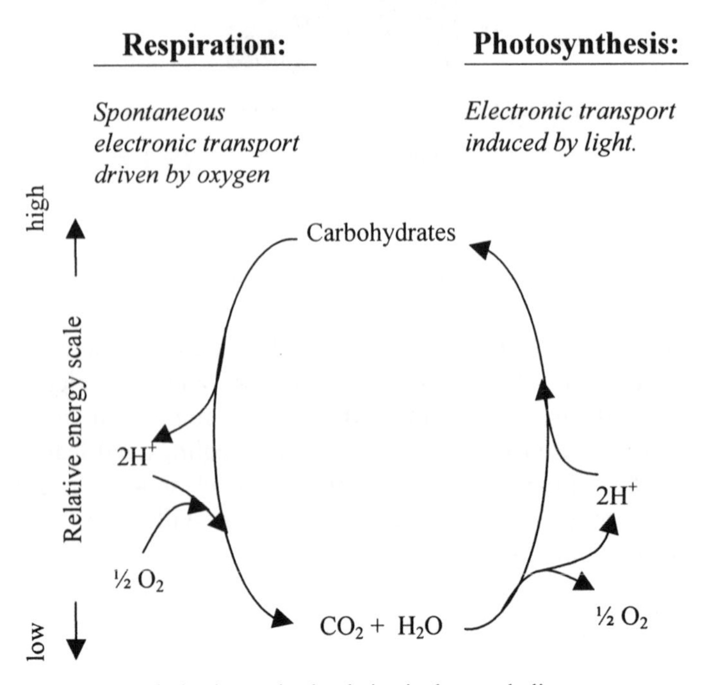

Figure 4. Whole electronic circulation in the metabolic process.

This molecule may be regarded as the final product of PS II. The electron deficit in [Chl a]$_2^+$ is compensated for by a cluster of four manganese atoms involved in the oxidation of water and they are derived to the positively charged chlorophyll radical by mean of a tyrosine residue (Figure 5).

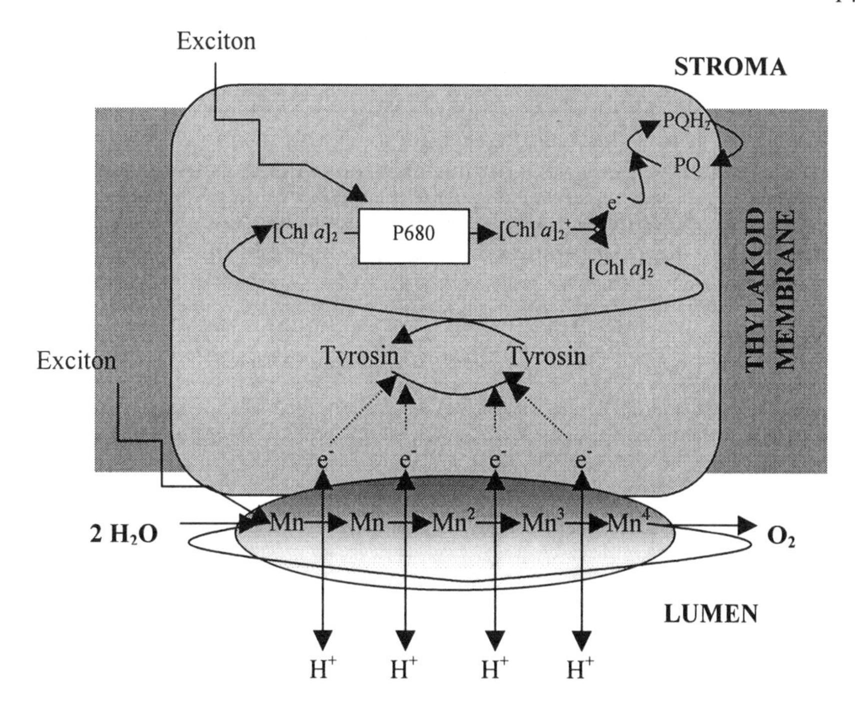

Figure 5 Scheme of photosynthetic electron transport in the photosystem II complex where the oxygen site production is shown.

To liberate one molecule of oxygen from water the reaction centres must withdraw four electrons and thus capture four excitons. The water-splitting machinery of the Mn clusters minimizes the formation of oxygen radical intermediates, especially at low light intensities (Heldt, 1997).

The oxidation of the NADH formed from degradation of substrates in the mitochondria matrix to NAD^+ provides two protons and two electrons to complex I and feeds the respiratory chain with the electrons. Via ubiquinone the electrons are passed to complex III (the cytochrome-b/c complex). Ubiquinone reduced by the NADH dehydrogenase complex or succinate dehydrogenase is oxidised by the cytochrome-b/c complex. Cytochrome-c is a mobile electron carrier. Due to its positive charge, reduced cyt-c diffuses along the negatively charged surface of the inner membrane to the complex IV (cytochrome-c oxidase, or cytochrome-a/a_3 complex) (Buchanan et al., 2000).

This is the terminal complex of the electron transport chain and reduces oxygen to water on the matrix side of the membrane. Cytochrome-a, in the complex, receives electrons from cytochrome-c and passes them to cytochrome a_3 with a Fe atom and a Cu atom bound to histidine. Sixth coordination position of the Fe atom is not saturated by an amino acid of the protein (Figure 6). This free coordination position, and the Cu atom, forms the binding site for the oxygen molecule, which is reduced to water by the uptake of four electrons (Heldt, 1997).

We will describe the use of the oxygen electrode for measuring photosynthesis and respiration in *Lactuca sativa* var. Great Lakes seedlings.

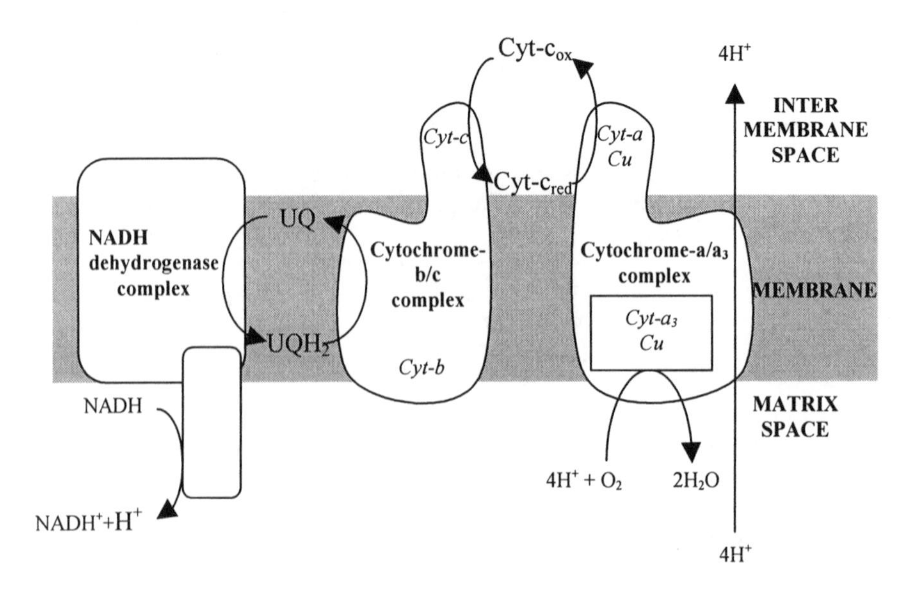

Figure 6 Scheme of the respiratory chain complexes in the mitochondrial inner membrane where the oxygen site consumption is shown.

MEASURE OF OXYGEN EVOLUTION

During photosynthesis, light energy is absorbed by chlorophyll with oxygen production while, during respiration, this molecule is consumed.

Accordingly, if a leaf, seedling or an algae suspension is enclosed in a chamber, provided with carbon dioxide (or bicarbonate as a source of CO_2) and then, illuminated or not, oxygen concentration will change. A 'Clark-type' electrode is able to detect polarographically the oxygen changes due to accumulation or consumption during photosynthesis or respiration in the chamber. Usually, Clark-type electrodes comprise a platinum cathode and a silver anode immersed in, and linked by an electrolyte (potassium chloride). A thin teflon or polythene membrane which is permeable to oxygen and a piece of paper (beneath the membrane) are usually placed over the electrodes surface (Figure 7) in order to protect them and to provide a uniform layer of electrolyte between both anode and cathode.

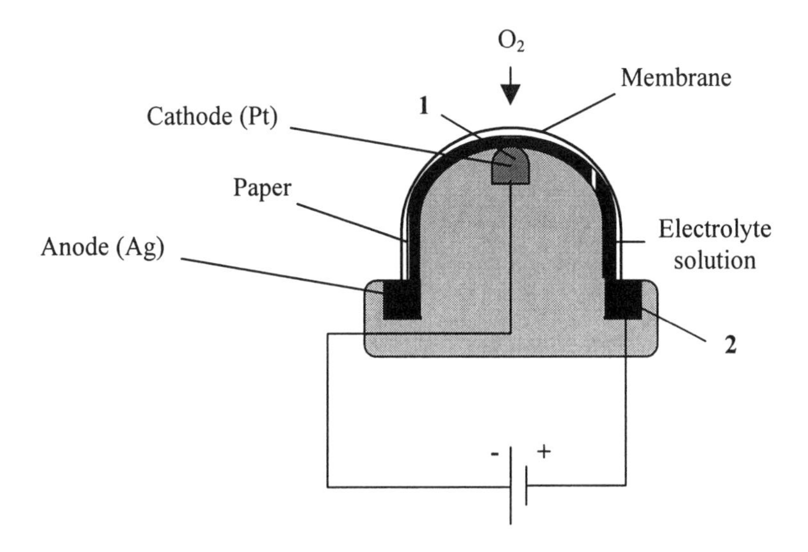

Figure 7 Scheme of the electrodes dome and place of oxygen electrode reactions.
1.- $O_2 + H_2O + 2e^- \rightarrow H_2O_2 + 2OH^-$; $H_2O_2 + 2e^- \rightarrow 2OH^-$
2.- $4Ag \rightarrow 4Ag^+ + 4e^-$; $4Ag^+ + 4Cl^- \rightarrow 4AgCl$

When a small voltage is applied across these electrodes (0.6-0.7 V), so that the platinum is made negative with respect to the silver, the current which flows is at first negligible and the platinum becomes polarised. Oxygen diffuses through the membrane and is reduced at the platinum surface, initially to hydrogen peroxide (H_2O_2). A thin layer of KCl solution closes the current circuit. The silver is oxidised and silver chloride deposited on the anode. The electrical current generated, by the reduction of oxygen at the cathode, is stoichiometrically related to the oxygen consumed and converted to a voltage output signal by a control box. The output signal is large enough to be monitored on a computer where the data are transformed in $\mu mol\ O_2 \cdot min^{-1} \cdot g\ (DW)^{-1}$.

As explained above oxygen solubility is temperature dependent and then the reaction chamber may be kept to constant temperature during the experiment and the final results corrected in function of this temperature.

Thirty Celsius degrees is the most usual temperature used in biological studies. Table 1 shows the oxygen content of air saturated water.

Table 1. Oxygen content of air saturated water

Temperature °C	**$[O_2]$ (µmoles·ml^{-1})**
0	0.442
5	0.386
10	0.341
15	0.305
20	0.276
25	0.253
30	0.230
35	0.219

In this system, photosynthesis and respiration are often expressed as a rate of oxygen uptake or release per unit of weight. This is convenient if it is used as a basis for comparison between different treatments. There are another possibilities such as to express the results as a function of chlorophyll or protein amount or number of cells in the suspension.

Electrochemical reactions in the electrodes always generate a residual current that is detected by the system and recorded in the computer even in the absence of oxygen. Discrepancies between zero oxygen and the electrical zero should be identified for accurate detection of variations in oxygen.

To know the electrical current generated by the system in absence of oxygen it would be necessary to remove the initial oxygen dissolved in the reaction chamber. To do this it is usual to add sodium dithionite to the solution in the chamber that consumes oxygen according to the following reaction:

$$Na_2S_2O_4 + O_2 + H_2O \longrightarrow NaHSO_4 + NaHSO_3$$

Dithionite is corrosive and it could cause damage to the membrane. It should be removed from the membrane as quickly as possible.

PRACTICAL MEASUREMENT

In order to describe the measure with more detail, we have selected the procedures used with a Hansatech D.W unit containing a Clark-type oxygen electrode based on a design by Delieu & Walker.

Electrode assembly (DW1/AD)

1. Switch on the temperature-controller bath fixing the temperature to 30 °C (temperature and flow key).
2. Put a 50 ml flask with distilled water in a magnetic stirrer to get oxygen saturated water.

Electrode disc preparation

1. The dome, which bears the platinum cathode, should be moisten with some drops of electrolyte (5% saturated KCl) flooding the metacrylate well which bears the silver anode.
2. Cut approximately 2.5 cm^2 of cigarette paper (avoid touching it with fingers to prevent finger marks and waste the gummed part) and place it over the cathode dome.
3. Cut approximately 2.5 cm^2 of teflon membrane and place it over the paper (avoid touching it with fingers and getting folds).
4. Ensure the membrane and paper in position, over the dome, with the O-ring provided for this purpose. Fix the membrane smoothly over the surface of the electrode dome avoiding wrinkle formation with the help of the applicator.
5. Ensure for smoothness and that there are not wrinkles. Otherwise it would be rejected and the procedure should be repeated.
6. Check that the paper is wet with the electrolyte securing an electric continuity between anode and cathode.
7. Aspire the electrolyte excess.
8. Fit a larger O-ring to the channel surrounding the electrode dome.
9. Place the electrode disc, dome upwards, on the bottom section of the reaction chamber and thread the assembly section over the

electrode on to the base (avoid over-tightening because this can cause damage to the membrane).

3. Connect the electrode to the circuit by plugging it through the control box to the computer.
4. With the computer switched on, click over the 'minirec' icon to get the recorder software.
5. Add two millilitres of oxygen saturated distilled water to the reaction chamber.
6. Switch on the magnetic stirrer (*main power on*) and introduce a little magnetic rod in the reaction chamber.
7. Ensure that *back-off* controls are setting to cancel, the gain control (*output*) to the x1 position and the variable gain control (*output*) should be set to the minimum position. Then switch on the control box, this should result in an output voltage of about 600-800 mV in a few minutes.

Electrode calibration (zero adjust)

1. When the recorder shows a constant and smooth slope in the graphic, add a spatula tip of sodium dithionite to the reaction vessel.
2. Insert the plunger and adjust it so that the water in the reaction vessel just enters the conical section (Figure 8). Conductivity will decrease very fast before reaching a stable plateau around 30-50 mV (background or residual signal) in some minutes.

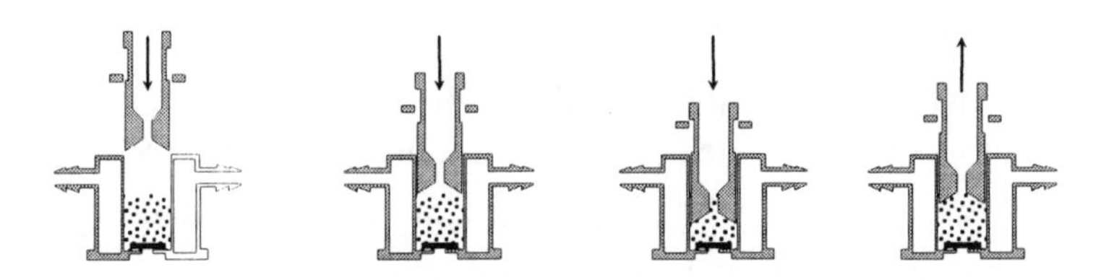

Figure 8. Scheme of the correct way to insert the plunger in the reaction chamber.

3. Set the electric difference between zero oxygen and the residual conductivity switching on the back-off key.

4. If the signal do not go to zero value, adjust it using the fine back-off key.
5. The content of the reaction chamber should be removed as quickly as possible by using an aspirator with a soft tip.
6. Clean the reaction vessel with distilled water for five or six times (remember that sodium dithionite is corrosive) to avoid damage to the membrane. Do not forget to clean the cover of the reaction vessel with distilled water as well.

Measuring

1. Put in the reaction vessel 1.5 ml of oxygen saturated water.
2. Introduce a small net in the reaction vessel to protect the sample from the magnetic stirring rod. Wait 5 minutes to get a stable reading and write down the water conductivity (*wc*).
3. Choose samples with homogeneous sizes for all the replicates. Introduce it in the reaction vessel with the help of the pincers.
4. Close the reaction vessel and tight with the locking nut adjusting the solution volume at the end of the dome of the cover.
5. Maintain the camera in darkness.
6. Wait 10 minutes by setting the electronic timer. Write down the second measurement of conductivity (*icond*).
7. Wait 5 minutes by setting the electronic timer. Write down the third measurement of conductivity (*fcond*).
8. Switch on the source of light, wait for one minute to avoid variations in the signal because of changes in physical conditions and write down the conductivity again (*icond*).
9. Wait 5 minutes by setting the electronic timer. Write down the new value of conductivity (*fcond*).
10. Move away the sample from the chamber. Remove the outer water excess over soft paper and weigh the sample to know the fresh weight (FW). Dry in an oven (100 ºC, 24 h), and weigh it again (DW).
11. Transform the data of conductivity to $\mu mol\ O_2\ min^{-1}\ g\ (DW)^{-1}$ consumed or produced using the following equation:

$$\mu\text{mol } O_2 \text{ min}^{-1} \text{ g (DW)}^{-1} = \frac{WOC \cdot Vol \cdot (fcond - icond)}{wc \cdot t^{-1} \cdot DW}$$

where:

WOC: Oxygen content of air saturated water at 30 °C (0.230 μmol·ml^{-1})
Vol: water volume in the reaction vessel
fcond: final reading of conductivity
icond: initial reading of conductivity
wc: water conductivity
t: minutes between readings
DW: dry weight of the sample

S.N°	Weight		*wc*	Respiration			Photosynthesis			NP
	FW	DW		*Icond*	*fcond*	ΔO_2	*icond*	*fcond*	ΔO_2	
1										
2										
3										
4										
5										

S.N°: Sample number; FW: fresh weight; DW: dry weight; ΔO_2: μmol/min·g (DW); NP: net photosynthesis ($P\Delta O_2$- $R\Delta O_2$)

REFERENCES

Bazzaz F.A. *Plants in Changing Environments. Linking Physiological, Population and Community Ecology*. Cambridge: Cambridge University Press, 1996

Buchanan B.B., Gruissem W., Jones R.L. *Biochemistry and Molecular Biology of Plants*. Rockville, Maryland: American Society of Plant Physiologists, 2000

Bryce J.H., Hill S.A. "Energy Production in Plant Cells." In *Plant Biochemistry and Molecular Biology*. P.J. Lea, R.C. Leegood, eds. Chichester: John Wiley and Sons, 1993

Garab G. *Photosynthesis: Mechanisms and Effects*. Dordrecht, The Netherlands: Kluwer Academic, 1999

Govindjee, Coleman W.J. How plants make oxygen. Sci Am 1990; 262:42-51

Hall D.O., Rao K.K. *Photosynthesis*. Cambridge: Cambridge University Press, 1999

Heldt H-W. *Plant Biochemistry and Molecular Biology*. Oxford: Oxford University Press, 1997

Hitchman M.L. "Measurement of Dissolved Oxygen." In *Chemical Analysis* Vol 49, P.J. Elving, J.D. Winefordner, I.M. Kolthoff, eds. New York: John Wiley and Sons, 1978

Maths P. *Photosynthesis: from Light to Biosphere*. P. Maths, ed. Dordrecht, The Netherlands: Kluwer Academic Publishers, 1995

Raghavendra A.S. *Photosynthesis: a Comprehensive Treatise*. Cambridge: Cambridge University Press, 1998

Van Gorkom H.J., Gast P. "Measurement of Photosynthetic Oxygen Evolution." In *Biophysical Techniques in Photosynthesis*. J. Amesz, A.J. Hoff, eds. Dordrecht, The Netherlands: Kluwer Academic Publishers, 1996

CHAPTER 10

FLUORESCENCE TECHNIQUES

Manuel J. Reigosa Roger[1] and Oliver Weiss[1,2]
[1]*Depto Bioloxía Vexetal e Ciencia do Solo. Universidade de Vigo. Spain*
[2]*Institut für Pflanzenbau. Rheinische Friedrich-Wilhelms-Universität. Germany*

INTRODUCTION

The investigation *in vivo* about the photosynthetic apparatus functioning allowed the development of techniques for the study of the chlorophyll fluorescence in intact photosynthetic organs. In this chapter we will see the possible use of an available instrumental technique in plant ecophysiology: chlorophyll fluorescence in continue excitement.

Besides of the uses that we will comment, this technique can be applied for the study of photosynthesis and for the study of the photosynthetic apparatus (Krause and Weis, 1984).

The *in vivo* chlorophyll fluorescence is affected by several phenomena, which affect photosynthesis, either if they are permanent or reversible. This kind of fluorescence could be measured by modern equipments, with non-invasive and non-destructive techniques, and is also easy to use in field experiments. Because of that any kind of factor affecting the photosynthesis could be individually studied, these techniques have been employed to measure the stress of the photosynthetic apparatus induced by different stress factors.

The following stress factors and phenomena have been specially studied using the chlorophyll fluorescence techniques: environmental pollution, water stress, photoinhibition, and low temperature. Another application is the 'quenching analysis' (Krause and Weis, 1988). With this technique even part of the mechanisms of the stress reaction, produced by the above mentioned stress factors, could be distinguished and, in certain cases, genotypic

M.J. Reigosa Roger, Handbook of Plant Ecophysiology Techniques, 155–171.

differences of the photosynthetic apparatus could be detected. In the last years this technique has been applied in a very wide range of fields and became a common used technique.

Applications

Many enthusiastic papers have been written about this technique and its possible applications (Krall and Edwards, 1992, Schreiber et al., 1995). The relationship found between the photosystem II activity and the CO_2 fixation in plant leaves could be a powerful parameter to estimate the efficiency of the photosynthetic apparatus. Today, the chlorophyll fluorescence became a potential tool for several basic studies and practical applications (Strasser et al., 1995; Strasser et al., 2000; Barth et al., 2001). The authors show different groups of applications for this technique:

- Measurement of chemical changes in the environment (chemical stress); Inhibitors, fertilisers, gases or any other chemical stress factors could be tested.
- Studies of physical parameters of the environment (physical stress), like light quality and intensity, photoinhibition (Krause and Somersalo, 1989; Koroleva et al., 1994) and temperature.
- Biosensing: analysing a system in terms of vitality, productivity, resistance to stress; studying the ecodynamics of a complex system, for example a tree or in horticulture, or forests (Ball et al., 1994).

Therefore, several applications are found for this technique. In agriculture it is widely used for testing herbicides (Lichtenthaler and Rinderle, 1988), other pesticides and growing regulators, or to test allelopathic effects of secondary metabolites on the plant metabolism (Devi and Prasad, 1996; Roschina and Melnikova, 1996), as also to evaluate genotypic differences in plant breeding. This technique was called to be 'productivity marker' (Planchon et al., 1989) in a study with barley. Also the resistance of cultures to drought, heat, cold, light, and salt stress has been tested by this technique (Strasser et al., 2000). For example, Prange (1986) found chlorophyll fluorescence to be a rapid and sensitive mean of detecting desiccation in potato leaves.

Chlorophyll fluorescence is also applied in quality control of commercialised products. For example, freshness and colour of vegetables, flowers and fruits, storage conditions or water quality (Merz et al., 1996).

Other possible applications could be (Strasser et al., 2000):

- Testing greenhouse conditions to optimise the production process.
- Testing effects of pollution on environmental conditions.
- Testing the effects of global changes on the behaviour of ecosystems.

The measurement of the *in vivo* fluorescence behaviour can be done simultaneously with measurements of gas exchange, allowing so an even deeper knowledge of what happens in the chloroplast.

The physiological aspects

When a plant leaf absorbs light a chain of reactions begins to work. This process is called photosynthesis. During photosynthesis, radiation energy is transformed into chemical energy by several physical and chemical mechanisms. The process starts in the chlorophyll with light absorption by the antenna molecules. In the next step the radiation energy is transferred as excitation energy and used in a reaction centre for chemically useful work, or dissipated mainly as heat, and less as emitted radiation, or in other words, as *in vivo* chlorophyll fluorescence (Strasser et al., 2000). So, trapped photons energy produces redox energy (thus inducing electrons flux and photochemical work), infrared radiation and fluorescence (Fig. 1).

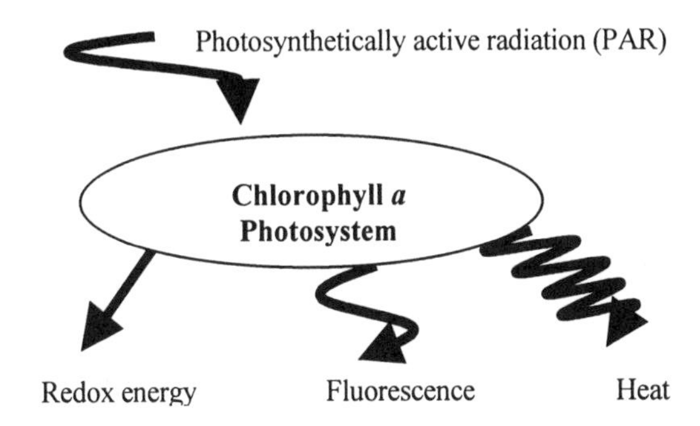

Figure 1. The main processes by which radiation energy is dissipated after excitation of a chlorophyll *a* (Reigosa, 1997)

Chlorophyll fluorescence and gas exchange measurements provide the 'experimental interface' between the plant response to the environment and the biochemical and biophysical mechanisms, which determine this response (McDermitt et al., 2000).

But, where does fluorescence occur? In Figure 2 the origin of chlorophyll fluorescence and where it takes place is shown, beginning in a green plant leaf and finishing at the antenna of the reaction centres located in the thylacoid membrane.

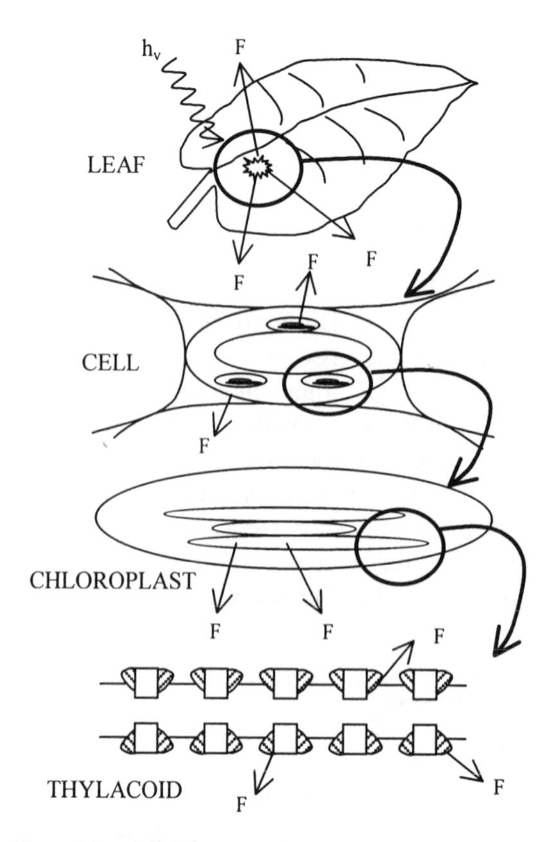

Figure 2. The origin of fluorescence from a macroscopic point of view to a more and more microscopic viewpoint (Adapted from Strasser, 1995)

Green plants contain two kinds of chlorophyll. They are called chlorophyll *a* and chlorophyll *b*. In white light they appear to be of green colour because they absorb light of the blue and red waveband of the spectrum (Fig. 3).

So, when light shines on a plant leaf only the energy content of the blue and red light will be absorbed by the chlorophyll. Some energy will be used for chemical reactions in photosynthesis. And all the energy, which is not used, will be dissipated as shown before.

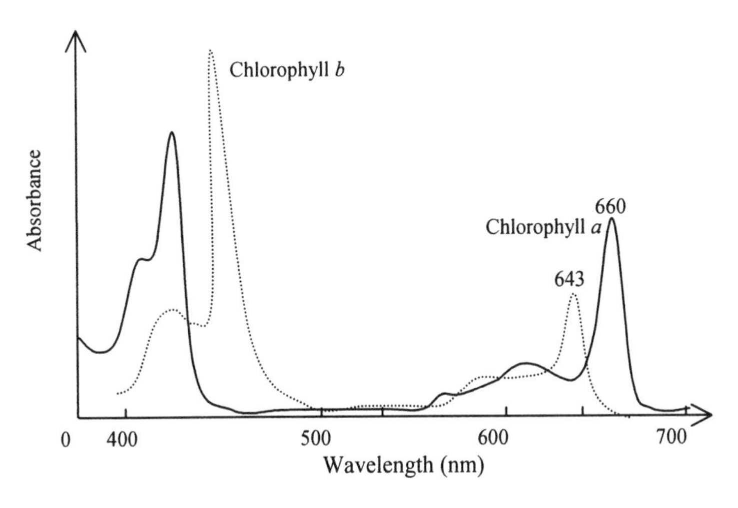

Figure 3. The absorption spectra of chlorophyll *a* (continuous line) and chlorophyll *b* (broken line) (Redrawn from Lichtenthaler and Rinderle, 1988)

The lost energy is from light of a lower energy content than the absorbed light. Light with a lower energy content is that kind of light of a longer wavelength (Planck's law). So, the fluorescence emission is always of a longer wavelength than the light absorbed by the leaves (Hansatech, 1996). This must be considered for an adequate detector-filter selection.

At physiological temperatures, in higher plants, most of the fluorescence comes from the chlorophyll *a* molecules, which are associated to the photosystem II (PS II, Fig. 4).

A fluorescence emission superior to a wavelength of 710 nm and associated to the photosystem I (PS I) is possible. In this sense the fluorescence activity of both photosystems can be compared by means of the fluorescence ratio.

$$\frac{\text{emission at 695 nm}}{\text{emission at 720 nm}}$$

However, the fluorescence answer is fundamentally confined to the photosystem II at the usual atmospheric temperatures (Krause and Weis, 1984). The fluorescence of this photosystem II can be measured with a photomultiplier or with a sensitive photodiode to the light in the region of 690 nm (red light in which the emission of *in vivo* chlorophyll *a* fluorescence takes place, see Fig. 3).

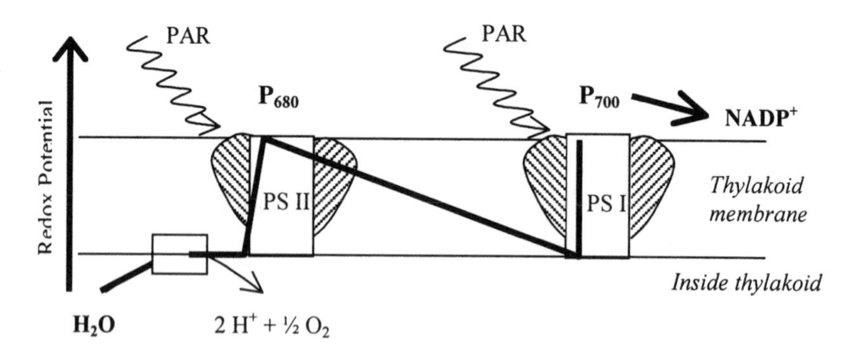

Figure 4. Outline of the Z-scheme or photosynthetic chain reaction of electron transport, disposition of molecules that allows the photochemical conversion, that means, the step of luminous energy to redox energy and charge of ATP. Hatched surfaces represent light harvesting complexes (LHC) of the photosystems.

To produce fluorescence the leaf must be illuminated with photosynthetically active radiation. That is the reason why the measurement of fluorescence emission requires the separation of any other radiation. Generally, this is done by means of a filter that prevents the reaching of longitudes shorter than 620 nm (see Fig. 5).

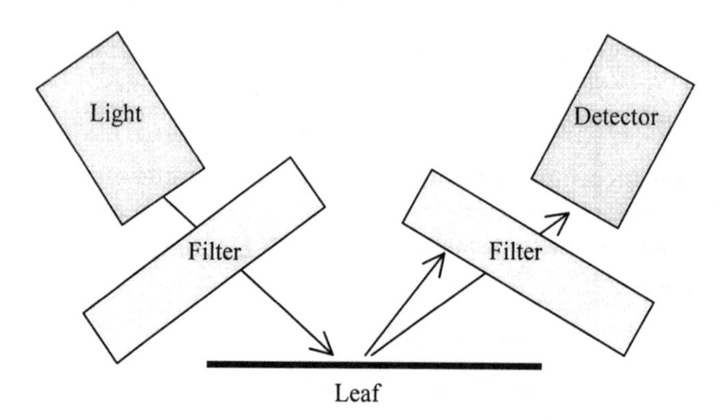

Figure 5. Outline of the operation for a non-modulated fluorometer. The detector filter needs to be a long-pass filter (Reigosa, 1997).

If a plant leaf is illuminated at constant intensity, the emission of fluorescence is continuously at a steady level. However, much more information will be obtained when the changes of fluorescence are studied. This occurs when the illumination levels are changing.

So, if a plant leaf is dark-adapted, that means kept for several minutes in darkness, and suddenly exposed to bright light, a typical fluorescence transient, also called Kautsky-effect, occurs (Fig. 6). This response is known as the curve of Kautsky, and a set of values denominated OIDPSMT has been identified in this curve. Therefore, sometimes it is called OIDPSMT curve, too. In this curve the fluorescence is represented in arbitrary units.

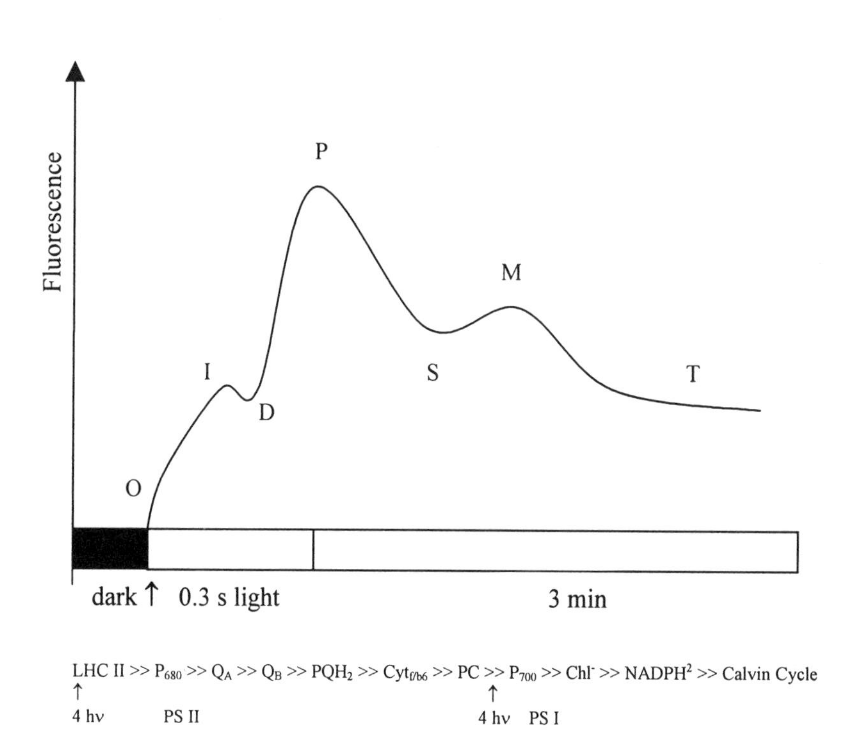

LHC II >> P_{680} >> Q_A >> Q_B >> PQH_2 >> $Cyt_{f/b6}$ >> PC >> P_{700} >> Chl^- >> $NADPH^2$ >> Calvin Cycle

↑ 4 hν PS II ↑ 4 hν PS I

Figure 6. Chlorophyll fluorescence induction kinetics (Kautsky effect: fast rise and slow decline) after illumination of a dark adapted plant leaf. The form of the curve is similar in all higher plants, although values change. OIDPSMT means: O = origin, I = inflection, D = dip level, P = peak, S = first steady state, M = maximum, T = terminal state. In the lower part of the figure the fluorescence induction curve has been aligned with the reactions in the electron transport chain; hν = photon; LHC II,I = light harvesting complex of PS II and I; P_{680} and P_{700} reaction centres of PS II and I; Q_A = primary plastoquinone acceptor of PS II; Q_B = secondary plastoquinone acceptor of PS II; PQH_2 = reduced plastoquinone; $Cyt_{f/b6}$ = cytochrome f/b6 complex; PC)= plastocyanin; Chl^- = intermediate electron acceptor of PS I. (Redrawn from Lichtenthaler and Rinderle. 1988: Bolhàr-Nordenkampf and Öquist. 1993)

Kautsky and Hirsch in 1931 were the first in describing this phenomenon.

They showed that the chlorophyll *a* fluorescence emission exhibits a fast rise to a maximum followed by a slow decline to reach finally after a few minutes a steady level. They also demonstrated that the slow fluorescence decline is related with a CO_2 assimilation increase.

Determined points on the Kautsky curve (Fig. 6)

O (origin): F_0 is the fluorescence value immediately reached after the illumination (< 1 ps). It represents the fluorescence when the practical entirety of the reaction centres of the photosystem II (PSII) are 'open'. These reaction centres are open when the primary acceptor (Q_A) is oxidised. During the darkness (if the period is long enough) these reaction centres get open. Then, if the illumination goes on, the fluorescence ascends until a first peak (F_M), well-known as phase P, going through an inflection phase (I) often after a stage (D) of slope of the curve. The increment from F_0 until F_M is interpreted as the increase of attributable fluorescence to the every time bigger competition among the photosynthetic process (or, rather, of the photochemical phase of the photosynthetic process) with those of exchange of heat and fluorescence (see Fig. 1). During this process more and more primary acceptors of the photosystem II (Q_A) get closed (in other words, reduced) (see Fig. 2). The increase of fluorescence until P usually lasts less than 2 seconds in moderate light and is denominated fast phase of the induction of fluorescence. The value of F_0 can be used as reference value for the rest of fluorescence values.

I (inflection): Where all the primary plastoquinone acceptors of PS II (Q_A) are reduced (0.5 ms);

D (dip level): Where all the secondary plastoquinone acceptors (Q_B) of PS II are reduced (100-600 μs)

P (peak): Where the plastoquinone pool is completely reduced (0.5-2 s). It's called 'closed traps' (F_M).

S (first steady state): At this stage an accumulation of $NADPH_2$ and ATP happens while the pH rises.

M (maximum): A secondary peak of maximum fluorescence usually exists after the peak P. Generally, it is interpreted as the moment in which the assimilation of the CO_2 begins: the Photosynthetic Carbon Reduction Cycle starts.

T (terminal steady state): After 180-300 s a constant level of chlorophyll fluorescence have been reached. It is interpreted as the moment in which a balance state is reached in the capture and assimilation of CO_2. This slope until a stable level of fluorescence from P is denominated as slow phase of the fluorescence induction.

q: The difference among the maximum fluorescence (P, F_M) and the fluorescence once reached the balance (T, F_T) is denominated 'fluorescence quenching'. Two types of quenching could be distinguished: photochemical quenching (q_P) and not photochemical quenching (q_{NP}). The first one is related to the energy transformation in the reaction centre of the photosystem II, the second one represents, after the chlorophyll excitement, the different energy return ways (mainly as heat and energy transfer to PS I).

Fluorescence parameters and plant stress

F_0: It is affected by any environmental stress, which causes alterations at the pigment level of PS II. For example, thermal damage of the PS II is characterised by a drastic increase in F_0. But freezing damage of the thylakoids does not affect F_0.

F_v: The difference $F_M - F_0$ is denominated variable fluorescence. Normally this value is lowered by environmental stresses, which cause damage on the thylakoids. Examples for this kind of stress are heat, freezing and photoinhibition.

F_M: The maximal fluorescence decreases after exposure of the leaf to high but not injurious temperatures. More severe heat treatment causes an increase of F_0 and a decrease of F_M. The PS II activity will be inhibited.

F_M/F_0: This ratio depends on the leaf water potential. Under drought conditions it will drop to 1. Under severe drought conditions no F_V was produced, but after watering F_M/F_0 were restored up to 3.0 (Hansatech, 1996).

F_v/F_M: This relation is typically in the range of 0.832 ± 0.004 (Krause and Weis, 1991) and it is proportional to the quantum yield of the photochemical phase of the photosynthesis. A decrease of this relation is a good indicator for damage of photoinhibition caused by the light when the plants undergo diverse types of environmental stresses, as drought, cold, freezing, excessive illumination or salinity. When the changes in this coefficient are examined, an increment of F_0 (usually attributed to the destruction of reaction centres of the photosystem II) can be separated from the decreases of F_v, which are usually explained as increments of the not photochemical quenching. It is noteworthy that photoinhibition, for example, stimulates both effects.

This ratio is used thoroughly as an indicator of stress induced on the photosynthetic apparatus. However, in the moment of its interpretation it is necessary to have certain cautions. In the first place, the definition of F_v depends on how well we are able to measure F_0. The point O of the Kautsky curve is defined by a discontinuity in the rate of change of fluorescence in regard to the time. In some equipment, the speed of opening of the diaphragm can make that the time corresponding to F_0 coincides with the opening moment. So, different fluorometers could give different values of F_0. In this possible error it is not necessary to consider the fluorometers of modulated pulse or modulated fluorometers, since for their mode of operation an error is not possible. The second consideration to keep in mind is that F_M will only be reached if the excitement energy is enough to reduce all the primary acceptors. To be sure of this, we should check the F_M values with different photon flows (see 'Light Saturation').

$T_{1/2}$.: half time between the point O and the point P (see Fig. 4 Kautsky curve). This half time is strongly affected by those processes that block the energy transfer from the reaction centres until the quinones.

Finally, in the interpretation of the results it is necessary to say that in a Kautsky curve, obtained with a non-modulated fluorometer, it is considered that the leaves have been previously adapted to the darkness. That means that all the reaction centres are absolutely open. Usually, periods from 15 to 30 minutes are considered as enough, but this depends on the species. In certain situations of stress this period could be longer. Therefore, it is necessary to check (by performing previous experiments) the necessary period of submission to darkness in the plants that we want to study. The current fluorometers have leafclips that allow carrying out a schematic measurement in the same leaves subjected to different darkness periods.

WORKING WITH A FLUOROMETER

A fluorometer of continue excitement, such as the Plant Efficiency Analyser (PEA) of Hansatech Instruments Ltd. (see Fig. 7), is adequate to obtain the Kautsky curve.

The following description is based on the experiences with a fluorometer 'Plant Efficiency Analyser (PEA)' from Hansatech Instruments Ltd, Pentney, Norfolk, England and operating instructions from Hansatech Instruments Ltd (1995).

The Plant Efficiency Analyser (PEA) measures the chlorophyll fluorescence, which has been emitted from green plants. It is not expensive, but lightweight and completely portable. This equipment has been designed to be easy to use, and its functioning is equally good in either laboratory or in the field.

The measurements of the fluorescence induction are made using dark-adapted intact plant leaves, needles or other samples including micro and macroalgae. For the last sample types, special accessories are necessary.

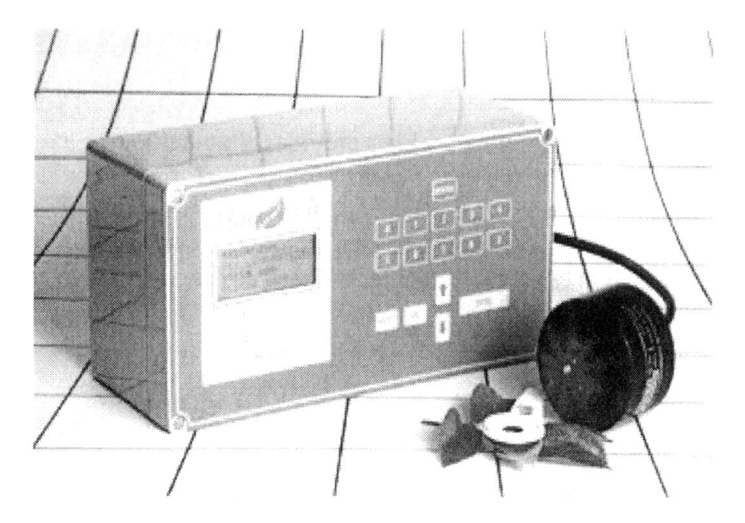

Figure 7. Plant Efficiency Analyser (PEA) for chlorophyll fluorescence measurement, photo courtesy of Hansatech Instruments Ltd., Pentney, Norfolk, England (www.hansatech-instruments.com).

The PEA system consists of the following items

Leafclips

Before any measurement the leaf or needle, which will be analysed, must be covered with a small, lightweight leafclip. The leafclip is a kind of pincer. It is made of white plastic to reduce the development of heat during the time where the leafclip is placed on the leaf. It consists of a locating ring for the sensor unit, a shutter plate to close the exposed leaf area in the centre of the locating ring, and a foam pad in order to minimise possible damages of the leaf structure.

When the leafclip is placed at the chosen leaf, the shutter plate must be closed so that light is excluded and dark adaptation starts. During dark adaptation any fluorescence yield is quenched. This process takes time. This period is variable and depends on the analysed plant species, the light levels prior to the dark adaptation, and if the plant is stressed. Normal time periods are about 30 minutes. Two strategies are offered to reduce waiting time: You can measure during a non-light period, or several leafclips can be simultaneously used to ensure sufficient dark-adapted leaves.

Sensor Unit

It houses an optical assembly, which provides illumination of the exposed leaf area and the detection of the consequent fluorescence signal. Therefore, the sensor unit consists of an array of 6 ultra bright LED, which are focused onto the leaf surface to provide even illumination over the exposed leaf area (4 mm diameter). They provide red light with a peak wavelength of 650 nm and a maximum intensity of 3000 µmol m^{-2} s^{-1}. The LEDs dissipate low heat radiation and they rise fast to full light intensity (microseconds) after switched on. So inaccuracies of F_0 measurements are eliminated.

The detector is a high performance Pin photodiode and associated amplifier circuit. An infrared long-pass filter (50% transmission at 720 nm) ensures that it responds maximally to the longer wavelength of the chlorophyll fluorescence signal and blocks the reflected shorter wavelength of the light source (LED light). The sensor unit plugs in the main control box via a 1.5 m cable. The sensor unit is located over the leafclip at the location ring. So, daylight will be excluded. During the measurement the leaf with the sensor unit connected should be held in its correct position with the hand.

Control box

It consists of three components: The LCD display, a waterproof membrane keypad and a side panel of connectors. PEA is totally solid-state, no moving parts or chemicals are needed. A rechargeable battery and a microprocessor control all instrument functions. Data and measurement settings could be stored for several days by backup device. But it is recommended to download data almost every day and don't store critical data for longer time periods.

Light level and recording interval are selectable from the keypad. During the initial fast rise the fluorescence signal is measured by a reading frequency of 100.000 readings s^{-1}, first of all to provide a good resolution of the F_0 value. After 2 milliseconds and until 1 s a slower frequency is taken (about 1.000 readings s^{-1}). After 1 s a rate of 1 reading s^{-1} is adopted. At the end of the recording time of data points are expressed automatically the fluorescence parameters F_0, F_M, F_v and the ratio F_v/F_M are calculated.

MAKING A MEASUREMENT

Plant material

It is worth to think about the plant material that will be measured for a while. Anyone who goes into a forest, writes Strasser et al. (2000), with a good idea, born in the lab, in order to analyse a tree, will find many practical problems. In a lab any material will be homogenous, clean and healthy. But standing in front of a tree a lot of crucial decisions have to be taken.

Which leaf, which branch should be taken? Every kind of leaf material, from green to brown, will be found. The answer is that many samples must be chosen. So, a fast screening method with lightweight equipment is required. Therefore, *in vivo* chlorophyll fluorescence is an excellent parameter

Experimental procedure

Two settings of crucial importance must be determined before starting the measurements. The first is the time that the plant leaf needs to be dark adapted. The other one is the determination of the saturating light level, or in other word how much light must be provided to get all reaction centres closed.

Determination of the dark adaptation time

When the measurements are taken in actinic light absence, the determination of the minimum dark adaptation period is important. During dark adaptation all reaction centres have to oxidise or to get 'open'. The time period may vary with the light level prior to the dark adaptation. So it is recommended that the determination be carried out on plants which have been in high light conditions. So, the dark adaptation period will be enough for all light conditions.

Hansatech recommends to place 20 leafclips on the leaves or needles and to close the shutter plates. Measure in intervals of 2 minutes, using the 100% light level. Build a table of dark time period and F_v/F_M. 'The ratio should plateau at the minimum satisfactory dark adaptation period' (Hansatech, 1996).

Determining the saturating light level

After the determination of the required time for dark adaptation place 10 leafclips on the leaves or needles and close the shutter plates for at least the determined dark adaptation time period (Hansatech, 1996).

Then make measurements at each clip using increasing light levels from 10 to 100%. The F_v/F_M should plateau at the saturation point. Generally, the 100% light level will not cause over-scaling problems.

Sample preparation and measurement

Attach a leafclip to a leaf and close the shutter plate to start the dark adaptation. Ensure that the base of the leafclip is not directly exposed to the light source, otherwise light will penetrate the clip and spoil the dark adaptation.

When the leaf is dark adapted fit carefully the sensor head over the location ring of the leafclip in order to seal out the light. Hold the clip and the sensor unit in your hand, open the shutter plate and press either the button on the sensor head or ENTER on the keypad of the console. The measurement starts. With the PEA full data files (FULL DATA – 75 full data sets) or only the fluorescence parameters (PARAMS ONLY – 8500 parameter sets) could be recorded. To measure only the fluorescence parameters a recording time of 3 s is enough. To achieve a complete OIDPSMT curve a recording time of 3-5 min should used. After the measurement data could be stored by pressing ENTER or the sensor switch.

Data analysis

For the data evaluation, software is supplied as well to download the data points as for graphical analysis of full data files or parameter files. These files could be converted to ASCII format and than imported into other software packages for further analysis.

A typical data sheet is shown in table 1. The data have been taken in three different maize populations, at 11:00 am and after a 30 min dark adaptation period.

Table 1. Fluorescence parameters measured at three plants of different maize varieties. Three measurement replicates are shown for each plant (unpublished data).

File	Light %	Sec	F0	Fm	Fv	Fv/Fm	Area
101	60	5	646	3332	2686	0.806	72500
102	60	5	651	3496	2845	0.814	68300
103	60	5	598	3463	2865	0.827	77800
201	60	5	659	2550	1891	0.742	92300
202	60	5	644	3183	2539	0.798	260000
203	60	5	643	2688	2045	0.761	62800
301	60	5	574	3035	2461	0.811	77400
302	60	5	585	2939	2354	0.801	80300
303	60	5	589	3453	2864	0.829	70400

The value 'area' means the area above the fluorescence curve between F_0 and F_M. It seems that it is proportional to the pool size of the electron acceptors Q_A of the PS II. If the electron transfer from the reaction centres to the quinone pool is blocked, for example in the case of a photosynthetically active herbicide like DCMU, this area will decrease dramatically (Hansatech, 1995).

Acknowledgements. We thank Adela Sánchez-Moreiras for her critical review of the manuscript and her invaluable help with the figures. O. Weiss thanks to the Konrad-Adenauer-Foundation for financial support.

REFERENCES

Ball M.C., Butterworth J.A., Roden J.S., Christian R., Egerton J.J.G. Applications of chlorophyll fluorescence to forest ecology. Aust J Plant Physiol 1994; 22:311-19

Barth C., Krause G.H., Winter K. Responses of photosystem I compared with photosystem II to high-light stress in tropical shade and sun leaves. Plant Cell Environ 2001; 24:163-176

Björkman O., Demmig B. Photon yield of O2 evolution and chlorophyll fluorescence characteristics at 77 K among vascular plants of diverse origins. Planta 1987; 170:489-504

Bolhàr-Nordenkampf H.R., Öquist G. "Chlorophyll Fluorescence as a Tool in Photosynthesis Research." In *Photosynthesis and Production in a Changing Environment: a Field and Laboratory Manual.* D.O. Hall, J.M.O. Scurlock, H.R. Bolhàr-Nordenkampf, R.C. Leegood, S.P. Long, eds. London: Chapman and Hall, 1993

Devi S.R., Prasad M.N.V. Influence of ferulic acid on photosynthesis of maize: analysis of CO_2 assimilation, electron transport activities, fluorescence emission and photophosphorylation. *Photosynthetica* 1996; 32

Hansatech Instruments Ltd. *An Introduction to Fluorescence Measurements with the Plant Efficiency Analyser (PEA).* Pentney, Norfolk, England: Hansatech, 1996

Hansatech Instruments Ltd. *Operating Instructions for Plant Efficiency Analyser - Special JIP Firmware Version.* Pentney, Norfolk, England: Hansatech, 1995

Kautsky H., Hirsch A. Neue Versuche zur Kohlensäureassimilation. Naturwissenschaften 1931; 19:96

Koroleva O.Y., Brüggemann W., Krause G.H. Photoinhibition, xantophyll cycle and *in vivo* chlorophyll fluorescence quenching of chilling-tolerant *Oxyria digyna* and chilling-sensitive *Zea Mays*. Physiol Plantarum 1994; 92:577-84

Krall J.P., Edwards G.E. Relationship between photosystem II activity and CO_2 fixation in leaves. Physiol Plantarum 1992; 86:180-87

Krause G.H., Somersalo S. Fluorescence as a tool in photosynthesis research: application in studies of photoinhibition, cold acclimation and freezing stress. Philos T Roy Soc B 1989; 323:281-93

Krause G.H., Weis E. "The Photosynthetic Apparatus and Chlorophyll Fluorescence. An Introduction." In *Applications of Chlorophyll Fluorescence*, H.K. Lichtenthaler, ed. Dordrecht, The Netherlands: Kluwer Academic Publishers, 1988

Krause G.H., Weis E. Chlorophyll fluorescence and photosynthesis: the basics. Annu Rev Plant Physiol Plant Mol Biol 1991; 42:313-49

Krause G.H., Weis E. Chlorophyll fluorescence as a tool in plant physiology. II. Interpretation of fluorescence signals. Photosynth Res 1984; 5:139-57

Lichtenthaler H.K., Rinderle U. The role of chlorophyll fluorescence in the detection of stress conditions in plants. Crit Rev Anal Chem 1988; 19: 29-85

McDermitt D.K., Garcia R.L., Welles J.M., Demetriades-Shah T.H. "Common Errors in Gas Exchange Measurement." In *Probing Photosynthesis,* M. Yunus, U. Pathre, P. Mohanty, eds. London - New York: Taylor and Francis, 2000

Merz D., Geyer M., Moss D.A., Ache H.-J. Chlorophyll fluorescence biosensor for the detection of herbicides. Fresen J Anal Chem 1996; 354:299-305

Planchon C., Sarrafi A., Ecochard R. Chlorophyll fluorescence transient as a genetic marker of productivity in barley. Euphytica 1989; 42: 269-73

Prange R.K. Chlorophyll fluorescence *in vivo* as an indicator of water stress in potato leaves. Am Potato J 1986; 63:325-33

Reigosa M.J. "Técnicas de Medida de la Fluorescencia *in vivo* en Ecofisiología Vegetal." In *Manual de Técnicas en Ecofisiología Vegetal.* N. Pedrol, M.J. Reigosa, eds. Vigo, Spain: Gamesal, 1997

Roshchina V.V., Melnikova E.V. Microspectrofluorometry: a new technique to study pollen allelopathy. Allelopathy J 1996; 3:51-58

Schreiber U., Bilger W., Neubauer C. "Chlorophyll Fluorescence as a Nonintrusive Indicator for Rapid Assessment of *in vivo* Photosynthesis." In *Ecophysiology of Photosynthesis.* E.D. Schulze, M.M. Caldwell, eds. Springer-Verlag, 1995

Strasser R.J., Srivastava A., Govindjee. Polyphasic chlorophyll *a* fluorescence transient in plants and cyanobacteria. Photochem Photobiol 1995; 61:32-42

Strasser R.J. Fluorescence Technique and its Use in Stressphysiology. Workshop, 1995

Strasser R. J., Srivastava A., Tsimilli-Michael M. "The Fluorescence Transient as a Tool to characterize and screen Photosynthetic Samples." In *Probing Photosynthesis,* M. Yunus, U. Pathre, P. Mohanty, eds. London - New York: Taylor and Francis, 2000

CHAPTER 11

MODULATED FLUORESCENCE

Oliver Weiss[1,2] and Manuel J. Reigosa Roger[1]
[1]*Depto Bioloxía Vexetal e Ciencia do Solo. Universidade de Vigo. Spain*
[2]*Institut für Pflanzenbau. Rheinische Friedrich-Wilhelms-Universität. Germany*

INTRODUCTION

Chlorophyll present in chloroplasts can absorb light in the waveband 400-700 nm, be it directly or driven by other pigment molecules. The light in those wavelengths is termed Photosynthetically Active Radiation (PAR), thus indicating that those photons can induce photosynthesis.

The amount of photochemical word driven by light is limited in any plant. This photochemical work depends of intrinsic factors (electron transport rates, photoprotection, number of chlorophyll molecules in the thylakoids, inefficiency of Rubisco, or up-regulation of several enzymes for example) but also by several environmental conditions, including any kind of stress. When excited, chlorophyll a in the photosystems can produce photochemical work, induce heating and release fluorescence. We term chlorophyll fluorescence to the reemission of photons previously trapped by chlorophyll. This emission of fluorescence is very complex. Of course, fluorescent measuring techniques have played an important role in the study of the photochemical process, but these basic studies are far beyond the objectives of this book. There are several excellent reviews and books about chlorophyll fluorescence and about the use of fluorescent techniques in the research of the photochemical phase of photosynthesis (see, for example, Amesz, 1995; Krause and Weis, 1991; Sauer and Debreczeny, 1995 or Strasser et al., 2000).

Chlorophyll fluorescence is very weak to be viewed, although several fluorometers have been developed to measure that emission of light and to

M.J. Reigosa Roger, Handbook of Plant Ecophysiology Techniques, 173–183.

distinguish it from other sources of light. Of course, the total amount of light trapped by chloroplasts will be equal to the sum of chlorophyll fluorescence + photochemical work (that is, photosynthesis) + heating. So, the emission fluorescence has been considered as a non-invasive probe of photosynthesis, and it is now very used especially when studying marine micro-algae. The technique of measure of fluorescence is very fast, cheap and the equipment does not influence to a great extent the life conditions of the leaves. The use of fluorescence techniques to estimate photosynthesis especially in the field has been proposed and it is now widely accepted due that it reveals all the photosynthetic work and not only the one related to Carbon, that can be affected by many circumstances (Osmond et al., 1999). However, there are still some doubts, for example because it relies on some not fully validated assumptions about photon distribution between photosystems, and so it must be validated by a simultaneous measure of CO_2 or O_2 exchange (Maxwell et al., 1998).

Here we will discuss the recent fluorescence *in vivo* measurements: the development of new equipment in the last year has opened a new field of research, with new possibilities both in the basic and applied research in basic plant physiology and in plant ecophysiology. Particularly, the measure in intact leaves is of special interest to plant ecophysiologists. Nevertheless some care should be taken: solid interpretations of fluorescence measured signals have been done after *in vitro* studies, done with isolated chloroplasts or even thylakoids. In the most favourable case, isolated protoplasts have been used (Krause and Weis, 1991). Those interpretations are not completely useful for plant ecophysiologists, while the evidence in the whole plant – intact leaf measurements rely on empirical observations and we still have not full comprehension of them.

FLUORESCENT TECHNIQUES. CONTINUOUS AND PULSE-MODULATED FLUOROMETERS

The fluorometers capable of *in vivo* measurements can be divided in two main categories: Continuous illumination and modulated fluorescence systems. Both techniques are non-destructive, usable in the lab and field conditions and very fast.

Continuous fluorescence systems use optical filters to separate the excitation light (provided by the apparatus) and the fluorescence signal produced by the chloroplasts. So, the leaf must be shielded to avoid ambient light. These systems allow a high frequency sampling of fluorescence induction kinetics, thus allowing a very fast and narrow sequencing of the fluorescent phenomena after the excitation. These fluorometers measure F_0, Fm, Fv/Fm, area over induction curves and OJIP analysis. The main uses of these instruments are analysis of photosystem II fluorescence and OJIP analysis. The use of non-modulated fluorometers is discussed in the previous chapter, but we include here some general considerations that can be also useful for the comprehension of modulated fluorescence techniques.

Fluorescence induction kinetics is normally studied after a dark adaptation of the leaf (or other photosynthetic part). This dark adaptation is a period in which the plant (or, normally, the part of the leaf where fluorescence will be measured) does not receive any light; this permits all the photosystem II reaction centres to be oxidised (we term it 'in open state'). When a weak light is applied, we suppose that reaction centres are still open, and the weak fluorescence that can be measured is termed F_0.

Pulse-Modulated fluorescence fluorometers use a pulsed light to induce a pulsed fluorescence in the leaf; that signal is electronically amplified and identified, so ambient light can be used. This allows measure of F_0 and Fm during induction and determination of quantum efficiency of photosystem II during photosynthesis under ambient light. The parameters that these pulse-modulated fluorometers measure include F_0, Fm, Fv/Fm, Photosystem II quantum efficiency, photochemical quenching, non-photochemical quenching, and, if they include a PAR analyser, Electron Transport Rate. So, these fluorometers are useful for quenching analysis. Pulse-modulated fluorometers can be used in combination with IRGA or oxygen electrode. So, a more complete overview of the photosynthetic process is attained (Reigosa, 1997).

White light cannot be used at least initially to study the fluorescent *in vivo* emission of chlorophylls, because it includes the wavelengths in which that fluorescence is emitted. However, pulse-modulated fluorometers allow the study of *in vivo* fluorescence of chlorophylls in illuminated plants by the use of a weak modulated light along with a system capable of separate that wavelength and phase. Figure 1 shows schematically how those fluorometers are built. Modulated light is usually of very low intensity, not being capable of inducing a fluorescent emission curve, although some modulated fluorometers also include a light with greater intensity just for generating that curve.

Quenching analysis is performed by the use of another light with saturating intensity, which is used to 'close' all reaction centres.

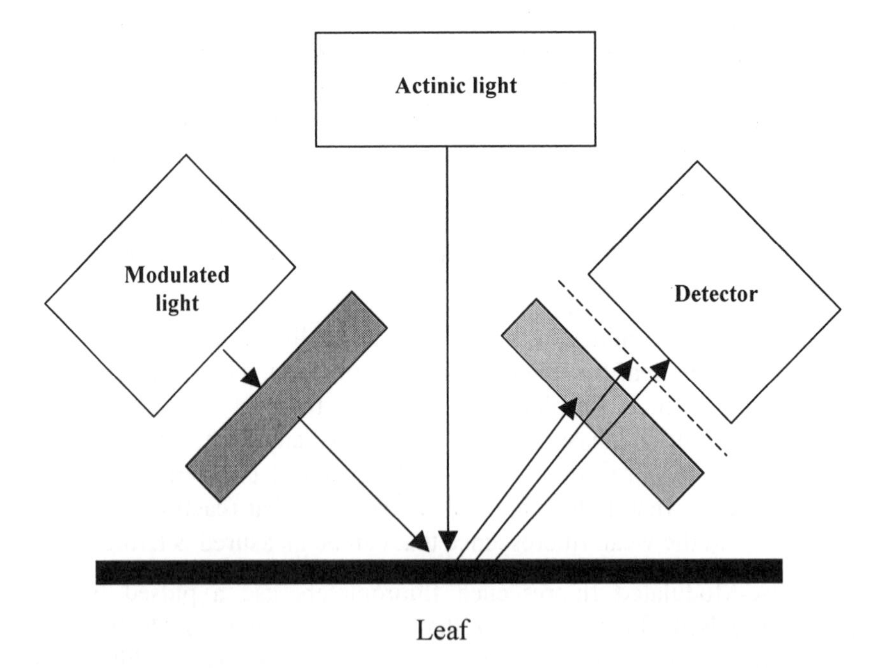

Figure 1. Scheme corresponding to a modulated fluorometer. The shadowed rectangles represent optical filters. The dashed line represents a signal-processing step.

Some modulated fluorometers allow, when actinic light is not used during dark adaptation, the same measurement than a non-modulated fluorometer allows about the Kautsky curve (see previous chapter). Illuminating using the weak modulated light (usually of a very low intensity, even less than 1 μmol photon m^2 sec^{-1}) permits measuring F_0. After some time, the leaf is illuminated using the actinic light and Kautsky curve will be induced. In any case, all non-modulated fluorometers are adequate for this task, and some of the modulated fluorometers lack some of the required capacities (Reigosa, 1997).

Pulse-modulated fluorometers are the only capable of performing quenching analysis. As Figure 7 shows, there are two components in chlorophyll fluorescence quenching:

Photochemical quenching (q_p). This is directly related to the oxidation state of the primary quinone type acceptor Q_A located after pheophytin.

Non-photochemical quenching (q_N o q_{NP}). This non-photochemical quenching is interpreted as caused *in vivo* due to three mechanisms:

- Energy-dependent quenching (q_E) due to the intrathylakoid acidification produced by electron transport.
- Quenching related to state 1 - state 2 transition (q_T), related to phosphorilation of light harvesting complex II.
- And photoinhibitory quenching (q_I) explained by the photoinhibition of photosynthesis (Giersch and Krause, 1991; Krause and Weis, 1991).

Considering these explanations, photochemical quenching is faster than non-photochemical quenching. Nevertheless, some considerations about the validity of this scheme are explained later.

Let us consider a typical curve of fluorescence when a leaf is submitted to the effect of a modulated fluorometer (Figure 2). The first event occurs when the plant receives modulated, low intensity light. This produces a weak fluorescence, which we interpret as F_0. After this, leaf receives a saturating light pulse of approximately 1 second, thus producing a peak in fluorescence: F_{max}. In this moment, we consider quenching (both photochemical and non-photochemical quenching) is zero. After this saturating light pulse, fluorescence falls slowly towards F_0, while the closed reaction centres re-oxidise (that is, 'open'). After going back to F_0, continuous actinic light is added, with a low intensity but sufficient to produce CO_2 fixation and reduction. Immediately, a series of saturating pulses of light are added, thus allowing the measure of both photochemical and non-photochemical quenching.

It is adequate to highlight that every fluorescence peak –(Fv) sat- produced by every light pulse is lower than the previous one. If all the quenching were due to photochemical quenching, all the peaks would have the same intensity (F_{max}), considering that the saturating light is rather sufficient to close all reaction centres. However, peaks decrease, so the common interpretation is the existence of non-photochemical quenching. Both coefficients are calculated according to the following formulae:

$$q_P = \frac{(Fv)\ sat - Fv}{(Fv)\ sat}$$

$$q_N = \frac{(Fv)\ max - (Fv)\ sat}{(Fv)\ max}$$

Most part of quenching produced in a healthy leaf under ambient light is thought to be photochemical quenching. On the contrary, when there is a process limiting energy conversion (limiting electron transport or CO_2 fixation and reduction for example) the predominant process will be non-photochemical quenching. For example, when CO_2 concentration is reduced sometimes total fluorescence remains unchanged, and there is a compensation with an increase in the non-photochemical quenching while photochemical quenching decreases. In that case, a non-modulated fluorometer would give little information about the real fluorescence processes.

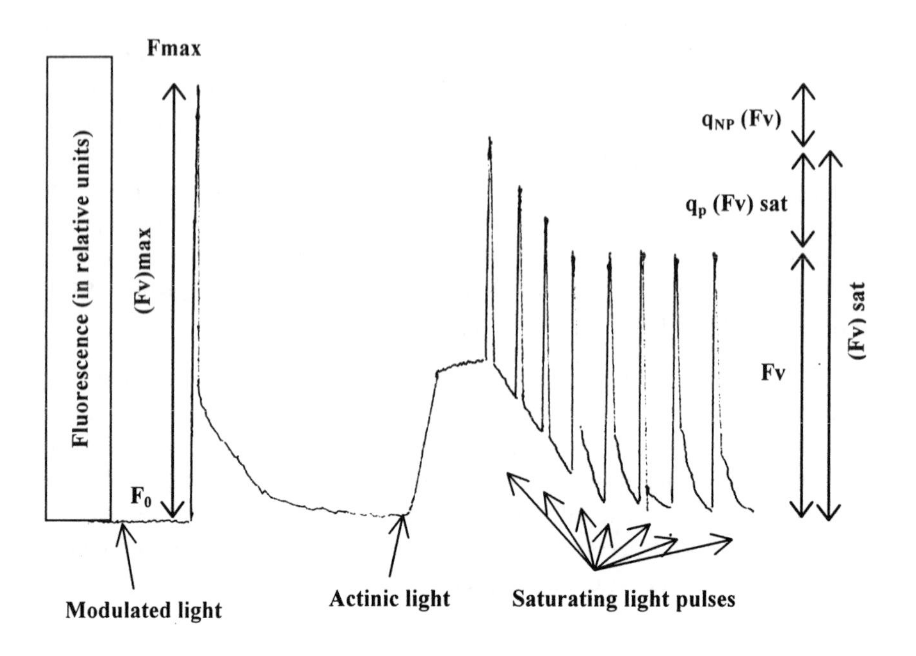

Figure 2. Time (length of this experiment was two minutes)

Pulse-modulated fluorometers can measure the excess light, indicated by a low value of photochemical fluorescence quenching (q_p), that is, can measure photoinactivation. This photoinactivation means that there is an increase in the pressure in the reaction centre of photosystem II and a reduced number of electron carriers between the two photosystems (Osmond et al., 1999).

Pulse modulated chlorophyll fluorometer measure in field and lab conditions fluorescence coming mainly from Photosystem II (Press et al., 1999). At physiological temperatures 95% of chlorophyll fluorescence derives from photosystem II.

Photosystem I is more difficult to study, but although generally less limiting photosynthesis rates, it could be important when plants are exposed to chilling stress (Press et al.,1999; Sonoike, 1966).

The term photoprotection refers to a light mediated decline in the efficiency of primary photochemistry. This photoprotection produces a decrease in the ratio Fv/Fm (Variable fluorescence / Maximal Fluorescence), and also a decline in F_0 (intrinsic fluorescence in weak light) with an increase in thermal dissipation. This photoprotection is usually rapidly reversible (Osmond et al., 1999).

The term photoinactivation means a light mediated decrease in the efficiency of primary photochemistry also (decrease in Fv/Fm like with photoprotection) but with an increase in the intrinsic fluorescence in weak illumination (F_0). When compared to photoprotection, this photoinactivation is slowly reversible. According to Park et al., 1995, this increase in F_0 can be correlated to a diminution in the number of functional reaction centres. This leads to an increase in the fluorescence lifetime in the antennae due to this decrease in reaction centres (Osmond et al., 1999).

Photoprotection and photoinactivation usually can occur simultaneously. Both promote an increase in Fv/Fm, but the differences in F_0 can be difficult to interpret, because they produce opposite effects and the lapses of time in which those effects arise are not necessarily the same (Osmond et al., 1999). New methods should be developed to distinguish in the field between photoprotection and photoinactivation, although some methods exist now to distinguish them in lab conditions (see Osmond et al., 1999).

ECOPHYSIOLOGICAL USES AND MISUSES

We have now a good understanding of the basic processes involved in fluorescence, related to primary processes of photosynthesis, parallel to our better knowledge of the structure of the photosystems and the behaviour of the electron transfer systems. Nevertheless, there is still some gap in the knowledge about photoinhibition, as well as about quenching processes. Nevertheless, this has not stopped us to use the new fluorometers capable of *in vivo* measurement of fluorescence and even separating non-photochemical and photochemical quenching (Bolhàr-Nordenkampf et al., 1989, 1991).

Fluorescence spectroscopy has been considered as a good complement to IRGA or far-red spectroscopy to study the regulation of photosynthesis in intact leaves (Krause and Weis, 1991).

Interpretation of chlorophyll fluorescence measurements is still open to criticisms in ecophysiological experiments (Proctor and Smirnoff, 2000). Of course, one of the classical utilities of the fluorescence curve measurements is in the study of the effects of stress on the photosynthetic systems. A non-modulated fluorometer helps us to study stress tolerance and also damages and genotype differences in the response to stress. When a modulated fluorometer is used, the information provided is richer, allowing the identification of some mechanisms of change in the photosynthetic apparatus (Ögren and Baker, 1985; Genty et al., 1990).

Interpretation of the physiological meaning of q_p and q_N should be very careful. According to Strasser et al. (2000), photochemical quenching refers to only one physiological state, referring precisely to the well-defined events of photochemical quenching. Nevertheless, q_N interpretation is less strict and accurate, because it refers to two physiological states. In this sense, it is not a specific expression of non-photochemical events, although it can be useful to understand this non-photochemical quenching. But some care should be taken when interpreting the experimental values of q_p and q_N.

According to Strasser et al. (2000), there are four basic groups of questions in which the fluorescent transient can be used. Some of them are applicable in non-modulated fluorescence measurements, so they appear also in the previous chapter. Here the following are applicable:

1. Chemical stress. Well-known plants under well-defined conditions to analyse the response to different chemicals. Examples of this type of experiments would be effects of allelochemicals, herbicides,

exogenous phytoregulators, fertilisers, CO_2, O_2, O_3, SO_2 or other gases concentrations, etc.

2. Physical stress. Well characterised plants under changing physical environmental parameters. Examples of this would be effects of quality or intensity of light or temperature.
3. Combined chemical and physical stress. Well-known plant materials can be used to analyse synergisms and antagonisms in the effects of chemical and physical co-stress. This is a very relevant and actual type of studies in plant ecophysiology.
4. Biosensing. In this kind of studies, the environmental conditions are maintained constant and fluorescence transient tests are used to:
   - 4.1 Describe a plant (or even an ecosystem) in terms of vitality, productivity, and response to stress.
   - 4.2 Study the relation structure-function in transgenic plants.
   - 4.3 Study complex systems (from trees to ecosystems).

Strasser et al (2000) even notice other possible or actual uses of fluorescent techniques (provided that a suitable experiment is designed), ranging from testing of agricultural productivity as influenced by style of agriculture, herbicides, pesticides, hormones, cultivars, environmental factors that can cause stress to even behaviour of plants at ecosystem level to the global climatic changes. Other cited uses are related to optimisation of culture conditions, testing the degree of fruit ripening or measuring the effects of pollution.

OTHER TECHNIQUES

The characteristics of modulated fluorescent allow the simultaneous use with other techniques. There are numerous possible combinations between IRGA and fluorometers; many of them are commercially available and are used frequently in plant physiology but specially in plant ecophysiology.

Photoacoustic detectors can also be combined with fluorometers. This photoacoustic technique is a non-invasive method based on the creation and detection of pressure changes –sound- created by absorption of modulated or pulsed light in a sample. This probe measures the energy of photosynthetic intermediates and allows distinguishing the individual activities of photosystems I and II (Malkin et al., 1998). Although less commonly used than the IRGA-Fluorometer combination, it can also be useful in lab and field conditions.

REFERENCES

Amesz J. "Developments in Classical Optical Spectroscopy." In *Biophysical Techniques in Photosynthesis*. J. Amesz, S.J. Hoff, eds. Series: Advances in Photosynthesis (Series Editor: Govindjee). Dordrecht, The Netherlands: Kluwer Academic Publishers, 1995

Bolhàr-Nordenkampf H., Hofer M., Lechner E.G. Analysis of light-induced reduction of the photochemical capacity in field-grown plants. Evidence for photoinhibition? Photosynth Res 1991; 27:31-39

Bolhàr-Nordenkampf H., Long S.P., Baker N.R., Öquist G., Screiber U., Lechner E.G. Chlorophyll fluorescence as a probe of the photosynthetic competence of leaves in the field: A review of current instrumentation. Funct Ecol 1989; 3:497-514

Genty B., Briantais J.M., Baker N.R. The relationship between the quantum yield of photosynthetic electron transport and quenching of chlorophyll fluorescence. Biochim Biophys Acta 1990; 87-92

Giersch C., Krause G.H. A simple model relating photoinhibitory fluorescence quenching in chloroplasts to a population of altered Photosystem II reaction centers. Photosynth Res 1991; 30:115-121

Krause G.H., Weis E. Chlorophyll fluorescence and photosynthesis: The basics. Annu Rev Plant Physiol Plant Mol Biol 1991; 42:313-49

Malkin S., Schreiber U., Jansen M., Canaani O., Shalgi E., Cahen D. The use of photothermal radiometry in assessing leaf photosynthesis: I. General properties and correlation of energy storage to P700 redox state. Photosynth Res 1991; 29:87-96

Maxwell K., Badger M.R., Osmond C.B. A comparison of CO_2 and O_2 exchange patterns and the relationship with chlorophyll fluorescence during photosynthesis in C_3 and CAM plants. Aust J Plant Physiol 1998; 25:45-52

Ögren E., Baker N.R. Evaluation of a technique for the measurement of chlorophyll fluorescence from leaves exposed to continuous white light. Plant Cell Environ 1985; 8:539-547

Osmond C.B., Anderson J.M., Ball M.C., Eggerton J.J.G. "Compromising Efficiency: the Molecular Ecology of Light-resource Utilization in Plants." In *Physiological Plant Ecology*. M.C. Press, J.D. Scholes, M.G. Baker, eds. Oxford: Blackwell Science, 1999

Park Y.I., Anderson J.M., Anderson J.M. Light inactivation of functional photosystem II in leaves of peas grown in moderate light depend on photon exposure. Planta 1995; 196:401-411

Press M.C., Scholes J.D., Baker M.G. *Physiological Plant Ecology*. M.C. Press, J.D. Scholes, M.G. Baker, eds. Oxford: Blackwell Science, 1999

Proctor M.C.F., Smirnoff N. Rapid recovery of photosystems on rewetting desiccation-tolerant mosses: chlorophyll fluorescence and inhibitor experiments. J Exp Bot 2000; 51:1695-1704

Reigosa M.J. "Técnicas de Medida de la Fluorescencia *in vivo* en Ecofisiología Vegetal." In *Manual de Técnicas en Ecofisiología Vegetal*. M.J. Reigosa, N. Pedrol, eds. Vigo, Spain: Gamesal, 1997

Sauer K., Debreczeny M. "Fluorescence." In *Biophysical Techniques in Photosynthesis*. J. Amesz, S.J. Hoff, eds. Series: Advances in Photosynthesis (Series Editor: Govindjee). Dordrecht, The Netherlands: Kluwer Academic Publishers, 1995

Sonoike K. Photoinhibition of photosystem I: its physiological significance in the chilling sensitivity of plants. Plant Cell Physiol 1996; 37:239-247

Strasser R.J., Srivastava A., Tsimilli-Michael M. "The Fluorescent Transient as a Tool to characterize and screen Photosynthetic Samples." In Probing Photosynthesis. Mechanisms, Regulation and Adaptation. M. Yunus, U. Pathre, P. Mohanty, eds. London: Taylor and Francis, 2000

CHAPTER 12

PLANT WATER STATUS

Luís González and Manuel J. Reigosa Roger
Depto Bioloxía Vexetal e Ciencia do Solo. Universidade de Vigo. Spain

INTRODUCTION

In this chapter some of the techniques commonly employed to measure plant water relations are explained.

Water plays an important role in the plant life. For every gram of organic matter made by the plant, the roots absorb approximately 500 g of water that is transported through the aerial parts and lost to the atmosphere. Most of cellular processes depend on the cell water status, then insignificant imbalances in the flow of water through the plant can cause water shortfall and severe malfunctioning of many metabolic processes (Koide et al., 1989).

Water is the most abundant and, at the same time, the most limiting resource, that plants need to grow and function for the best physiological efficiency or productivity in terms of dry weight. In any study involving biological activity, it is not sufficient to report the moisture content of the soil or plant in question. This has little or no meaning in terms of how much of the water is available to the cell. The relationship between tissue water content and tissue water potential and its turgor, osmotic and matric components is required to understand how plants respond to water stress.

$$\Psi = \Psi_p + \Psi_s + \Psi_m + \Psi_g$$

Greek letter psi (ψ) symbolises water potential of plant solutions. Water potential is proportional to the work required to move 1 mol of pure

M.J. Reigosa Roger, Handbook of Plant Ecophysiology Techniques, 185–191.

water at ambient pressure and temperature to another state at the same temperature. Inside plant cells water potential is negative, because pure water has a higher potential than the water inside the cell.

Term ψ_p corresponds to the turgor pressure and denotes the positive hydrostatic pressure within cells. The value may be negative in the xylem and in the walls between cells, where a negative hydrostatic pressure, can develop.

ψ_s is the osmotic potential and characterises the effect of solved solutes on water potential. Solutes reduce the free energy of water.

Expression ψ_m designates matric potential and is used to account for the reduction in free energy of water when it exist as a thin surface layer absorbed on the surface of some particles.

ψ_g represents the potential for water movement thus depends on height. Because of the value depends on the density of water and the acceleration due to gravity, a vertical distance of 10 m translates into a 0.1 MPa change in water potential.

There is a strong relation between the different components of the water potential in a cell (Figure 1). The component of pressure can be positive or negative depending on the force is in favour of the system or against the system. Usually, if the water potential of the environment that surrounds the cell causes that the water moves temporarily toward the exterior of the cell so that the turgor potential goes down below zero, the cell contracts separating the wall (plasmolysis) and the turgor potential remains to zero. The plasmolysis is usually destructive for the cells because the cellular membranes can crack at plasmodesmata level during the plasmolysis.

Water potential can be decreased without an accompanying decrease in turgor. This process is called osmotic adjustment where the change in tissue water potential results from changes in osmotic potential. Osmotic adjustment is a net increase in solute content that is independent of the volume changes that results from loss of water. Most of the adjustment can usually be accounted by increases in concentration of a variety of common solutes (sugars, organic acids and ions).

The appropriate experimental approach to water relations will depend on the objectives and will be variable. If the objective is to describe the water status of a plant community through time, it may be sufficient to measure the leaf water potential at dawn and in the early afternoon (maximum and minimum leaf water potential) during the days that the experiment would need. The number of replicates will depend on the species and the special features of the assays. Where the design of the trial consider many treatments, the measure days should be chosen when the environmental conditions vary shortly between the first plants recorded and the last ones, assuring a randomly distribution of the plants among the different treatments. If the experiment make necessary recording different related measurements (for example, photosynthesis, fluorescence, stomatal conductance, or leaf elongation) to minimise errors arising from comparing one variable on one plant and another from other plant, it will be better taking a complete set of measurements on one plant as close together in time as possible. Use of the same upper fully-expanded leaf is recommended as well (Slavík, 1974).

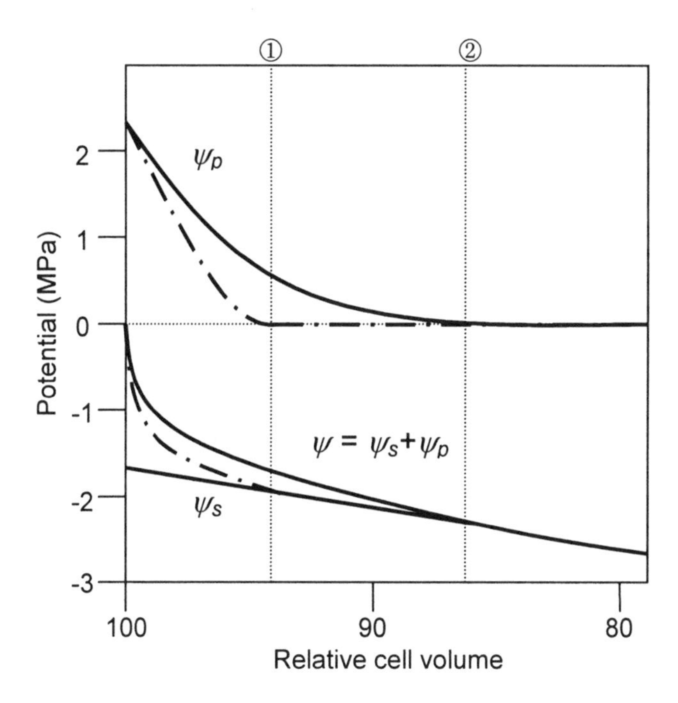

Figure 1. Modified Höfler diagram, showing the relationship between the components of water potential where ① is incipient plasmolysis point for a rigid cell and ② incipient plasmolysis point for a young cell. After Taiz and Zeiger, 1998

Usually plants take up the water from the soil and loss it through the stomata to the air environment. The movement of water from the soil to the atmosphere occurs in a continuous column of water from the soil solution to the atmosphere through the xylem. The column of water in plant vascular tissue is maintained by the cohesive and adhesive properties of water molecules and any break in the water stream prevents water movement through the xylem. The direction of water flow is determined by the water potential gradient between the soil solution and water vapor in the atmosphere (Figure 2), which is the driving force for transport water to the plant (Taiz and Zieger, 1998).

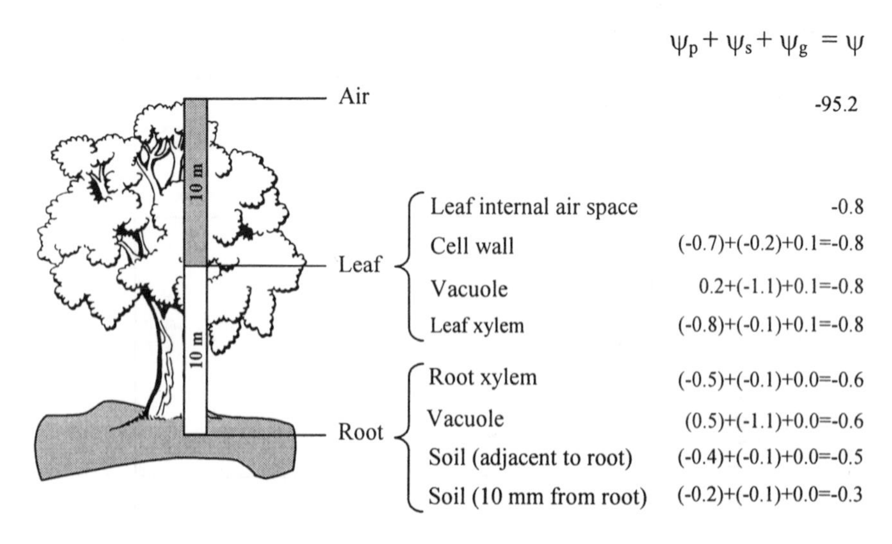

Figure 2. Representative values for water potential and its components at various points in the soil-plant-atmosphere continuum. After Nobel, 1991

In this sense water potential governs transport across cell membranes (Figure 3). Because plant cells have rigid walls, a change in protoplasm volume and water potential is usually accomplished by a change in turgor pressure, with little change in the osmotic potential of the cell. Components of water potential depend on the type of plant and vary with growth conditions and location within the plant, Values of osmotic potential can vary significantly between species and growth environment (Table 1).

Well-watered plants usually have a high osmotic potential (-0.5 to -1.2 MPa). The upper limit for cell osmotic potential is set probably by the minimum concentration of dissolved ions, metabolites, and proteins in the cytoplasm of living cells. At the other extreme, halophytes (plants that grow in saline environments) typically have a very low values of osmotic potential (lower than -2.5 MPa). By these means plants under water, salt stress or living in arid climates reduces cell water potential enough to extract water from the soil solution and to avoid excessive loss by transpiration (Tyree and Jarvis, 1982).

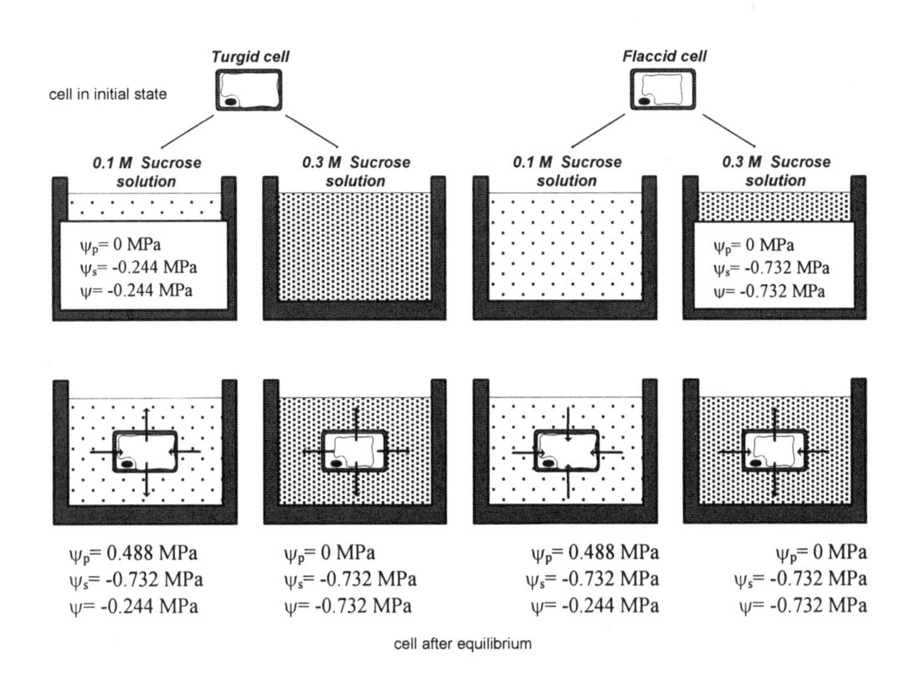

Figure 3. Osmotic behaviour of plant cells with different turgor pressure at 20 °C.

Table 1. Characteristic values for osmotic potential in leaves of different types of plants.

Plant Type	Osmotic potential (MPa)
Halophytes plants	-3 to -20
Deciduous trees	-2 to -5
Coniferous trees	-1.5 to -3
Shrubs	-1.4 to -2.5
Herbs of dry forest	-1.1 to -3.0
Herbs of moist forest	-0.6 to -1.4
Herbs of the alpine zone	-0.7 to -1.7

Turgor pressure within cells of well-watered plants usually range from 0.1 to 0.8 MPa, depending on the osmotic potential inside the cell (Figure 3). Positive turgor pressure drives growth by stretching of the cell walls and gets to maintain mechanical rigidity of cells and tissues so important in young non lignified tissues.

Water potential often influences many physiological aspects of the plant (Figure 4) thus affecting growth and development.

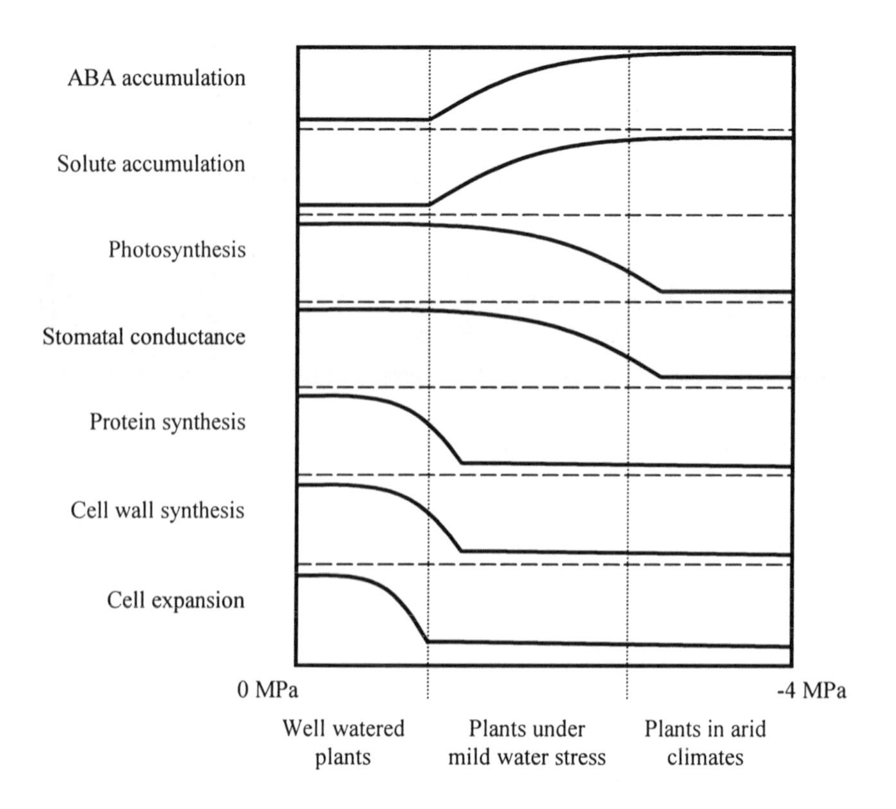

Figure 4. Sensitivity of various physiological processes to diverse growing conditions and water potential. After Hsiao, 1979.

REFERENCES AND FURTHER READINGS

Davies W.J., Gowing D.J.G. "Plant Responses to Small Perturbations in Soil Water Status." In *Physiological Plant Ecology*. M.C. Press, J.D. Scholes, M.G. Barker, eds. Oxford: Blackwell Science, 1999

Hsiao T.C. Plant responses to water deficits, efficiency, and drought resistance. Agr Meteorol 1979; 14:59-84

Koide R.T., Robichaux R.H., Morse S.R., Smith C.M. "Plant Water Status, Hydraulic Resistance and Capacitance." In *Plant Physiological Ecology. Field Methods and Instrumentation.* R.W. Pearcy, J. Ehleringer, H.A. Mooney, P.W. Rundel, eds. London: Chapman and Hall, 1989

Kramer P.J., Boyer J.S. *Water Relations of Plants and Soils.* San Diego: Academic Press, 1995

Nobel P.S. *Physicochemical and Environmental Plant Physiology.* San Diego: Academic Press, 1991

Slavík B. *Methods of Studying Plant Water Relations.* Berlin: Springer-Verlag, 1974

Taiz L., Zieger E. *Plant Physiology.* Sunderland: Sinauer Associates, 1998

Tayree M.T., Jarvis P.G. "Water in Tissues and Cells." In *Encyclopedia of Plant Physiology, New Series, Vol. 12B, Physiological Plant Ecology.* O.L. Lange, P.S. Nobel, C.B. Osmond, H. Ziegler, eds. Heildelberg: Springer-Verlag, 1982

CHAPTER 13

DETERMINATION OF WATER POTENTIAL IN LEAVES

Luís González
Depto Bioloxía Vexetal e Ciencia do Solo. Universidade de Vigo. Spain

INTRODUCTION

Solutes distribution and their movement in the plant take place, predominantly, dissolved in water. The energy status of the water from the root to the leaves will be, therefore, who drives this movement and the thermodynamic parameter commonly used to describe the energy status of the water in the plant is the water potential (Ψ) that is defined as:

$$\Psi = \frac{\mu_w - \mu^o}{V}$$

where,

μ_w: Chemical potential of water at same point in the system at constant temperature and pressure (free energy per mole)

μ^o: Chemical potential of pure water at the same temperature and at atmospheric pressure.

V: Partial molal volume of water.

Water potential expresses the free energy variation of the water in a point, because of a variation of water mol that enter or leave that point when the other parameters are constant (temperature, pressure, etc.). Therefore, it is the work that would be necessary to move an unit of mass

M.J. Reigosa Roger, Handbook of Plant Ecophysiology Techniques, 193–205.

of bound water to the tissues of a plant, to take it from that state of union to a reference status corresponding to the pure water at same temperature and the atmospheric pressure. As the value zero it is adopted for this reference potential, all the potentials that characterise the bound water to a system are negative because it is necessary to provide a work to take this water to a potential equal to zero. The units of the water potential are those of pressure, being commonly used MPa (megapascals). Alternatively bars can be used (1MPa = 10 bars).

For a plant cell, water potential may be expressed as the sum of three components:

$$\Psi = \Psi_p + \Psi_s + \Psi_m$$

where:

ψ_p: Difference in hydrostatic pressure with the reference or turgor potential (due to the cell wall pressure).

ψ_s: Osmotic potential (due to the pressure of dissolved solutes).

ψ_m: Matric potential (due to the pressure of matrices).

There is a strong relation between the different components of the water potential in a cell (Figure 1). The component of pressure neither can be positive when the force is in favour of the system or negative if it is against the system. The turgor potential is a result of hydrostatic pressure in cells, which occurs when the cellular pressure balances the difference in water potential between the environment around cells and the cytoplasm. Usually, if the water potential of the environment that surrounds the cell causes that the water moves temporarily toward the exterior of the cell so that the turgor potential goes down below zero, the cell contracts separating the wall (plasmolysis) and the turgor potential remains to zero. The plasmolysis is usually destructive for the cells because the cellular membranes can crack at plasmodesmata level during the plasmolysis.

The osmotic component is consequence of the solutes dissolved in the system of study and therefore it reduces the free energy of the water, being always negative. The term osmotic pressure ($\pi = -\psi_s$) can also be used, although ψ_s is preferred in plant physiology.

The matric potential is similar to the osmotic potential except that the reduction in the activity of the molar fraction of the water is due to forces in the surfaces of the solids. Usually, we must consider it if the tissue hydration is very low or there are large quantities of mucilage or colloids in the tissues. We can contemplate matric potential as special form of osmotic potential. Instead of relatively small ions regulating osmotic potential, large polar macromolecules or polar surfaces composed of such molecules affect matric potential.

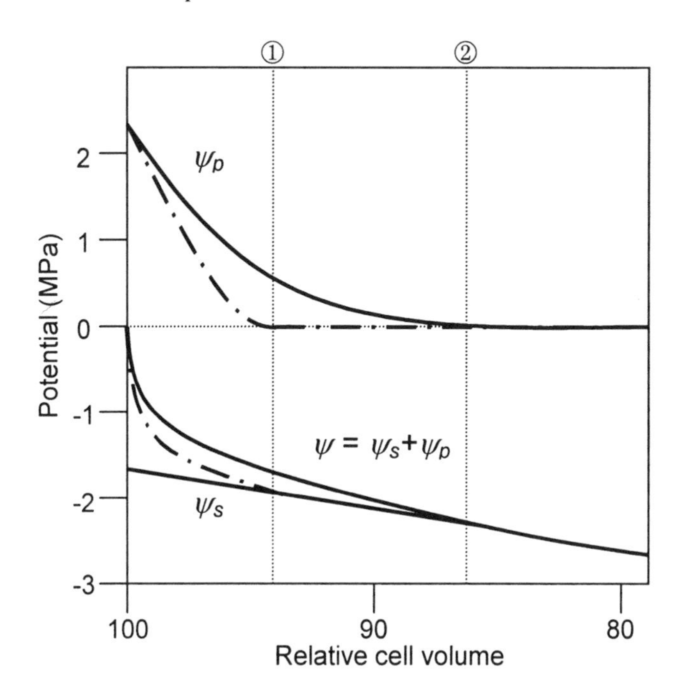

Figure 1. Modified Höfler diagram, showing the relationship between the components of water potential where ① is incipient plasmolysis point for a rigid cell and ② incipient plasmolysis point for a young cell. After Taiz and Zeiger, 1998

When a cell losses water, turgor pressure diminishes sharply with the initial decrease in cell volume from 100 to 90%. The shape and volume range depend on the elasticity of the cell wall, therefore of the species and of the cell kind as is showed in figure 1 by the point line. If the cell wall is very rigid the potential tension make it smaller very fast with a minor loss of water. At this range, most of the decrease in water potential is due to decrease in turgor potential while osmotic potential decrease smoothly as a result of water loss by the cell and consequent increased concentration of cell solutes. In contrast, at the range of volume between 90 to 85% the slope of the curve of turgor pressure is smaller than the slope of the curve of the osmotic potential, as a consequence the descent in water potential in this region of loss of volume is due mainly to the osmotic potential.

Water potential is a good overall indicator of plant health because of cell growth, photosynthesis and crop productivity are all strongly influenced by water potential and its components (Repellin et al., 1997; Pardossi et al., 1998 and Morales et al., 1998; Taiz and Zieger, 1998). Then, the use of an accurate, reliable and simple method for evaluating the water status of a plant would be an efficient tool to know the physiological condition of a plant. The pressure chamber is one of the easiest ways of following the water status in plants under natural conditions.

This chamber consists on a pressure vessel, a compressed gas source (nitrogen) and a pressure gauge that allow us to make the measurements (Figure 2).

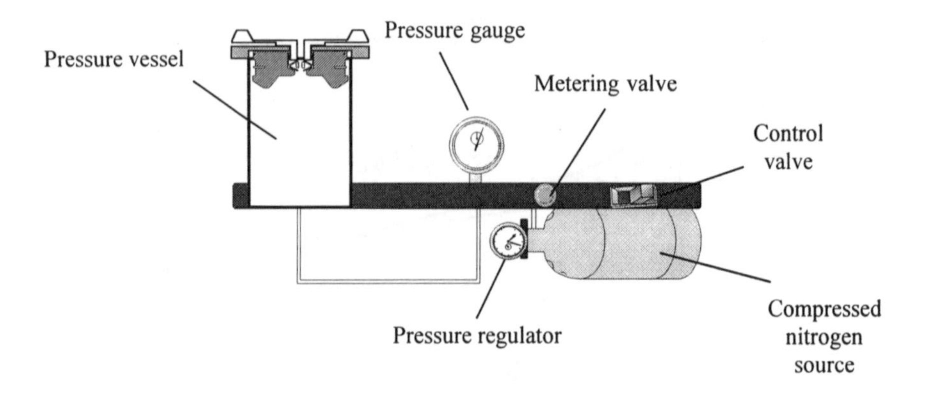

Figure 2. Diagram of a pressure chamber used to determine water potential.

PHYSIOLOGICAL ASPECTS

At equilibrium the water potential is the same across the heterogeneous phases of the cell (through the vacuola, cytoplasm and cell wall). However, the components of the water potential should be clearly different through these phases. For water present in the symplasm (vacuola and cytoplasm), the dominant components are usually the turgor and the osmotic potential. There are four main classes of osmotically active solutes that can significantly affects tissue osmotic potential: nonprotein amino acids, carbohydrates, inorganic ions, and organic acids (Nilsen and Orcutt, 1996). Although there are other molecules that can also affect osmotic potential, nonprotein amino acids and carbohydrates are compatible with the protoplasm whilst inorganic ions and organic acids can reach high concentrations only in the vacuole producing damage in metabolic processes or cellular metabolism when they are free in the protoplasm.

In a cell at equilibrium the turgor and osmotic potential are probably uniform through the membranes of the vacuole and cytoplasmic organelles but under determined circumstances we should not discard the reduction in the osmotic potential brought about by the effect of certain solutes. For apoplasmic water (water in the cell wall and intercell spaces), the dominant component is usually the turgor potential while the osmotic and matric potential contribute fundamentally in the region immediately adjacent to the charged wall surface, being the contribution to the turgor and the osmotic potential specially important in some halophytes.

Inside a tissue, the symplasmic values of the pressure and osmotic potential should vary significantly from one cell to other even at the equilibrium. Given this variation, the most appropriate parameters to describe the water relationships of a tissue at equilibrium will be the averages values of the pressure, osmotic and matric potential.

$$\Psi = \overline{\Psi}_p + \overline{\Psi}_s + \overline{\Psi}_m$$

Comparing with the symplasmic values of the pressure and osmotic potential, the matric potential is insignificant. Although large polar molecules or surfaces will interact with the polar nature of the water, forming hydrogen bonds and, therefore, reducing the kinetic energy of the water, matric potential may become important only in cells with high concentrations of mucilage because of the high proportion of surface area to volume in these systems. Then, it is usual to express water potential as

$$\Psi = \overline{\Psi}_p + \overline{\Psi}_s$$

In a similar way we should define the apoplasmic components of the water potential in a tissue. Besides the water in the living cellular walls we should keep in mind, the water in the walls and lumen of dead cells, (i.e. vessel elements, tracheids and fibers). For any study, at the equilibrium the water potential in the apoplasm of a tissue equals to the water potential in the symplasm of the tissue.

A gravitational term is often incorporated on the whole plant. This component of the water potential is consequence of changes in the potential energy due to differences of height.

$$\Psi = \Psi_p + \Psi_s + \Psi_g$$

The gravitational potential value is increased with the height at a rate of 0.0098 $MPa \cdot m^{-1}$, being, therefore, valueless except in tall trees, thus:

$$\Psi = \Psi_p + \Psi_s$$

Following the development of precise techniques for the measure of water potential and their components in higher plants that grow under field conditions, numerous studies have analysed the ecophysiological significance of the variation of these parameters for plants that grow under different environments (Hinckley et al., 1980; Parker et al., 1982; Barnes, 1985 and Davis and Mooney, 1986). These studies have contributed to show the presence of significant variations among species, particularly in relationship to the availability of water in the territory. Variations have also been demonstrated in these parameters within the own individuals in response to ontogenic, diurnal and seasonal factors (Nilsen et al., 1983, 1986; Calkin and Pearcy, 1984, Pavlik, 1984 and Myers et al., 1987).

MEASURE OF WATER POTENTIAL

Although techniques using thermocouple psychometers were broadly used, at the moment the most extended method to measure the water potential in higher plants is the pressure chamber technique (Scholander et al., 1965) because it does not need a precise temperature control and it is a fast method for estimating the water potential of large pieces of tissue.

The pressure chamber measures the apoplasmic value of the hydrostatic pressure (turgor potential). If we consider the value of the osmotic and matric potential as insignificant, the apoplasmic turgor potential is similar to the water potential and, if the tissue is at equilibrium the apoplasmic water potential will be analogous to the symplasmic water potential. Although most of the plants that grow under natural conditions present values of the osmotic potential close to zero (higher to -0.05 MPa) it is not the same state with halophyte plants or with those that grow under specific conditions. Then, the apoplasmic osmotic potential would be checked in a given study expressing sap from the xylem for over-pressurisation and measuring the osmotic potential of this solution with an osmometer.

The technique is simple, rapid and the equipment is not difficult to use. A leaf, small branch or a shoot is cut with a razor blade (or any similarly sharp instrument) and it is placed in the pressure chamber with the stem or petiole protruding through the rubber-sealing gasket (Figure 3).

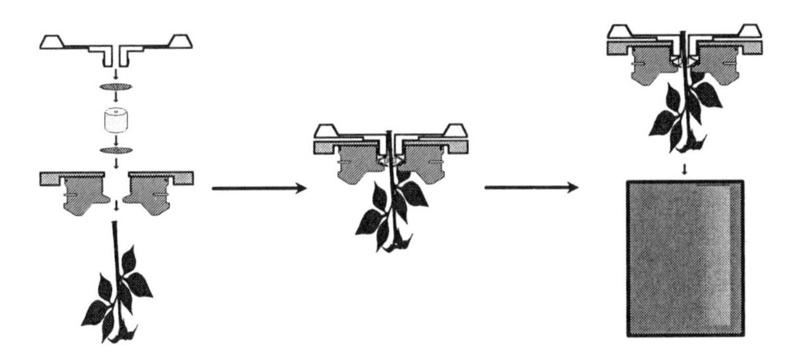

Figure 3. Diagram illustrating the use of the pressure chamber to measure the water potential of a shoot.

When the water column inside the plant is broken for the excision of the organ, the water is moved away back in the xylem capillary for an opposite tension. Consequently, the cut surface appears dry (Figure 4). The petiole or stem should not be re-cut, as this will result in an overestimated value of water potential. Then, the pressure is increased inside the chamber adding air from a compressed gas source. The pressure is increased until the water coming from the xylem appears in the cut surface. The accurate measure of this end point is facilitated by the use of a magnifying glass. As it has been reflected previously, the balance pressure in the chamber at this end point, taken as a negative value, is equal in magnitude to the apoplasmic value of the turgor potential in the tissue, which equals the symplasmic value of the water potential under the same conditions. If we know the osmotic potential for the xylem sap from other measurements, we may calculate the xylem water potential, which is assumed to be close to the water potential of the whole organ.

Plants with extensive aerenchyma make pressure chamber measurements difficult because it seems difficult to difference between air bubbles, xylem or phloem sap. It takes extensive experience with the tissue and the apparatus before being confident of the results.

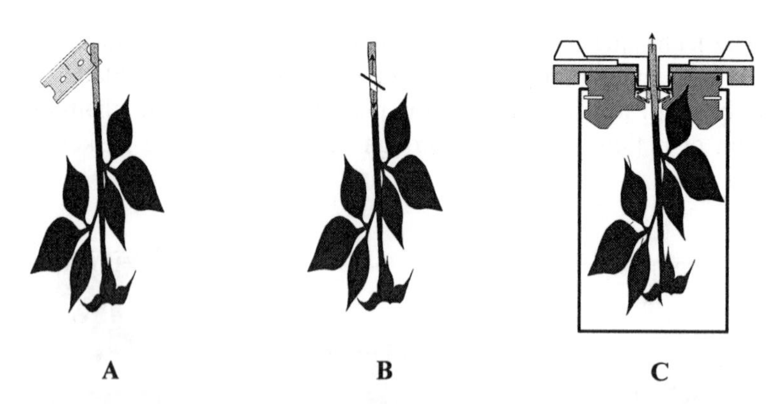

Figure 4. State of the water column within the xylem before cut the plant material (A), after cut the shoot (B) and after the chamber is pressurised (C).

PRACTICAL PROCEDURE OF MEASURING LEAF WATER POTENTIAL

In order to explain the measurement of water potential we have chosen the plant water status console is an equipment from Soilmoisture Equipment Corp. based on the pressure chamber designed in 1965 by Scholander.

Measurement of water potential

1. Close the flow valve turning the key clockwise (metering valve on the console).
2. Turn control valve to OFF position.
3. Open the compressed gas tube valve turning the key counterclockwise.
4. Select the plant material to assess the water potential of the plant.

5. Introduce a wet paper inside the chamber to avoid dry atmosphere. Take care not to cover the gas entry in the centre of the bottom chamber.

6. Place a metallic support washer in the hole of the specimen holder (cover of the chamber), on it place the sealing grommet and insert the plant through the cover, the washer and the grommet. Introduce another metallic washer on the rubber (Figure 5).

7. Put in the sealing knob and turn clockwise until sealing sleeve press sufficiently around the plant material to hold it firmly.

8. Insert the plant material into the pressure vessel and push down and turn the specimen holder 45° clockwise to lock it under the cams of the pressure vessel.

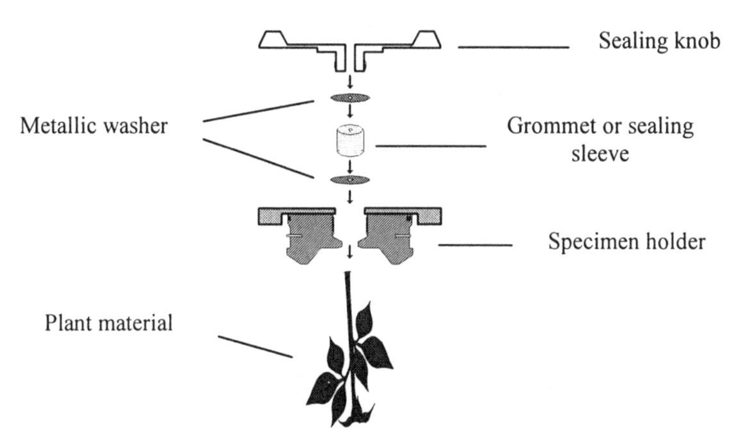

Figure 5. Scheme of the specimen holder of the pressure chamber

9. To be sure that the sealing sleeve is sealing well around the plant material, tightens the sealing knob further if it was necessary and take care do not choke the plant.

10. Turn the control valve on the console to the PRESSURIZE position.

11. Open the metering valve slowly, by rotating in counterclockwise and check that the needle of the read-out gauge moves slowly (approximately. 5-10 psi·sec).

12. As pressure is built up in the pressure vessel you can hear air escaping from the chamber, simply tighten the sealing knob further to exert higher sealing pressure. Additionally, you can take the reading more exactly with the aid of a magnifier and a supplementary light source. Caution! Never put the eye directly over the pressure vessel during a run.
13. Observe carefully the cut end of the petiole of the sample being tested.
14. The equilibrium pressure for the individual sample is reached when sap starts to flow from the exposed cut end of the sample. This is the moment to take a reading.
15. Stop immediately the increment of pressure inside the chamber by rotating the control valve to the OFF position. At any time the build up of pressure can also be stopped or reduced by turning clockwise the metering valve. This option is extremely useful when working close to the balance pressure to reach more accuracy in the pressure reading.
16. Take note down of the pressure reading at that point. This equilibrium value (negative) is referred to as plant water potential.
17. Exhaust the pressure in the pressure vessel by turning the control valve to the EXHAUST position.
18. To remove the sample the sealing knob must be loosen and the closing cap of the specimen holder turned 45° in a counterclockwise and take it out by pulling up.
19. Release the plant material, the washers and the sealing sleeve.
20. When all the samples were run, shut off the compressed gas tube.

Collect sap

1. Repeat previous steps from 1 to 13.
2. When the equilibrium pressure is reached sap starts to flow from the end of the sample.
3. Stop the increment of pressure inside the chamber by rotating the control valve to the OFF position.

4. Blot with smooth tissue the cut surface to avoid collect the first released sap because it may be contaminated by solutes released from damaged cells at the cut surface.
5. Overpressurise the sample by turning the control valve smoothly to the ON position.
6. When sap start to flow again put a glass capillary on the cut surface and allow the sap go up through it.
7. Put the content of the glass capillary in an Eppendorf tube.
8. Once your have collected enough sap, stop the increment of pressure inside the chamber by rotating the control valve to the OFF position.
9. Repeat previous steps from 17 to 20.

Measurement of osmotic potential

Osmotic potential of the vacuolar sap of tissues may be determined basically obtaining the sap from the tissues and then determining its osmotic potential by an indirect method. The expressed cell sap is assumed to have the same osmotic potential as the vacuolar solution.

Total concentration of osmotically active dissolved particles in a solution without considering particle size, density, configuration or electrical charge is known as the expression osmolality.

The addition of solute particles to a solvent (water in plants) changes the free energy of the solvent molecules and this allow to us indirect means (vapor pressure, freezing point or boiling point) for the measurement of osmotic potential.

The measurement of osmotic potential, can only be made indirectly by comparing one of the solution colligative properties (vapor pressure and freezing point are the most common) with the corresponding cardinal property of the pure water.

In the laboratory we can made the measurement of the osmotic potential by using osmometers based in the depression of the freezing point or modern osmometers based in the measurement of vapor pressure depression.

Osmometer of vapor pressure has a significant advantage over the older methods: measurement can be performed without the change in the physical state of the sample and this technique avoids measurement artefacts that often occur when the sample to be tested must be altered physically.

REFERENCES AND FURTHER READINGS

Barnes P.W. Adaptation to water stress in the big bluestem-sand bluestem complex. Ecology 1985; 66:1908-1920

Calkin H.W., Gibson A.C., Nobel P.S. Xylem water potentials and hydraulic conductances in eight species of ferns. Can J Bot 1985; 63:632-637

Davis S.D., Mooney H.A. Tissue water relations of four co-occurring chaparral shrubs. Oecologia; 1986; 70:527-535

Hinckley T.M., Duhme F., Hinckley A.R., Richter H. Water relations of drought hardy shrubs: osmotic potential and stomatal reactivity. Plant Cell Environ 1980; 3:131-140

Kramer P.J., Boyer J.S. *Water Relations of Plants and Soils*. San Diego: Academic Press, 1995

Morales M.A., Sánchez-Blanco M.J., Olmos E., Torrecillas A., Alarcon J.J. Changes in the growth, leaf water relations and cell ultrastructure in *Argyranthemum coronopifolium* plants under saline conditions. J Plant Physiol 1998; 153:174-180

Myers B.J., Robichaux R.H., Unwin G.L., Craig I.E. Leaf water relations and anatomy of a tropical rainforest tree species vary with canopy position. Oecologia 1987; 74:81-85

Nilsen E.T., Orcutt D.M. *Physiology of Plants under Stress*. New York: John Wiley and Sons. 1996

Nilsen E.T., Sharifi M.R., Rundel P.W., Jarrell W.M., Virginia R.A. Diurnal and seasonal water relations of the desert phreatophyte *Prosopis glandulosa* (honey mesquite) in the Sonoran Desert of California. Ecology 1983; 64:1981-1993

Nilsen E.T., Sharifi M.R., Rundel P.W., Virginia R.A. Influences of microclimatic conditions and water relations on seasonal leaf dimorphism of *Prosopis glandulosa* var. Torreyana in the Sonoran Desert, California. Oecologia 1986; 69:95-100

Nobel P.S. *Physicochemical and Environmental Plant Physiology*. San Diego: Academic Press, 1991

Pardossi A., Malorgio F., Oriolo D., Gucci R., Serra G., Tognoni F. Water relations and osmotic adjustment in Apium graveolens during long-term NaCl stress and subsequent relief. Physiol Plantarum 1998; 102:369-376

Parker W.C., Pallardy S.G., Hinckley T.M., Teskey R.O. Seasonal changes in tissue water relations of three woody species of the *Quercus-Carya* forest type. Ecology 1982; 63:1259-1267

Passiura J.B. The meaning of matric potential. J Exp Bot 1980; 31:1161-1169

Pavlik B.M. Seasonal changes of osmotic pressure, symplasmic water content and tissue elasticity in the blades of dune grasses growing *in situ* along the coast of Oregon. Plant Cell Environ 1984; 7:531-539

Repellin A., Laffray D., Daniel C., Braconnier S., Zuily-Fodil Y. Water relations and gas exchange in young coconut palm (*Cocos nucifera* L.) as influenced by water deficit. Can J Bot 1997; 75:18-27

Ray P.M. On the theory of osmotic water movement. Plant Physiol 1960; 35:783-795

Roger T.K., Robichaux R.H., Morse S.R. "Plant Water Status, Hydraulic Resistance and Capacitance." In *Plant Physiological Ecology*. R.W. Pearcy, J. Ehleringer, H.A. Mooney, P.W. Rundel, eds. London: Chapman and Hall, 1989

Scholander P.F., Hammel H.T., Bradstreer E.D., Hemmingsen E.A. Sap pressure in vascular plants. Science 1965; 148:339-346

Slavík B. *Methods of Studying Plant Water Relations*. Berlin: Springer-Verlag, 1974

Steudle E. "Pressure Probe Techniques: Basic Principles and Application to Studies of Water and Solute Relations at the Cell, Tissue and Organ Level." In *Water Deficits: Plant Responses from Cell to Community*. J.A.C. Smith, H. Griffiths, eds. Oxford: BIOS Scientific Publishers, 1993

Taiz L., Zieger E. *Plant Physiology*. Sunderland: Sinauer Associates, 1998

Tyree M.T., Jarvis P.G. The measurement of the turgor pressure and the water relations of plants by the pressure bomb technique. J Exp Bot 1972; 23:266-282

CHAPTER 14

DETERMINATION OF RELATIVE WATER CONTENT

Luís González and Marco González-Vilar
Depto Bioloxía Vexetal e Ciencia do Solo. Universidade de Vigo. Spain

INTRODUCTION

Tissue water content may be expressed in several ways, including the amount of water per unit dry or fresh weight and per unit weight of water at full hydration. Fresh weight seems to be the less accurate of them to measure tissue water content because is highly influenced by changes in tissue dry weight (Turner, 1981).

The relative water content (RWC) stated by Slatyer in 1967 is a useful indicator of the state of water balance of a plant essentially because it expresses the absolute amount of water, which the plant requires to reach artificial full saturation. Thus there are a relationship between RWC and water potential (Figure 1). This relation varies significantly according to nature and age of plant material.

The RWC express the water content in per cent at a given time as related to the water content at full turgor:

$$RWC = \frac{fresh\,weight - dry\,weight}{saturated\,weight - dry\,weight} \times 100$$

The procedure to get full saturation needs to reach constant weight in the tissue. Young leaves, which are still suffering expansion, absorb water

M.J. Reigosa Roger, Handbook of Plant Ecophysiology Techniques, 207–212.

for a substantially longer period (usually several days). Young tissue is respiring up and consumes part of its dry weight. This can result in a significant error. To avoid this problem it is necessary to reduce the time for saturation of the tissue by reducing the sample size as much as you can manage it (weigh, dry...) and keep away the sample from physiological activity by physical inhibition of growth and respiration during determination. Saturation of the tissue portions at 4°C inhibits satisfactorily the growth.

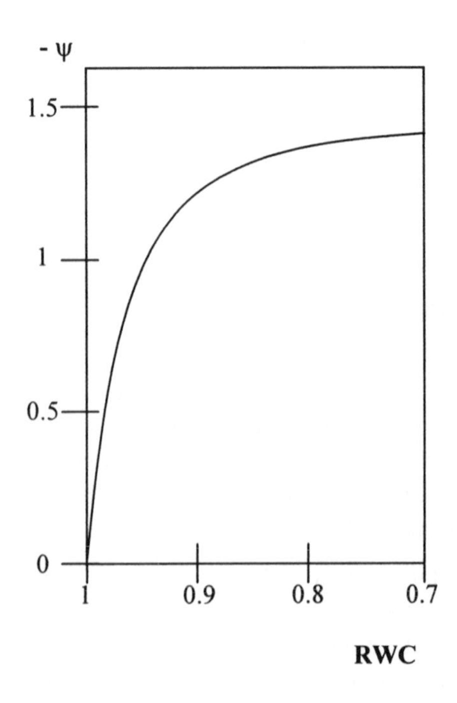

Figure 1. Relationship between tissue water potential and relative water content.

PHYSIOLOGICAL ASPECTS

Water makes up most of the mass of plant cells. In each cell, cytoplasm makes up only 5 to 10% of the cell volume and the remainder is a large water-filled vacuole.

There is a strong correlation between alterations in leaf protoplast volume and changes in leaf photosynthetic activity. Sometimes decreases in tissue water content may be more important than decreases in water potential or pressure potential in terms of influencing growth. Then, it is not enough to study water relations in plants to know only water content but is often important to measure both, tissue water content and water potential for plants growing under field conditions. Plants in natural environments can support different environmental conditions that interact with its genetics characteristics. Diploid *Dactylis glomerata* showed a smaller RWC than tetraploid citotype (González-Vilar et al., unpublished). Values for tetraploid plants were close to 80 and diploid plants presented values round 70 with significant differences at 0.05 level. This could imply that these plants suffered marked metabolic changes with ceasing of photosynthetic activity, respiration increment and proline and abscisic acid accumulation.

Small discs or tissue pieces are used to determine a great variety of physiological processes in plants (photosynthesis, enzymatic activity, pigments content...) although it should be taken in account the possible heterogeneity of the leaf to get a good correlation between RWC and some physiological processes.

In the process of tissue saturation there are two phases of water uptake. Initially there is a rapid uptake of water to satisfying the water deficit of the tissue. Later there is a slower uptake of water caused by growth and other physiological factors, difficult to define, how it takes place. Young tissues are more sensible to this second step (Figure 2).

Samples with a long saturation time, like young tissues, need some kind of data correction for growth.

RWC is a measure of the relative cellular volume that shows the changes in cellular volume that could be affecting interactions between macromolecules and organelles. As general rule, a RWC among 100-90 % is related to closing of the stomata pore in the leaf and a reduction in the cellular expansion and growth. Contents of 90-80 % are correlated with changes in the composition of the tissues and some alterations in the relative rates of photosynthesis and respiration. Levels of RWC below 80 % imply usually water potential of the order of -1.5 MPa or less, and this would produce changes in the metabolism with ceasing of the photosynthesis, increment of the respiration and proline and abscisic acid accumulation.

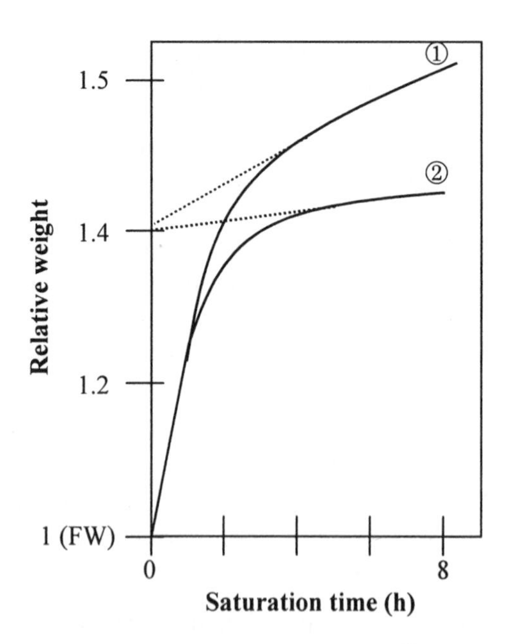

Figure 2. Saturation curves of discs cut out from young ① and mature ② tissue. FW: fresh weight

MEASUREMENT OF RELATIVE WATER CONTENT

For the previously exposed reasons the 'piece extrapolation routine' is the most suitable method to assess the plant RWC. The following procedure was modified from Sharper et al. 1990.

1. Weigh and label with a number 0.2 ml Eppendorff tubes and label with the same numbers 1.5 ml Eppendorff tubes. (Take into account the number of replicates. It will depend on the species but for this technique you must use at less five replicates for sample).
2. Fill with cold de-ionised water the 1.5 ml tubes.
3. Remove the youngest developed leaf of the plant. (If the sample is collected in the field or greenhouse and is not processed immediately you must place it in sealed plastic bags and transfer it to a cooled insulated container to prevent growth and reduce evaporation of water from the cut tissue).
4. Put four sections (0.5 cm^2) of a same leaf in a tube (choose homogeneous pieces and avoid apical and basal parts of the leaf.

Those fractions would have really different expansion rate). Seal the tubes and place it in ice to prevent growth and evaporation.

5. Weigh the tubes with tissue sections inside to give a value for tissue fresh weight (FW).
6. Once every sample has been weighed, transfer the piece of leaf to a 1.5 ml Eppendorff tubes containing de-ionised water with the same number of the small tube.
7. Tubes should be placed in ice, and left for 4 hours in a fridge to allow the tissue taking up water.
8. Take away the tissue sections from the de-ionised water. Remove carefully the excess of water on the leaf surface with tissue paper.
9. Transfer every sample to their original tubes (0.2 ml) and reweigh. This gives a measure of fully turgid fresh weight (TFW).
10. Open the tubes, and place them in a stove to 70 °C during at less 48 hours.
11. After drying reweigh the tissue to obtain a value for dry weight (DW).

Relative water content (RWC) can then be calculated using the following equation:

$$RWC = \frac{FW - DW}{TFW - DW} \times 100$$

REFERENCES AND FURTHER READINGS

González-Vilar M., González L., Reigosa M.J. Ecophysiological responses to light intensity and water stress in two citotypes of cocksfoot (*Dactylis glomerata* L.) in Galicia, NW of Spain. Submitted.

Larcher W. *Physiological Plant Ecology*. Berlin: Springer-Verlag, 1995

Lawlor D.W. "The Effects of Water Deficit on Photosynthesis. In *Environment and Plant Metabolism. Flexibility and Acclimation*. N. Smirnoff, ed. Oxford: BIOS Scientific Publishers, 1995

Sharp R.E., Hsiao T.C., Silk W.K. Growth of the maize primary root at low water potentials. II. Role of growth and deposition of hexose and potassium in osmotic adjustment. Plant Physiol 1990; 93:1337-1346

Taiz L., Zieger E. *Plant Physiology*. Sunderland: Sinauer Associates, 1998

Turner N.C. Techniques and experimental approaches for the measurement of plant water status. Plant Soil 1981; 58:339-366

CHAPTER 15

DETERMINATION OF WATER POTENTIAL IN SOILS

Luís González
Depto Bioloxía Vexetal e Ciencia do Solo. Universidade de Vigo. Spain

INTRODUCTION

The amount of water that soil pores contain is of fundamental importance in determining soil biological activity, availability of this water, soil aeration, supply of soluble nutrients, pH and Eh of the soil solution and regulates osmotic potential (Killham, 1994).

One of the most important features in plant-soil water relations is that there is not relationship between the amount of water in soil and the availability to the plant. For example, in a comparison between a sandy soil and a clay loam we expect that a sandy soil has a low water content but most of this water will be available to the plant roots. On the other hand, the clay loam may have a higher water content but a roughly similar amount of water available to the plant roots because it has smaller pores from which water is more difficult to extract (Kramer and Boyer, 1995).

The tension required to withdraw water from a soil at certain moisture content is referred to as the matric potential of the soil and it is usually measured in megapascals. Graphs of matric potential against water content are termed 'moisture release curves' and they provide information on available soil moisture (Mullins, 1991, Klute 1986).

To study water potential in soils considering matric tension alone is an inaccurate generalisation because, as the soil pores dry out, the remaining water will tend to become increasingly saline. Thus, on the soil system there is other potential that affect the water potential: osmotic

M.J. Reigosa Roger, Handbook of Plant Ecophysiology Techniques, 213–222.

potential of soil water. There is also a gravitational component to soil water potentials, but this becomes totally negligible on the first 10 cm of soil surface. Then, plants must overcome both components for plant uptake of water.

$$\psi = \psi_m + \psi_s$$

Where: ψ = total soil water potential
ψ_m = matric potential
ψ_s = osmotic potential

Tensiometers, sensors for electrical resistance (fiberglass or gypsum), heat dissipation sensors, psychrometers or neutron probe are the most common equipments to measuring matric or water potential in soils.

Usually, plant root cannot extract sufficient water to maintain full turgor when water potential is close to -1.5 MPa. and the plant will wilt. When rewatering, most plants will recover fully all the physiological traits. This point is called 'reversible wilting point'. Approximately at -1.5 MPa, the roots of most plants can no longer extract any water and the plant will die. This is referred to as the 'permanent wilting point' (Figure 1).

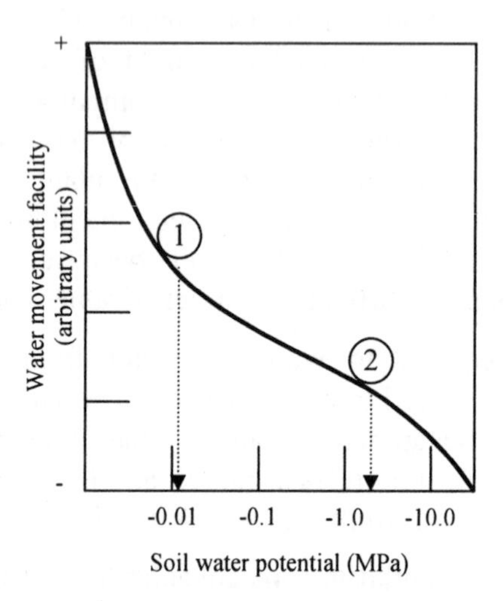

Figure 1. Water mobility through the soil as a function of the water potential of a typical clay soil. The shape of this curve is representative of many soils. ①, field capacity, and ②, permanent wilting point.

In the range from the field capacity (the water potential of a water-saturated soil after it has been allowed to drain under gravity) to the permanent wilting point, the osmotic component of soil water potential is generally very small and, in the most of cases, is not necessary to consider it. Over this range, only saline soils and highly fertilised soils will have an appreciable osmotic component of water potential

Water available for plant growth is the amount of water held in the soil between field capacity and permanent wilting point and is generally termed as 'available water capacity'.

In function of this and because matric potential can have only zero or negative values (pressure exerted against the soil system), the terms soil water potential or matric suction (which represent the same quantity but take the opposite sign) are often used as convenient alternatives (Reeve and Carter, 1991). The units used to express the energy levels of soil water are diverse but usually is preferred MPa as SI unit. Some authors use bar (1 MPa = 10 bar) or the less extended pF scale, which is the logarithm of the soil water suction expressed in centimetres of water to avoid the use of very large numbers. Table 1 shows the correspondence of values to different soil water potentials.

Table 1. Conversion table for different soil water potentials. FC: Field capacity, PWP: permanent wilting point.

Soil water potential (ψ)			
MPa	bar	pF	
-0.001	-0.01	1.01	
-0.005	-0.05	1.71	
-0.010	-0.10	2.01	
-0.020	-0.20	2.31	
-0.030	-0.30	2.49	
-0.040	-0.40	2.61	FC
-0.050	-0.50	2.71	
-0.060	-0.60	2.79	
-0.070	-0.70	2.85	
-0.080	-0.80	2.91	
-0.090	-0.90	2.96	
-0.100	-1.0	3.01	
-0.200	-2.0	3.31	
-1.500	-15.0	4.18	PWP

PLANT WATER ABSORPTION

The water potential of soil water solution refers to the potential energy of water in the soil with respect to a defined reference state. The physiology meaning of soil water potential is that it controls water flow in the soil: from the soil into the plant root. The movement of water through soils takes place in the numerous channels between the soil particles and despite the different pore sizes the water movement should be considered that take place in a continuum, with a uniform flow averaged over space.

Water moves through soils predominantly by bulk flow driven by a pressure gradient, although diffusion also plays some role in water movement. As a plant absorbs water from the soil, it depletes the soil water content in the vicinity of the roots thus establishing a pressure gradient with respect to neighbouring regions of soil that have higher water potential values. Because the pore spaces in the soil are interconnected, water moves to the root surface by bulk flow through these channels down in the pressure gradient.

In very dry soils the water potential is so low that plants cannot get back turgor pressure. As we say above, water potential value where these occur is called permanent wilting point. At this point the water potential of the soil is less than or equal to the osmotic potential of the plant. Permanent wilting point does not depend only of the soil, it is variable to many species because cell osmotic potential varies with plant species (Taiz and Zieger, 1998).

In most of the soils matric potential is the major factor that determines the availability of water to plants and, as we stated previously, it refers to the tenacity with which the soil matrix holds water. Then, differences in the value of matric potential between different parts of the soil also provide the driving force for the flow of soil water.

Most of the water collected by plants travels from soil solution into the roots systems. There are some cases in which another way to enter water in the plant exists but this is not relevant in terms of most of the plants. Water diffuses into root cells because root cells have more negative water potential than the soil solution (Figure 2). Therefore, any factors that reduce the water potential of the soil solution (high matric interaction into the soil, high salinity into soil solution or freezing soil solution) will inhibit the movement of water into the roots. For those reasons, optimum physiological activity of the plant is usually related to physical and chemical properties of the soil.

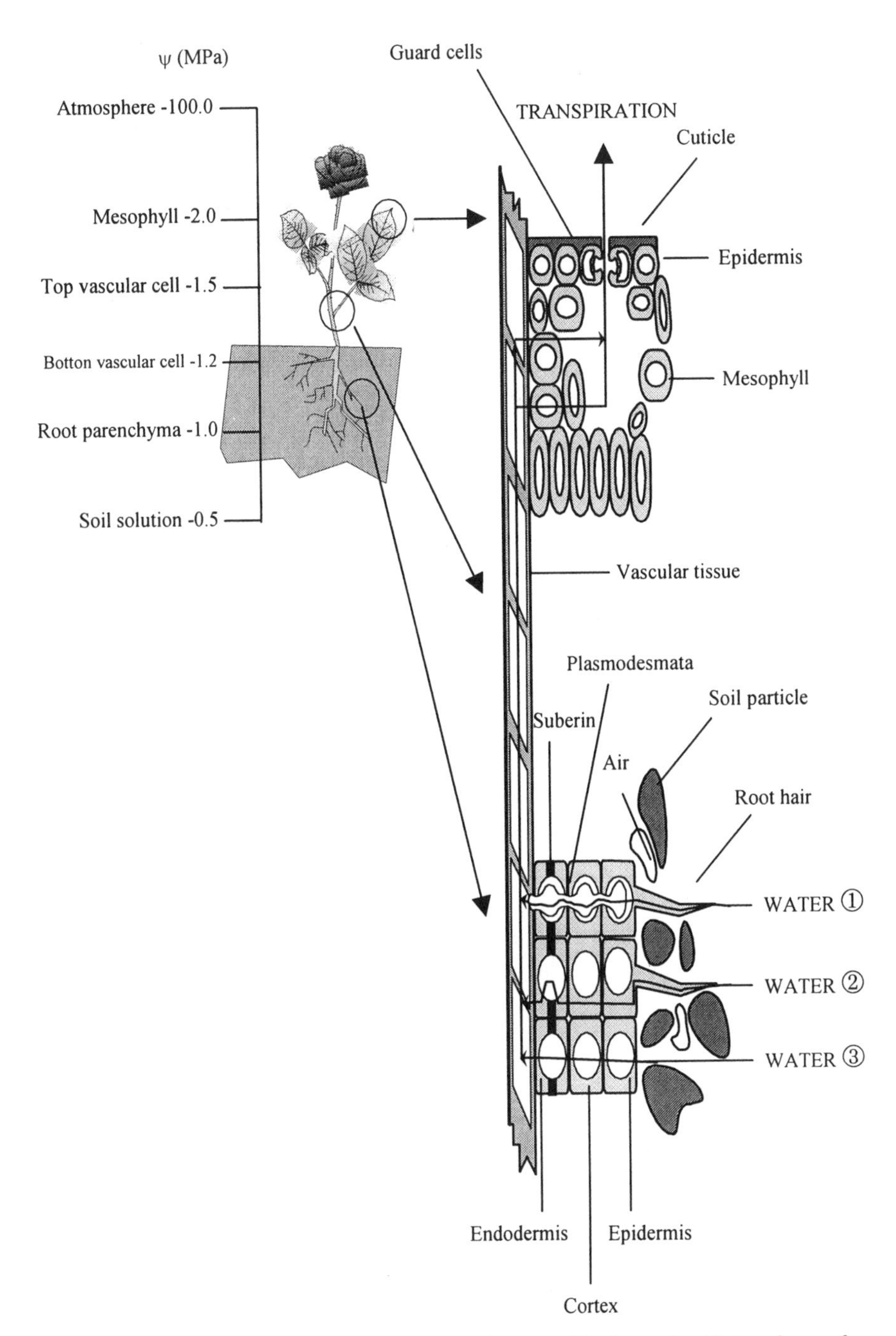

Figure 2. Diagram of the soil-plant-water continuum with the main driving forces for water flow from the soil through the plant to the atmosphere and several pathways for water uptake from the soil solution by the root: ① vacuolar route, ② apoplastic route and ③ symplastic route.

Thus, the relative dimensions of soil particles determine the size of a soil pore and small soil pores only allow settling the smallest soil bacteria eliminating other useful soil biota. The degree to which the soil pore space is water filled is of fundamental importance in determining soil biological activity as well. When the soil dries out, this water is restricted to thin films around the soil particles and the pores are occupied by air, thus becoming unavailable to live bacteria and protozoa useful for plant life. In addition, to facilitate different niches and compartmentalisation for different components of the soil biota, the capacity of the soil pores to maintain the water determines the accessibility of this water for the plants and the soil aeration, as well as the supply of soluble nutrients, and regulation of the osmotic potential, pH and Eh of the soil solution. Table 2 shows that there is a clear relationship between the soil matric potential and the neck diameter of the largest water-filled pores (Killham, 1994):

$$\psi_m = -0.3 \cdot d^{-1}$$

where ψ_m = matric potential (MPa)

d = neck diameter (μm) of largest water-filled pores.

Differences in water potential are the driving force to move water from the soil to the growing plants and this is associated with pore size: pore size greater than 30 μm provide water to plants freely under gravity and between 0.3 to 30 μm plants can withdraw it by absorbing roots. Hence, pore size distribution is critical in water supply to the roots. Ideally, a soil should have a reasonable number of large pores to facilitate root extension, but also a large number of small pores (0.3-30 μm) to sustain water supply to roots during periods of drought (Killham, 1994).

Table 2. Water potential related to soil water distribution.

Soil water distribution	**ψ (MPa)**	**Approximate size (μm)**
Water-filled pores	>-0.01	>30
	-0.1 to -0.01	3 to 30
	-1.0 to -0.01	0.3 to 3
	<-1.0	<0.3
Water films	-1 to -0.1	<0.003
	<-1.0	Few molecules thick

MEASUREMENT

As we justify above, if we assume that in most of the soils for the range of soil moisture from field capacity to the permanent wilting point, the osmotic component of the water potential is insignificant. At this way, the most comfortable method of measuring the water potential is by means of an accurate measure of the matric potential.

There are different methods to measure matric potential. Tensiometers are one of the most accurate equipment used. It consist of a porous cup attached via a liquid-filled column to a manometer or a porous cup whose content in water varies with the matric potential in a reproducible way. When the probe is inserted into the soil, a physical property of this material, that varies with the content of water, is measured and transformed in soil moisture (figure 3). Soil moisture is related to matric potential using a calibration curve.

With the probe placed into the soil, the matric potential within the porous material rapidly equilibrates with the neighbouring soil. Matric potential value is obtained by measuring the electrical resistance between the measuring rods, which is a function of the water content of the porous material where they are embedded. Unfortunately, the electric conductivity is also a function of temperature and of solute concentration in the soil solution. Temperature effects can be corrected, but the sensors cannot be used in saline soils.

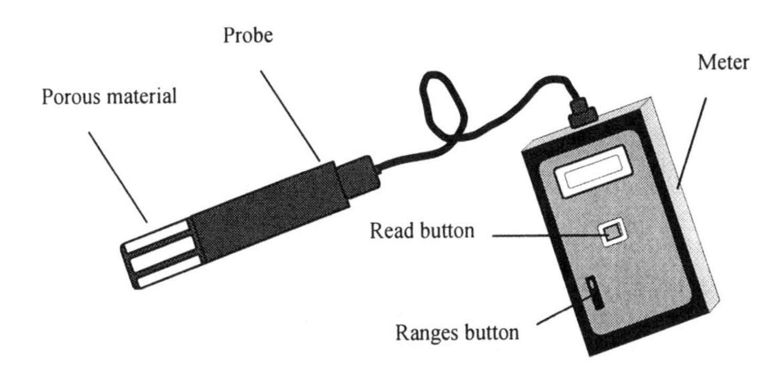

Figure 3. Equitensiometer components

When the soil is too heterogeneous (density, composition, rates of percolation, runoff, evaporation), or the distribution of roots is not very uniform the matric potential of the soil will vary depending on local situations. It is necessary, therefore, to take different measures in the soil region that is studied.

The porous material is calibrated by equilibrating it at a set of known matric potentials; therefore this method requires a separate calibration for each soil. Continuous use of the probe will likely produce significant changes in the porous material and then in its capacity to equilibrate the matric potential. Thus the only guarantee of consistent results is to recheck at regular intervals the calibration curve.

AN EXAMPLE OF TECHNIQUE EQUIPMENT AND PERFORMANCE

The equitensiometer probe type EQ2 (Theta-meter) was development by Delta-T Devices and it is associated to a moisture meter type HH1 that provide instant readout of soil moisture data. With the appropriate transformation the equitensiometer from Delta-T is a matric potential sensor.

1. Wet thoroughly the equitensiometer probe water.
2. Insert the probe into the soil at horizontal or slanting angle.
3. Select one of the two possibilities of measurement in the Theta-meter in order to you need to know: matric potential (select 'direct probe output') or soil moisture (select 'mineral soil moisture' or 'organic soil moisture' in function of the type of soil).

   3.1. Matric potential:

      3.1.1. Select 'direct probe output' in the HH1 Theta meter. When this range is chosen, electric potential is displayed directly.

      3.1.2. Press the READ button in the Theta meter for 5 seconds. Then, conductivity is displayed (this value is valid only to 20 °C).

      3.1.3. Check the soil temperature. If it is strong different from 20 °C would be necessary to apply the following temperature correction:

Correct value (mV) = measured value (mV) + [temperature (ºC) - 20]

3.1.4. Use the following table to transform the data in matric potential values.

Table 4. Interpolation of the calibration values for equitensiometer type EQ2.

MV	-MPa	MV	-MPa	MV	-MPa
445.0	0	297.6	0.11	169.2	0.22
435.0	0.01	273.2	0.12	167.3	0.23
426.0	0.02	248.8	0.13	165.4	0.24
416.5	0.03	224.4	0.14	163.5	0.25
407.0	0.04	200.0	0.15	161.6	0.26
397.0	0.05	194.6	0.16	159.7	0.27
387.0	0.06	189.2	0.17	157.8	0.28
377.5	0.07	183.8	0.18	155.9	0.29
368.0	0.08	178.4	0.19	154.0	0.30
358.5	0.09	173.0	0.20	152.8	0.31
322.0	0.10	171.1	0.21	151.6	0.32

3.2. Mineral soil moisture:

3.2.1. Select this range to know the soil moisture ($m^3 \cdot m^{-3}$) of a mineral soil.

3.2.2. Press the READ button in the Theta meter for 5 seconds. Theta meter HH1 converts the output signal into a volumetric moisture fraction reading. It is necessary take care because accuracy decreases above the value 0.46 $m^3 \cdot m^{-3}$; '1' is displayed when very large errors would occur.

3.3. Organic soil moisture:

3.3.1. Select this range to know the soil moisture ($m^3 \cdot m^{-3}$) of a organic soil.

3.3.2. Press the READ button in the Theta meter for 5 seconds. Theta meter HH1 converts the output signal into a volumetric moisture fraction reading. It is necessary take care because accuracy decreases above the value 0.54 $m^3 \cdot m^{-3}$ and '1' is displayed when very large errors would result.

REFERENCES AND FURTHER READINGS

Gardner C.M.K., Bell J.P., Cooper J.D., Dean T.J., Hodnett M.G. "Soil Water Content." In *Soil Analysis. Physical methods*. K.A. Smith, C.E. Mullins, eds. New York: Marcel Dekker, 1991

Kilham K. *Soil Ecology*. Cambridge: Cambridge University Press, 1994

Klute A. "Water Retention: Laboratory Methods." In *Methods of Soil Analysis. Part 1. Physical and Mineralogical Methods*. A. Klute, ed. Madison: American Society of Agronomy, 1986

Kramer P.J., Boyer J.S. *Water Relations of Plants and Soils*. San Diego: Academic Press, 1995

Mullins Ch.E. "Matric Potential." In *Soil Analysis*. K.A. Smith, Ch.E. Mullins, eds. New York: Marcel Dekker, 1991

Nobel P.S. *Physicochemical and Environmental Plant Physiology*. San Diego: Academic Press, 1991

Reeve M.J., Carter A.D. "Water Release Characteristic." In *Soil Analysis*. K.A. Smith, Ch.E. Mullins, eds. New York: Marcel Dekker, 1991

Roger T.K., Robichaux R.H., Morse S.R. "Plant Water Status, Hydraulic Resistance and Capacitance." In *Plant Physiological Ecology*. R.W. Pearcy, J. Ehleringer, H.A. Mooney, P.W. Rundel, eds. London: Chapman and Hall, 1989

Rundell P.W., Jarrell W.M. "Water in the Environment." In *Plant Physiological Ecology*. R.W. Pearcy, J. Ehleringer, H.A. Mooney, P.W. Rundel, eds. London: Chapman and Hall, 1989

Slavík B. *Methods of Studying Plant Water Relations*. Berlin: Springer-Verlag, 1974

Steudle E., Frensch J. Water transport in plants: Role of the apoplast. Plant Soil 1996; 187:67-79

Taiz L., Zieger E. *Plant Physiology*. Sunderland: Sinauer Associates, 1998

CHAPTER 16

DETERMINATION OF TRANSPIRATION USING A STEADY-STATE POROMETER

Manuel J. Reigosa Roger and Adela M. Sánchez-Moreiras
Depto Bioloxía Vexetal e Ciencia do Solo. Universidade de Vigo. Spain

INTRODUCTION

In the terrestrial ecosystems the water deficit is generally considered as the most important global cause for the plant stress. The adaptation (in an evolutionary sense) of the plants to the terrestrial medium was carried out with the necessary development of the adequate structures. This development implicated the necessity of regulating the water loss via transpiration and the water transport through the plants. This regulation is carried out directing the most part of the gas exchange through the stomata. Stomatal aperture and closure are regulated establishing a balance between the photosynthesis (possible only if atmospheric CO_2 is captured) that plants try to maximise, and the stomatal transpiration that should be minimised as an inevitable and negative consequence of the plant adaptation to the terrestrial medium (Field et al., 1989).

Water availability is the first restrictive factor of the most important plant growth in the terrestrial ecosystems, and therefore the great importance of measuring the different parameters related with the plant water status. Furthermore, in a transpiring leaf the symplasmic and apoplasmic water potential values will be not equal (Turner, 1981), by that these values should be checked in these types of plant experiments (Koide et al., 1989):

1. The tissue water potential (measured by psychrometric techniques or by pressure-chamber techniques as that from Scholander et al., 1965)

M.J. Reigosa Roger, Handbook of Plant Ecophysiology Techniques, 223–233.

2. The symplasmic water potential (mainly measured with some modified psychrometric and pressure-chamber techniques, but also with pressure- probe techniques), and
3. The apoplastic water potential (measured by osmometric techniques applying overpressurization to the xylem and measuring the value of the expressed water).

But also measurements about water availability in the soil are especially important. It can be measured by very diverse techniques including the neutron probes, pressure-chamber on ceramic plate, humidity measurement and ratio, etc.

In Plant Ecophysiology, the data availability about the atmospheric water status has also a great importance, usually obtained by the relative humidity values and by the evapotranspiration calculations (calculated theoretically or, preferably, by lysimeters). The determination of the water content, the hydraulic resistance of the different tissues and the water transport through the xylem are also relevant.

But the leaf transpiration, fundamentally the stomatal transpiration, highlights between all these factors. This measurement is a summary of the previously pointed values, especially if its measure is completed with data about edaphic water availability and atmospheric humidity that are usually measured with the common techniques to measure transpiration.

During transpiration, water evaporates primarily in leaves, in a process that is driven energetically almost entirely by the net absorption of radiation. Therefore the sun is the most important source for this process (Buchanan et al., 2000). But not only this environmental factor has a marked importance in transpiration measurements. Other environmental factors can influence not only the physical process of evaporation and diffusion but also the stomatal aperture and closure on the leaf surface (Salisbury and Ross, 1992).

The stomatal resistance (R_E) is, together with the atmospheric resistance adjacent to the plant, one of the restrictive factors for the transpiration rate.

The environmental dependence of this measurement makes very important the exhaustive knowledge of all the ambient conditions, because a change in one of them can change all the plant water processes. Therefore, transpiration and stomatal resistance, and inside its relation with the photosynthetic efficiency, are usual measurements in salt (Elhaak et al., 1997), light (Shiraishi et al., 1996), nutrient (Sharma, 1995), contamination (Noormets et al., 2001), water, or thermal (Shiraishi et al., 1996) stress experiments.

In 1986 Einhellig proposed the stomatal interference as a mechanism of action of some allelochemicals. A too long or unbalanced stomatal closure can limit the CO_2 availability, and interference in the photosynthetic process can affect the stomatal aperture induced by the light.

Good correlation between the allelopathic action and the effects on stomatal function and photosynthesis were also previously described (Einhellig and Kuan, 1971; Patterson, 1981; Plumbe and Willmer, 1986). Patterson (1981) studied the effects of phenolic acids on *Glycine maxima* and could observe a decrease in the photosynthetic rate and also in the stomatal conductance. In the same way, in 1986 Bhatia et al. observed stomatal closing in the presence of very low concentrations of salicylic acid.

In recent years, more allelopathic effects on transpiration were investigated. Jose and Gillespie (1998) studied the effects of different concentrations of juglone on transpiration and stomatal conductance (and other physiological processes) of corn and soybean seedlings. They could observe in this experiment significant inhibitory effects on all measured physiological events. Effects on transpiration and stomatal conductance could be also correlated with plant growth in both species.

By the other hand, allelopathic effects of hydroquinone (isolated from *Antennaria microphylla*) on water relations and photosynthesis of leafy spurge (*Euphorbia esula*) could be tested by Barkosky et al. one year later. Plants grown in presence of hydroquinone had considerably lower transpiration rates than control plants. They could conclude that interference on stomatal function with a disruption of water balance is one inhibitory mechanism of *A. microphylla* on leafy spurge.

But not always a variation in the stomatal conductance implicates a similar variation in the transpiration rate. In this way, Manthe et al. (1992) could observe an unequal behaviour in the stomata response and the transpiration values in plants of *Vicia faba* exposed to the salicylic acid addition. Low concentrations of this allelochemical caused a decrease in the stomatal aperture but higher concentrations could cause an inhibition in the transpiration. It is possible that the plant cell is able to balance in some way, by detoxification of the phenolic acid, the alterations produced in the plant metabolism.

THE TECHNIQUE FOR TRANSPIRATION MEASUREMENT

As previously said, transpiration measurement is one of the most determinant measures in the knowledge of the leaf energy balance and plant water status (Pearcy et al., 1989). This importance and its relation with the CO_2 exchange made that several researchers developed different techniques for measuring and calculating it. Many techniques and several humidity sensors were developed for industry but the reliability and repeatability of their measurements was not enough for research measurements. So, psychrometric systems, infrared gas analysers (usually used for CO_2 exchange measurement but with the possibility of having a water vapour detector, as LI-6200 and LI-6400. See Chapter 8), and other known sensors were used for this measurement.

In plant ecophysiology, transpiration measurement is very important, but a basic condition must be taken in account: the plant must not suffer damage during the transpiration measurement, and so the number of techniques available for measuring transpiration is limited. If additionally, the instrument must be able to take the measurement as well in laboratory as in the field, the most widely used instruments for measuring stomatal conductance 'the diffusion porometers' are the only choice. Three main diffusion porometers are in our hands to obtain transpiration and stomatal conductance values (Pearcy et al., 1989):

- **Transient Porometers**. These porometers, primarily developed by Wallihan in 1964, are based on the diffusion of water vapour from a leaf placed in a small chamber to the humidity sensor inside the chamber. The porometer will measure the time required to reach a determined humidity into the chamber or the humidity value that is detected by the sensor in a prefixed time.

  This chamber has the inconvenience that for a good measurement the chamber must totally enclose the leaf. Therefore thin leaves (as those from conifers) can not be measured by this method. Furthermore, we obtain a transpiration value, but to calculate the stomatal conductance is necessary an empirical calibration by comparing the measurement of the leaf into the chamber with that of a plate of pores in the place of the leaf. This makes that the precision of the values obtained with this method is broadly questioned.

  Transient porometers are cheap and easy to use, but its use is limited by the temperature dependence, the complicated calibration or the possible errors.

- **Constant Flow Porometers**. Parkinson and Legg (1972) and Day (1977) developed a diffusion porometer based in a constant flow of dry air through the chamber where the leaf is placed. So, the humidity value, at which a steady state is reached, is recorded.

  This method is exposed to several differences between the environmental conditions outside the chamber and the microenvironmental conditions at which the leaf is exposed into the chamber. Furthermore, if a change in the vapour pressure occurs, the stomata can react in front of it and the actual plant status will be different from the starting plant status.

  Instruments based on a constant flow system able to measure CO_2 exchange are also available (Schulze et al., 1982). This type of porometer has inside an infrared analyser that allows the measurement of the differences in the CO_2 decrease and H_2O increase. These apparatus are interesting when simultaneous measurements of photosynthesis and transpiration are needed in the experiment, or when the environmental humidity is really high. In any case, they are more precise than the transient porometer but also more expensive and more difficult to use.

- **Null Balance Porometers**. The first null-balance porometer appeared in 1972. Beardsell et al. developed a diffusion porometer based in a dry airflow through the chamber to balance the humidity generated into the chamber by a transpiring leaf reaching a steady-state humidity. With this method it is necessary taking care when measuring the air and leaf temperatures, because differences can occur with the dry airflow. That is one of the advantages of this porometer, that it does not assume as equal the leaf and air temperatures, avoiding important errors in the measurement.

  As it happens with constant flow porometers, the conductance value is calculated in base of the mole fraction gradients of water vapour and the measured transpiration rates (Pearcy et al., 1989). As well in constant flow porometer as in null balance porometer errors in the humidity measurement can occur and it is possible to quantify them (Campbell, 1975). By contrary, errors in transient porometer are more difficult to know.

  This system (null-balance porometer) is perhaps the most complex of the three presented here, but it is also the most flexible, able for field and laboratory measurements and with a quite easy calibration, which can be made at constant humidity values being also more precise.

In any case, it is necessary to take in account that the extrapolation of the measurement obtained in these porometers to the whole plant transpiration is not easy and must be made with equations that are now still in development. For making it easier, the user and the apparatus must be able to maintain undisturbed the leaf during the measurements, to avoid abrupt changes of temperature trying to take the measurement at shade, protect the plant against wind and not changing the external conditions during the measurement (Pearcy et al., 1989).

Furthermore, the necessary number of replicates is high and we must be capable to take them in a short time period to avoid external environmental changes. Some scientists have developed models where the environmental conditions are directly implicated (Küppers and Schulze, 1985).

LI-1600. A NULL-BALANCE POROMETER

LI-1600 (Li-Cor model 1600, Steady State Porometer) is a diffusion porometer operating in the null-balance principle.

As previously said, the null-balance porometers are based in the measurement of the dry air flow rate necessary to maintain a constant relative humidity inside a chamber where a transpiring leaf is placed. So, we can know the water loss by this leaf. According to this principle, this apparatus gives direct values of stomatal resistance (s cm^{-1}) calculated from the relative humidity, leaf and air temperature and flow rate measured values. Stomatal resistance values are also corrected for boundary layer conductance (Li-Cor, 1989).

The equation used by LI-1600 to obtain the leaf transpiration rate (mg cm^{-2} s^{-1}) is as follows:

$$E = (\rho_c - \rho_a)\frac{F}{A}$$

where: E is the leaf transpiration rate, expressed in mg cm^{-2} s^{-1}

P_c is the water vapour density in the chamber, expressed in mg cm^{-3}

P_a is the water vapour density in the dry air flow entering in the chamber, assumed as a 2% of relative humidity

F is the volumetric flow rate in the chamber, expressed in cm^{-3} s^{-1}

A is the leaf area, expressed in cm^2

As shown in this equation, the calculated leaf transpiration rate directly depends on the volumetric flow rate into the chamber. This flow rate depends also on the chamber temperature and the barometric environmental pressure.

Considering above statements, it is necessary to use an instrument able to obtain these measurements and calculations implying on it the as many environmental parameters as possible. In this way, LI-1600 allows inclusion of atmospheric pressure data, measurements of chamber and leaf temperatures or relative humidity. Light intensity is also measured although it is not used in the calculations.

A LI-1600 porometer is represented schematically in Fig.1. It consists of two main parts:

1. **Console:** It has some control elements that make possible the transpiration measurement. The pump draws air from inside the console and pumps it to the air tank where the airflow is dispersed in two directions. A part escapes via the Internal Flow Controller (that controls and adjusts the relative humidity in the chamber to the set point), and the other part goes to the Null Adjust Valve (that limits the maximum flow rate entering in the chamber and therefore, forces a major air flow to escape from the system).

   This air dispersion is essential in the system because leaves with very different transpiration rates must be measured and control on the flow rate entering in the chamber will be necessary to maintain the relative humidity in the set point. For example, if the leaf placed in the chamber has a very low transpiration rate, only a small dry air flow rate will be necessary into the chamber to balance the leaf transpiration rate; if the air entering in the system is too high the relative humidity can drop below the setpoint and the measurement will be not real. Therefore, the system has the option of venting off excess flow.

   Once that the airflow passes through the null adjust valve, it will pass through two desiccants (see Fig 1.), one in the console and the other one in the sensor head. So, we ensure that the airflow entering in the chamber is dry.

2. **Leaf Chamber:** Transpiration depends on many factors. Therefore, the leaf which will be measured, must be normally enclosed in a chamber where the major number of possible external factors can be controlled. In the same way the size of the selected chamber must permit the measurement of very different leaves, with very different sizes but also with very different stomatal resistances. Therefore a high variety of chambers are available for measuring (with different shapes and sizes). The cylindrical chamber of a LI-1600, for example, is made by nickel plated aluminium to minimise water absorption. The sensors inside the chamber check for several environmental parameters (see Fig. 1), for example leaf and air temperatures, relative air humidity and also the PAR (photosynthetically active radiation).

   Once that the just dry air flow enters in the chamber it is mixed with the water vapour loss by the transpiring leaf, and a dry air flow rate will be necessary to balance the relative humidity into the chamber. This value is the key for the later stomatal resistance and transpiration calculations.

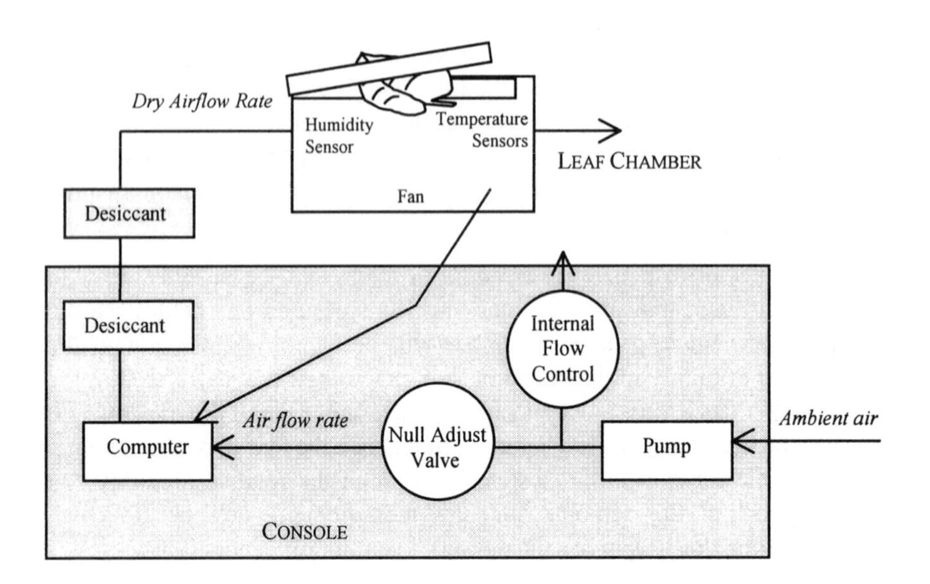

Figure1. Scheme of null-balance porometer to measure transpiration, for example used by the LI-1600 Steady State Porometer

Measuring

The first step to use the LI-1600 is to check that the desiccant from the console is completely dry, changing it if necessary. Every week the desiccant from sensor head must be changed, too.

The chamber must be installed at this time. The selected leaf chamber and the aperture (size and form) will depend on the plant species that will be measured. Is not the same to measure grass leaves that conifer needle.

Once that the sensor head is cleaned without presence of pollen or dust (it can increase the water vapour absorption with unusually increased values), the apparatus is ready to work.

For starting the work with the LI-1600 a previous habituation to the environmental parameters is necessary (30 min). Therefore, we must be careful and to install the instrument in the definitive measuring place. Direct sun must be avoided to protect the chamber against abrupt temperature changes. At this time, checking of chamber and leaf temperatures, flow, relative humidity, PAR, etc. can be done.

Once finished the acclimatisation and after calibrating relative humidity sensor if necessary, and setting the null point humidity to ambient relative humidity (null points different from ambient values can also be used), and the dry air flow rate entering in the chamber, data recording can start. Along this time, we can display the parameter value of our interest.

The LI-1600 needs generally a mean time of 10 to 20 seconds to reach a steady-state situation on a leaf with a diffusive resistance of 5 s cm^{-1}. With higher values the instrument needs a little longer time.

Once that the instrument is at or near the null point relative humidity and the diffuse resistance value is stable, data can be stored in a data memory and transferred to a PC with a data transfer program for later analysis.

REFERENCES

Barkosky R.R., Butler J.L., Einhellig F.A. Mechanisms of hydroquinone-induced growth reduction in leafy spurge. J Chem Ecol 1999; 25:1611-1621

Beardsell M.F., Jarvis P.G., Davidson B. A null-balance diffusion porometer suitable for use with leaves of many shapes. J Appl Ecol 1972; 9:677-690

Bhatia D.S., Vandana J., Malik C.P. Effects of salycilic acid and tannic acid on stomatal aperture and some enzyme changes in isolated epidermal peelings of *Euphorbia hirta* L. Biochem Physiol Pfl 1986; 181:261-264

Buchanan B.B., Gruissem W., Jones R.L. *Biochemistry and Molecular Biology of Plants.* Rockville, Maryland: American Society of Plant Physiologists, 2000

Campbell G.S. Steady state diffusion porometers. Measurements of stomatal aperture and diffusive resistance. Wash Univ Agric Res Cent Bull 1975; 809:20-23

Day W. A direct reading continuous flow porometer. Agr Meteorol 1977; 18:81-89

Elhaak M.A., Migahid M.M., Wegmann K. Ecophysiological studies on *Euphorbia paralias* under soil salinity and seawater spray treatments. J Arid Environ 1997; 35:459-471

Einhellig F.A. "Mechanism and Modes of Action of Allelochemicals." In *The Science of Allelopathy.* R.A. Putnam, Ch-Sh Tang, eds. New York: John Wiley and Sons, 1986

Field C.B., Ball J.T., Berry J.A. "Photosynthesis: Principles and Field Techniques." In *Plant Physiological Ecology.* R.W. Pearcy, J.R. Ehleringer, H.A. Mooney, P.W. Rundel, eds. London: Chapman and Hall, 1989

Jose S., Gillespie A.R. Allelopathy in black walnut (*Juglans nigra* L.) alley cropping. II. Effects of juglone on hydroponically grown corn (*Zea mays* L.) and soybean (*Glycine max.* L. Merr.) growth and physiology. Plant Soil 1998; 203:199-205

Koide R.T., Robichaux R.H., Morse S.R., Smith C.M. "Plant Water Status, Hydraulic Resistance and Capacitance." In *Plant Physiological Ecology.* R.W. Pearcy, J.R. Ehleringer, H.A. Mooney, P.W. Rundel, eds. London: Chapman and Hall, 1989

Küppers M., Schulze E.-D. An empirical model of net photosynthesis and leaf conductance for the simulation of diurnal courses of CO_2 and H_2O exchange. Austr J Plant Physiol 1985; 12:512-526

Li-Cor Inc. *Li-1600. Steady State Porometer.* Lincoln: Li-Cor, 1989

Manthe B., Schulz M., Schnabl H. Effects of salicylic acid on growth and stomatal movements of *Vicia faba* L.: evidence for salicylic acid metabolization. J Chem Ecol 1992; 18:1525-1539

Noormets A., Sôber A., Pell E.J., Dickson R.E, Podila G.K., Sôber J., Isebrands J.G., Karnosky D.F. Stomatal and non-stomatal limitation to photosynthesis in two trembling aspen (*Populus tremuloides* Michx.) clones exposed to elevated CO_2 and/or O_3. Plant Cell Environ 2001; 24:327-336

Parkinson K.J., Legg B.J. A continuous flow porometer. J Appl Ecol 1972; 9:669-675

Pearcy R.W., Schulze E.-D., Zimmermann R. "Measurement of Transpiration and Leaf Conductance." In *Plant Physiological Ecology.* R.W. Pearcy, J.R. Ehleringer, H.A. Mooney, P.W. Rundel, eds. London: Chapman and Hall, 1989

Plumbe A.M., Willmer C.M. Phytoalexins, water-stress and stomata III. The effects of some phenolics, fatty acids and some other compounds on stomatal response. New-Phytol 1986; 103:17-22

Salisbury F.B., Ross C.W. *Plant Physiology.* Frank B. Salisbury, ed. Belmont, California: Wadsworth, 1992

Scholander P.F., Hammel H.T., Bradstreet E.D., Hemmingsen E.A. Sap pressure in vascular plants. Science 1965; 148:339-346

Schulze E.-D., Hall A.E., Lange O.L., Walz H. A portable steady-state porometer for measuring the carbon dioxide and water vapor exchanges of leaves under natural conditions. Oecologia 1982; 53:141-145

Sharma P.N. Water relations and photosynthesis in phosphorus deficient mulberry plants. Indian J Plant Physiol 1995; 38:298-300

Shiraishi S.I., Hsiung T.C., Shiraishi M., Kitazaki M. Changes in the photosynthetic rate, transpiration rate, stomatal conductivity and water use efficiency of *Vitis* varieties grown under different temperature and light conditions. Sci Bull Fac Agr Kyushu Univ 1996; 51:33-38

Turner N.C. Techniques and experimental approaches for the measurement of plant water status. Plant Soil 1981; 58:339-366

Wallihan E.F. Modification and use of an electric hygrometer for estimating relative stomatal apertures. Plant Physiol 1964; 39:86-90

CHAPTER 17

STOMATA IMPRINTS: A NEW AND QUICK METHOD TO COUNT STOMATA AND EPIDERMIS CELLS

M.H.Meister and H.R.Bolhàr Nordenkampf

Division of Horticultural Plant Physiology and Primary Production. Institute of Ecology and Conservation Biology. University of Vienna. Austria

Abstract

Comparing several techniques for creating epidermal replicas, we found that imprints on cellulose-di-acetate and on polymethyl-metacrylate can be easily performed within a few seconds, which proves to be a considerable advantage especially in field trials. The result is a permanent impression of the epidermis' surface, perfect for long-term storage. In the case of extremely sunken stomata, pleated leaf surfaces, or coniferous needles, we additionally used a cyanacrylate adhesive. Reliable and reproducible results could be achieved for use when analysing the imprints by a drawing microscope or an image analyses program.

INTRODUCTION

Stomata are found in all plants above the evolutionary level of the saprophyte generation of mosses and the sporophyte generation of ferns, respectively. Those of mosses are mainly non-functional and stomata of pteridophyta show rather sluggish movements. On the other hand, stomata of gymnosperms and angiosperms are highly versatile. They are predominantly controlled by water availability and CO_2 concentration, in

M.J. Reigosa Roger, Handbook of Plant Ecophysiology Techniques, 235–250.

a combination of various endogenous responses to changes in the environment (Raschke 1975, Martin et al. 1983).

Even though stomata are most commonly associated with leaves, they can also occur in inflorescences, fruits, herbaceous stems, petioles and tendrils of gymnosperms and angiosperms, whereas no holoparasites are reported to express any.

In the course of leaf development, stomata create their own inhibitory field which may result in a nearly equidistant spacing. Nevertheless, in fully developed leaves stomata pores are usually non-randomly scattered within the epidermis. In some species stomata complexes are clustered or restricted to certain morphological areas e.g. they occur in bands adjacent to the venation (Willmer, 1983). The guard cells are generally crescent-shaped, but dumb bell-shaped in sedges and grasses. The arrangement of stomata depends on the formation of guard mother-cells and their polarity caused by unequal cell division, which is thought to be genetically controlled and influenced by environmental factors (Bünning, 1965; Barlow and Carr, 1984).

Previously it has also been demonstrated that guard-cell size and stomata density develop corresponding to environmental conditions during leaf growth (Napp-Zinn, 1984; Jones, 1985). Although local climatic conditions will influence the development of stomata patterns in nearly all plants, dispersal, number and arrangement of subsidiary cells and shape of guard-cells are attended to in comprehensive studies on families and genera (Metcalfe and Chalk 1950; Mauseth, 1988), see Table 1.

Stomata frequency and arrangement, as well as expansion of the leaf area can be affected by water availability, light intensity, temperature and CO_2 concentration. However, in case of a larger leaf area, accompanied with a lower stomata density (number per mm^2), the number of stomata per leaf as well as the stomata index (number of stomata per number of epidermis cells) will remain unchanged (Salisbury, 1949).

To permit an effective control through the central slit (aperture) between the guard-cells above the aperture, the front cavity and the outer cuticular ledges minimise forced transpiration, e.g. during wind, whereas the nozzle-shaped back cavity is thought to guide the gasses to and from the substomatal cavity (Esau, 1965; Fahn, 1982).

Table 1. Stomata density of selected species after Bolhàr-Nordenkampf HR & Draxler G (1993) and Woodward FI & Kelly CK (1995).

	STOMATA FREQUENCY per square millimeter leaf surface		
Species	**Upper surface**		**Lower surface**
Grasses:			
Avena sativa (oats)	25 - 48		27 - 35
Phragmites communis (riparian)	426 - 536		493 - 754
Saccharum officinarum (sugar cane)	59 - 167		176 - 351
Lolium perenne (English rye grass)	70		9
Oryza sativa (rice)		931	
Zea mays (corn)	52 - 94		68 - 158
Herbaceous plants:			
Arachis hypogaea (peanut)		15 - 20	
Helianthus annuus (sunflower)	27 - 326		90 - 408
Phaseolus vulgaris (bean)	8 - 16		132 - 184
Solanum tuberosum (potato)	10 - 20		230
Trifolium pratense (clover)	207		335
Succulent leaves:			
Agave sp.		8 - 72	
Portulaca oleracea	0		21
Sedum acre	21		14
Shrubs and trees:			
Acer pseudoplantanus		396	
Pistacia lentiscus		287	
Citrus sinensis (subtropic)	0 - 90		620
Juniperus communis		229	
Castanea sativa		597	
Fagus sylvatica (beech, temperate)			
sun exposed leaves		130 - 295	
shade exposed leaves		94 - 177	
Hedera helix (ivy, shade, temperate)	0		105
Ligustrum sp. (subtropic)	0		830
Myrtus communis (subtropic)	0		158
Populus alba (temperate, riparian)	0		315
Quercus rubra (oak,temperate, dry)	0		680
Quercus triloba (oak, subtropic, dry)	0		1192
Sambucus nigra (elder, temperate)			
sun exposed leaves		42 - 260	
shade exposed leaves		12 - 147	
Salix alba (temperate, riparian)	114		(?)
Floating leaves:			
Nymphea alba (water lily)	400		0

Consequently, stomata density and size (length of aperture) may strongly affect photosynthetic and transpiration rates per unit leaf area, and therefore cause great differences in the efficiency in the use of light and water (Leunig, 1995; Jarvis et al., 1999).

Thus, especially as a basis for causal analyses of data from stress physiology research, e.g. water stress, elevated CO_2 or ozone concentrations, studies of functional leaf anatomy are presently becoming more important (Radoglou and Jarvis, 1990). Together with the structural differentiation of epidermis cells and stomata pores, the size of the mesophyll (palisade, spongy) and the amount of phloem and xylem elements will determine the performance of plants in a certain environment (Körner et al., 1989; Bolhàr-Nordenkampf and Draxler, 1993).

Studies on stomata density of entire leaves can be performed only, using a sophisticated incident light microscope, whereas surface parallel sections or epidermis strips can be investigated easily using a calibrated grid in the eyepiece of any microscope (Cousson, 2000).

Due to the high natural variation in shape, size and frequency of stomata complexes, big sample-sizes have to be dealt with to trace up statistical significance. Additionally, studies on ecophysiology require field methodologies which are easy to perform.

Based on the requirements mentioned above, we started to search for a technique with the two key features: Firstly, it should allow a quick performance of a large set of measurements to meet statistical requirements, and secondly, be simple enough for usage in the field.

METHODS

Cellulose-di-acetate

Vazzana et al. (1988) reported a new technique which creates imprints of the leaf surface on pieces of Cellulose-di-acetate (CdA; Rhodoid®, Rhone-Poulenec via Weber Metaux, Paris, F) softened by acetone. Impressions can be sputtered and investigated in the electron-microscope. For examinations under the light-microscope, slide-shaped rectangles should be cut from 2.0 mm thick sheets, which is costly.

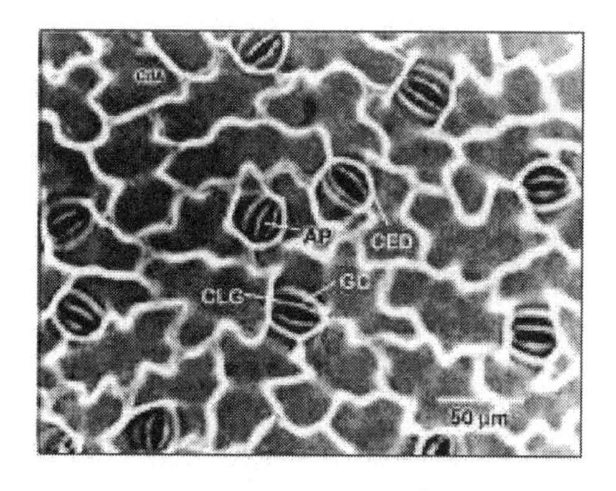

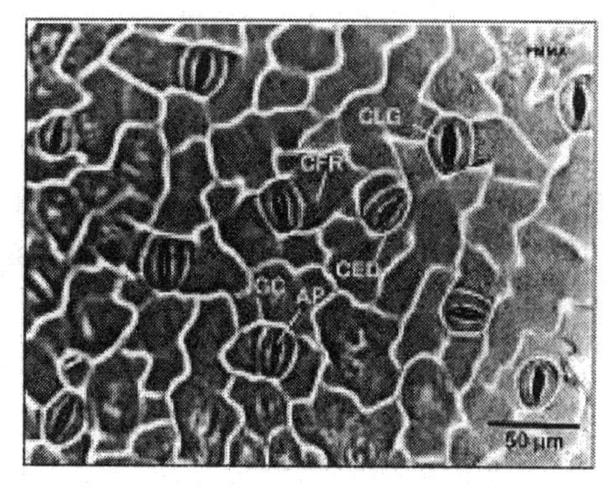

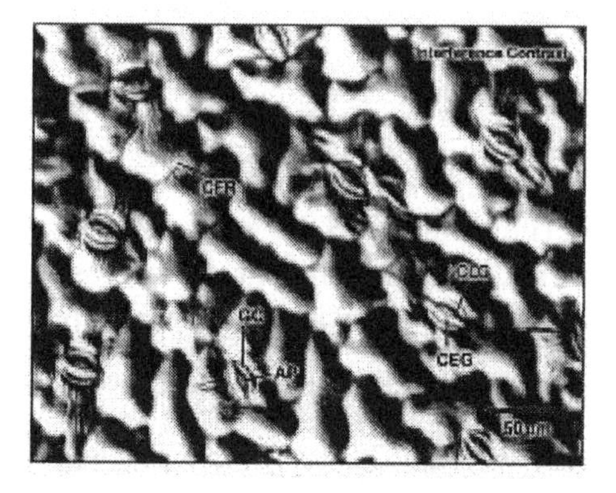

Figure 1.Beta vulgaris var. altissima 'Zumo': amphistomatous leaf, abaxial (lower) side, stomata in slightly risen position with lateral cuticular furrows, undulated epidermal cell walls. AP aperture (central slit), CED cuticular edge of epidermal cell, CFR cutilar furrows, CLG cuticular ledge of front cavity, GC guard cells of stoma.

a. Imprint on Cellulose-di-Acetate softened by Acetone, stomata and undulated cell walls clearly visible, aperture and central slit become visible.
b. Imprint on PolyMethylMetAcrylate softened by Methyl-Ethyl-Cetone, outlines more distinct compared to a. cuticular structures become visible.
c. Imprint on PMMA seen in Nomarski Differential Interference Contrast, aperture (central slit) not clearly shown, but cuticular furrows and domed epidermis cells are visible.

To reduce costs, irregular-sized pieces can be cut from sheets using pruning shears or sharp pincers. Thinner material tends to bend after application of acetone, which makes microscopic work more difficult. For further investigation photographs were taken and analysis could either be performed on projected negatives or on prints (e.g. Bettarini et al., 1998).

Imprints on CdA clearly show complexes of stomata as well as guard-cells forming the pore in the outer cuticular ledge (Fig. 1a). Besides the clear impression of the aperture (central slit) the outlines of epidermal cells can usually be easily recognised as well. Therefore, determinations of stomata frequency and counting of epidermal cells can be performed, permitting calculation of the stomata index. This also applies to measurements of cell-size and length of the aperture.

Cellulose-di-Acetate sheets may differ in their consistency. Probably different trademarks are more reliable. We used Ultraphan® -1.0 mm (Lonza, Weil am Rhein, Germany) quite successfully, which was easy to cut by a cutting-blade (Stanley) or a guillotine. With four imprints on one sheet, bending did not start immediately. The CdA-slides can be successfully stored for months.The main disadvantage of CdA is the availability of thick, crystal-clear sheets (1-2 mm), as many companies have stopped production.

Polymethyl-metacrylate

Difficulty with CdA availability was the main reason why we started test-series with perspex (Poly-Methy-MetAcrylate), the latter being offered by many suppliers and at comparably low price. Microscopic slides were cut from a 1 mm thick PMMA sheet using a Stanley-knife or a guillotine.

Ethyl-methyl-cetone (easily flammable, irritant) was used instead of acetone to soften the surface of the slide. A drop of solvent was placed on the surface of the Perspex® slide. While the surface was left to soften (10-20 seconds), leaf-disks (diameter 6-7 mm) were punched out with a cutter (cork borer) or square pieces cut with a razor-blade. Imprints of leaf surface were simply created by gently impressing a leaf disk onto the softened surface (finger tip). In some cases this procedure was good enough for counting both stomata and epidermal cells (Fig. 1b), whereas sunken stomata and the shape of epidermal cells were not clearly visible. Therefore, this simple technique was modified to achieve imprints of higher quality. In case of sunken stomata (Fig. 2a,b) another drop of solvent was placed on the slide before the leaf-disk was impressed on the softened surface by a lead-cylinder (diameter 10 mm, 30g). Depending on leaf-anatomy the load was increased to 0.5 kp for several seconds. The leaf-disks were removed after 3-10 minutes, depending on the amount of solvent used.

Stomata and epidermis cells (Fig. 2d) were counted by means of a drawing microscope. Impressions were studied, depending on cell size mostly under the fields of view 0.09 mm^2 (Neofluar 25/0.60 Zeiss, Oberkochen) and 0.0625 mm^2 (EF 40/0.64). One imprint can serve for many counts, if small areas are chosen.

Microscopic slides made from PMMA with 3-4 imprints do not bend and can be stored for many years.

Direct microscopic examination of imprints is to be preferred over photographs as it permits to gain information from different optical layers (Fig. 3) and by advanced microscopic techniques (Fig: 1c), e.g. Nomarski differential interference contrast (Gay and Hauck, 1994). This also applies to image-analysing systems which can be successfully used only with excellent imprints of regularly shaped stomata and epidermis cells (LUCIA M, Version 3:0).

Measurements on the width of stomata are doubtful if not verified by other techniques, because the aperture might change during the generation of the imprint. Measurements of cell-size can also be performed on a digitising-board (Körner et al., 1989). For field-studies, a counting-grid mounted in the eyepiece of a microscope can be used.

During numerous studies of stomata imprints in the light-microscope, some improvements of the method were established:

Leaf surfaces should be even and smooth, and cleaned from organic or inorganic particles. Arthropods (e.g. spider mites) were removed using adhesive tape, which can also be used to remove any hairs or trichomes. Dense layers of epidermal appendages and especially epiphytes should also be removed. Imprints turned out to be blurred and difficult to analyse if leaves were infected with fungi or the epidermis was heavily destroyed by spider mites.

Thin leaves (e.g. clover) should be used fully water saturated only. Thicker leaves (e.g. sugar-beet) with rough surfaces gave better results if used in wilted condition. All these treatments, including the pressure by the load when creating the imprints, will either change and flatten the micro-relief of the epidermis, or open folds of pleated leaves (Fig. 4).

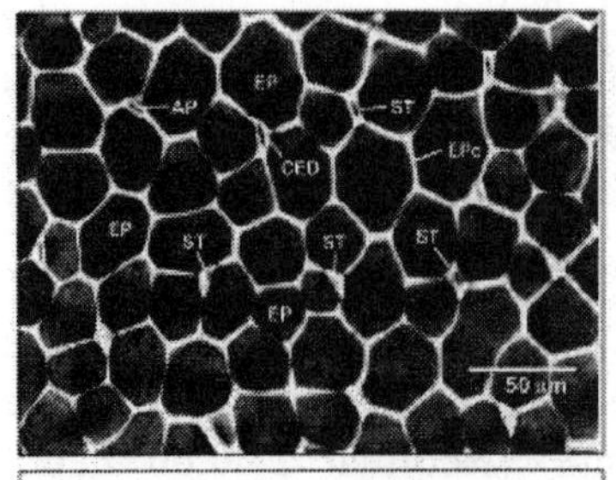

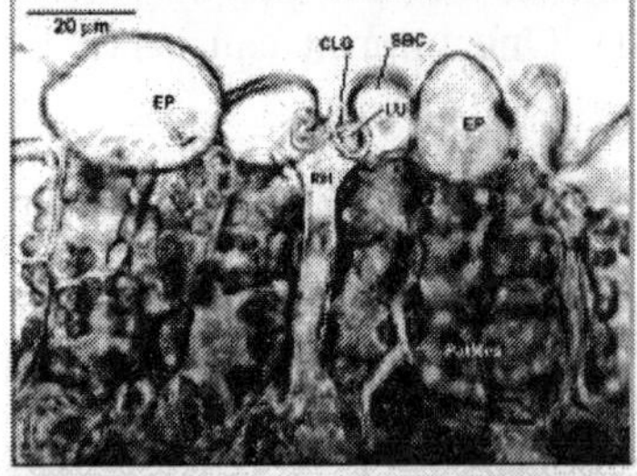

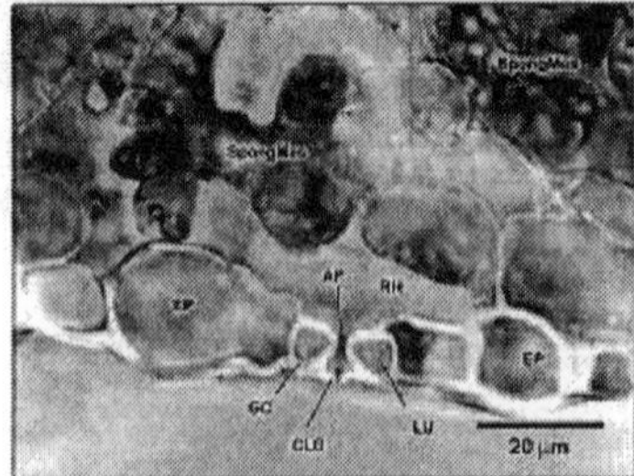

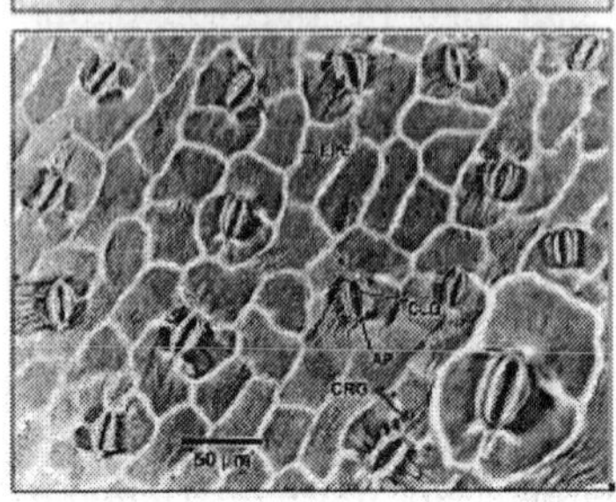

Figure 2. Trifolium repens "Milkanova": amphistomatous leaflet. AP aperture (central slit), CED cuticular edge of epidermal cell, CLG cuticular ledge of front cavity, EPc epidermal cell walls, GC guard cells of stoma, LU cell lumen, RH respiration hole, SBC subsidiary cell, ST stoma.

a. Upper (adaxial) side, imprint (PMMA), stomata in sunken position, rectangular or triangle shaped pore, hexagonal and heptagonal epidermal cells, no of stomata.mm^{-2} = 244 ± 48.

b. Upper (adaxial side), cross-section (~100 µm), living tissue in water, (50/0.85,oil- imm.) stoma in sunken position, over-arching epidermal cells, chloroplast aligned on cell walls of palisade-mesophyll (parenchyma).

c. Lower (adaxial) side, cross-section (~100 µm), living tissue in water, (50/0.85, oil-imm.), stoma surrounded by thicker epidermal cells, spongy mesophyll with chloroplasts and large inter cellular spaces.

d. Lower (adaxial) side, imprint (PMMA), stomata slightly sunken, front cavity edges and aperture (central slit) clearly shown, cuticular ridges run down to the pore, insert 1.23 times lager, no of stomata.mm^{-2} = 156 ± 23.

For life-size studies of the epidermal surface, an incident-microscope with image analyses can be used. Nevertheless, if no further treatment of the leaf is required and the pressure is well adjusted by the procedure described above, epidermis cells will be fixed immediately. This makes the method suitable for drought-stress experiments or research involving changes of epidermal and stomatal cell-size during ontogenesis (Fig. 4a-d).

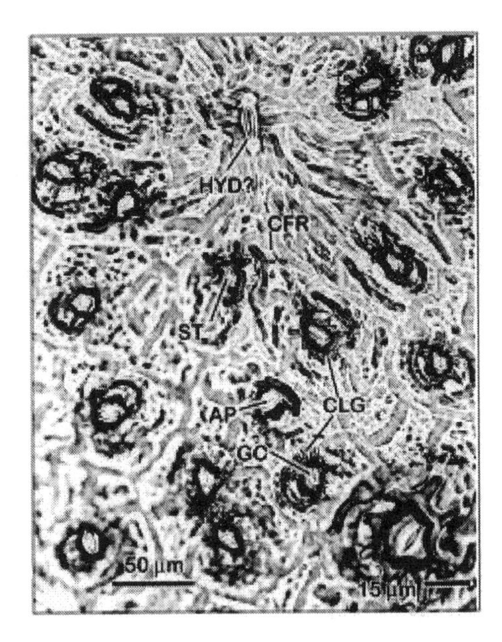

Figure 3. x Fatshedera lizei: hypostomatous leaves, lower (abaxial) side, guard cells of stomata in a sunken position, three subsidiary cells, slopes of outer ledges with cuticular ridges leading up to the pore, optical layer of epidermal cells not shown, insert 1.78 times enlarged.
Cyanacrylate adhesive combined with an imprint on PMMA:
AP aperture (central slit), CED cuticular edge of epidermal cell, CFR cutilar furrows, CLG cuticular ledge of front cavity, GC guard cells of stoma, ST stoma, EPc epidermal cell walls, HYD hydathode?.

PMMA slides were also tried extensively on herbarium plants. More than 50 tropical species were analysed for taxonomic purposes (see also Vazzana et al., 1988; Bettarini et al., 1998). Moreover, it is important to notice that plant material having been stored in 40% alcohol was easy to handle and gave imprints of good quality.

Imprints on PMMA slides combined with replicas by cyanacrylate adhesive

The technique described above cannot always be used when coming across pleated laminas, as in some grasses (Fig. 4). It needs some skill and patience to obtain replicas by means of acrylic varnish or fast-drying nail varnish. Therefore, we created replicas using liquid cyanacrylate adhesive

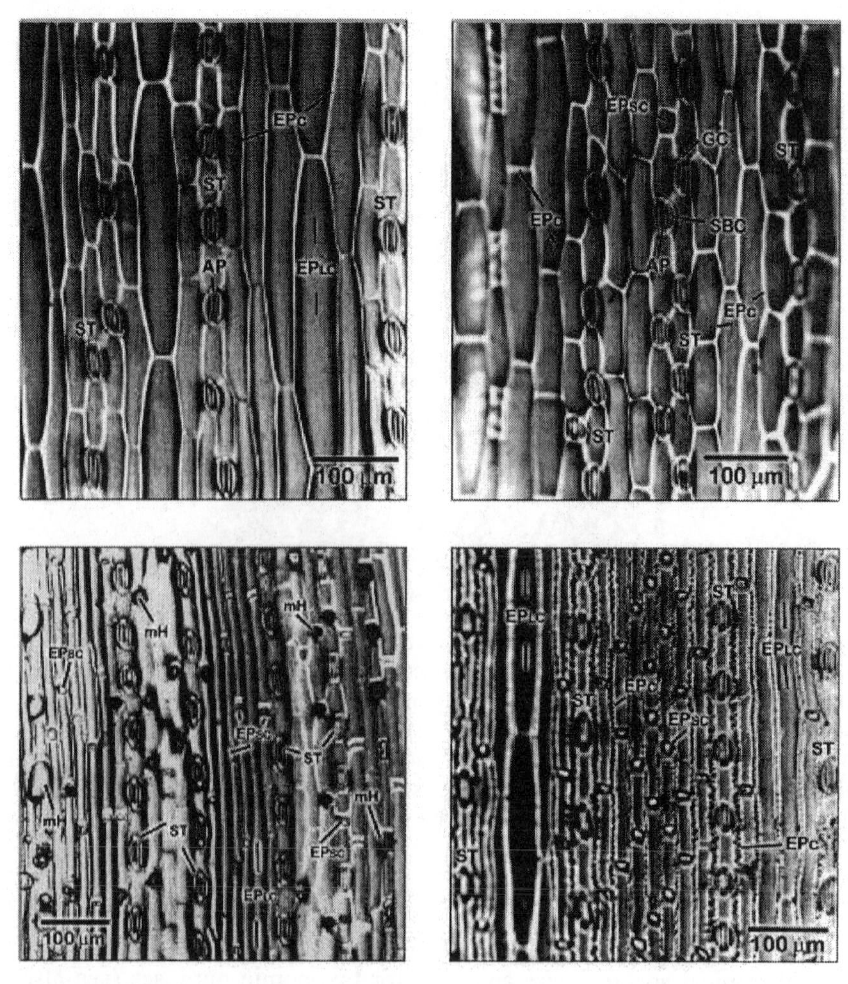

Figure 4. Triticum aestivum: amphistomatous pleated leaf, stomata gramineae-typ with two lateral triangle shaped subsidiary cells, no inter-stomatal short cells. AP aperture (central slit), EPlc long-cells, EPsc short-cells, GC guard cells of stoma, mH micro hair, SBC subsidiary cell, ST stoma.

Young plant well watered:

a. upper (adaxial) surface: slightly pleated, stomata frequently in two adjacent rows, some micro-hairs (not shown), no of stomata.mm^{-2} = 40 ± 4

stomata apparatus: length 61.5 ± 3.3 µm, width 33.6 ± 4.8 µm,

b. lower (abaxial) surface: nearly flat, stomata in single rows separated by long-cells, a few short-cells, straight epidermal cell walls, no of stomata.mm^{-2} = 27 ± 3.

Plant after 3 weeks of draught stress:

c. upper surface: pleated, single rows of stomata separated by narrow long-cells, many short -cells with silica bodies, numerous micro-hairs in rows, epidermal cells and stomata are smaller compared to those from young, well watered plants, no of stomata.mm^{-2} = 82 ± 9, stomata apparatus: length 56.1 ± 5.4 µm, width 25.9 ± 2.7 µm.

d. lower (abaxial) surface: slightly pleated, single rows of stomata, in between either rectangular epidermis cells in turn with short-cells with silica-bodies or extremely stretched long-cells, markedly sinuous epidermal cell walls, no of stomata.mm^{-2} = 63 ± 5.

and fixed them in the softened surface of a PMMA slide. A droplet of super-glue was put on the leaf surface, and after infiltration (3-5 sec) the leaf was inverted and pressed into the softened surface (10 sec) of a PPMA slide. Increasing the pressure produced better results but is likely to harm cells, and thus change actual size. The distinct structures of the epidermis are best recognised using a Perspex® imprint, whereas tiny details often emerge better in glue replicas.

Replicas done by a combination of super glue and glass slides have the disadvantage that dried glue tends easily to splinter off

Plants with inter-costal areas surrounded by the protruding venation on the lower surface will create another problem: Even though the liquid adhesive infiltrates the leaf tissue in these small areas, air bubbles might remain.

Recently modified tubes help to avoid the problem that the adhesive starts to harden the tube after use (Sekunden Kleber blitzschnell®, UHU®, Bühl, Germany; Patex Blitz Kleber®, Henkel, Vienna, Austria; super glue LSK3®, Duro super glue 5®, Loctite, Rocky Hill, CT, USA).

EXAMPLE

Research mainly focused on stomata distribution using imprints. To give a short overview of the possibilities of imprints, 'Aralia Ivy` was selected. Surface parallel sections and cross sections of leaves were also studied to evaluate results.

Methods

x *Fatshedera lizei* Guill. (hort. ex Cochet) (Araliaceae) is a genus hybrid of *Fatsia japonica* x *Hedera helix*. Leaves are long-lived, show three to five pointed lobes; stems have secondary growth in girth and are perennial. Samples were taken from plants cultivated in open-top chambers, which were set up to study morphological, anatomical and metabolic adaptations to rising CO_2 concentration in the atmosphere.

Small pots (180 cm^3, 'D' plants) were used to simulate restricted root growth and large pots (750 cm^3, 'U' plants) were used for unlimited root-growth over the investigated period of 20 weeks.

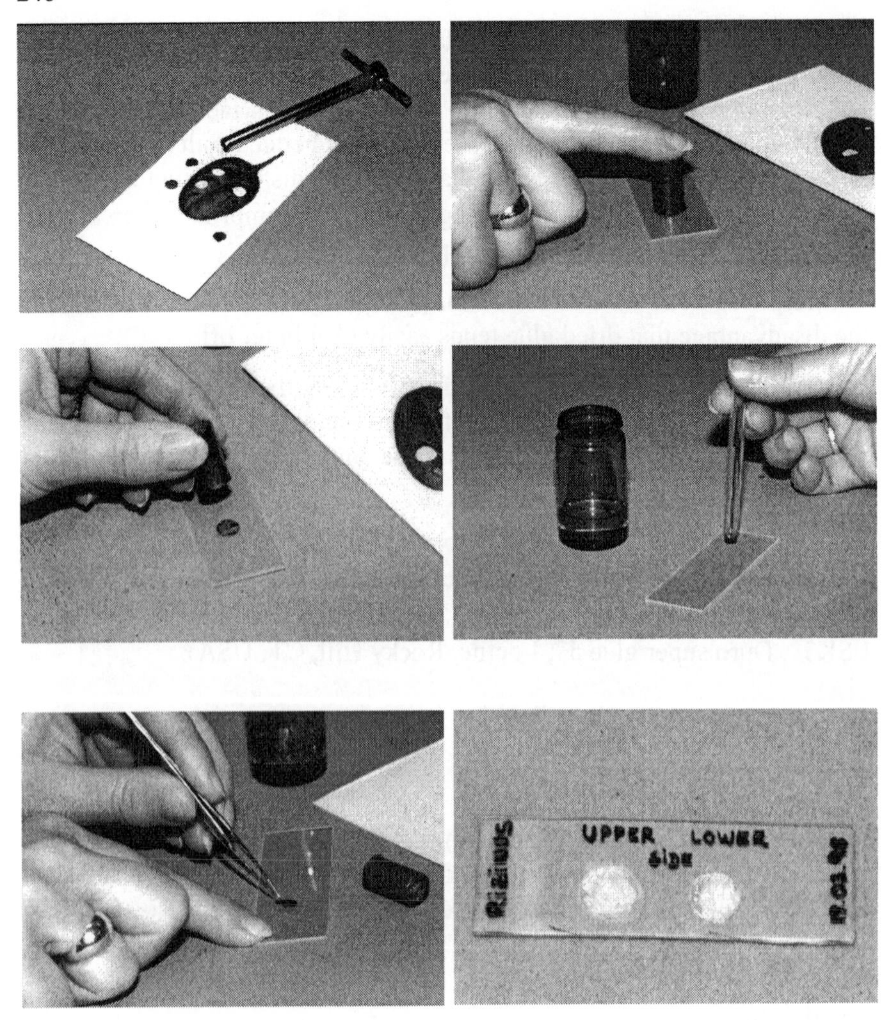

Figure 5. Sequential steps of the technique

As the micro-relief of the abaxial epidermis is very rough, we used a combination of liquid adhesive with a surface-softened PMMA slide. Imprints showed stomata and shades of epidermal cell walls. Thus, stomata pores and epidermis cells could only be counted, while their shape and area had to remain undetermined. Leaf disks were punched out from the central region of fully developed leaves, avoiding protruding major veins.

Leaf area was determined using the leaf-area-meter WinDIAS (Delta-T Devices, Burwell, UK).

Microscopy and statistics

A research microscope was used to study smaller details of imprints and leaf cross-sections (Microphot-SA, PlanApo 20/0.75, Plan 50/0,85-oil Tokyo Japan). Images were either preserved as photographs or videos. Adobe Photoshop 5.0® was used to level the different optical layers of the microphotographs.

Data were tested by calculating confidence-intervals based on Student`s *t*-test (Sigma Plot® 4.0, Jandel, Chicago, USA). These intervals were calculated to include the arithmetical mean with a 90% probability. Differences in plants due to the experimental set-up, growth conditions and the selection of leaf pieces for imprints were the mean reasons to calculate probability-errors of 10% ($p = 0.1$).

Result and discussion

For most species previously grown in carbon-dioxide-enriched conditions, it seemed proven that leaf conductance (approx. 95% stomatal) is reduced (Morison, 1987; Radoglou and Jarvis, 1990; Bunce, 1992; Bunce, 1999; Bettarini et al., 1998; Tognetti et al., 1998; Oren et al., 1999). This reduction can only partly be considered to be due to changes in stomata density or stomata index, as results are not consistent (Woodward, 1987; Körner, 1988; Woodward and Bazzaz, 1988; Radoglou and Jarvis, 1992; Gay and Hauck, 1994; Ticha et al., 1997).

Leaves of *Fatshedera lizei* are hypostomatous and stomata are bordered by three subsidiary cells with rather straight cell walls, which differ from the rest of the epidermis cells, the latter showing multiply lobed and serrated (dented) cell walls. The lower surface (abaxial) showed, besides a rough micro relief, some stemmed asterisk-shaped hairs and domed epidermis cells. The guard cells are sunken but the outer ledges of the large front cavity rise above the subsidiary cells. The slopes show cuticular ridges leading to the top of the front cavity where the pore is situated (Fig. 3).

Plants of 'Aralia Ivy' grown in enriched CO_2 atmosphere developed larger leaves (+U- plants by +12.5% ,+D-plants by +14.5%). Restricted root growth (D-plants) caused smaller leaves (+D versus +U –18.4%, D versus U –21.1%). Therefore, D-plants in ambient air showed a significant reduction ($p = 0.1738$) in leaf area of 29% when compared with +U-plants. Stomata frequency followed an opposite trend, with D-plants in ambient air increasing by 15% compared to +U-plants. This apparent increase was not significant (103.2 ± 8.1 mm^{-2} versus 89.8 ± 6.7 mm^{-2}, $p = 0.1791$) and resulted purely from the expansion in leaf area, as demonstrated by an unchanged number of stomata per leaf. The insignificant ($p = 0.2404$) rise in stomata index of +D-plants was due to a slightly increased number of epidermis cells. Bettarini et al. (1998) described the same phenomenon in *Conyza canadensis*.

A paper, dealing with *Zea mays, Glycine max* and *Liquidambar styraciflua,* reported on a significant increase in stomata index under a CO_2 enriched atmosphere (Thomas and Harvey, 1983), a phenomenon that was hardly ever noticed.

Conclusions

Imprints of leaf surfaces on CdA and PMMA-slides proved to be a technique, which is not time-consuming and enables straightforward investigations especially in field trials. With most plants investigated, good replicas could be produced if the technique was adjusted to the micro-relief and the surface-structures of the epidermis. Present results show that such imprints are perfectly suited for studying changes in stomata-density and stomata-index, especially under conditions of CO_2 enrichment, and when combined with other stress-factors. Additionally, silica-cells, complex cuticular structures like ridges and furrows, and sometimes appendages (hairs) can be easily studied applying the new method.

Thus, this newly developed technique for reproduction of leaf surfaces on CdA and PMMA microscopic slides can greatly facilitate the interpretation of physiological data, and we hope that the new method will give a greater incentive to future work in the field.

ACKNOWLEDGEMENTS

The authors would like to thank Ao.Prof. Dr. Irene Lichtscheidl, Alfons Svoboda and Ass.Prof. Dr. Wolfgang Postl for technical support. Our sincerest thanks to Dr. Brigitte Grimm for her great help with the manuscript. We are pleased to acknowledge the help of Dr. Antonio Raschi with the first CdA imprints.

The work was supported by a grant from the Austrian 'Ministry for Science, Traffic and the Arts' (GZ 30.545/I-IV/8b/95).

We are grateful for all the information we got from the supplier 'Wettlinger-Kunstoffe Hges.m.b.H' (Vienna). Due to the company ('Wettlinger-Kunstoffe Hges.m.b.H', Vienna) stocking a large amount of CdA-Ultraphan or PMMA-Perspex, they offer to cut microscopic slides in any quantity required (wettlinger@apanet.at, Att. R.Drahonsky).

REFERENCES

Barlow P.W., Carr D.J., eds. *Positional Controls in Plant Development.* Cambridge: Cambridge University Press, 1984

Bettarini I., Vaccari F.P., Miglietta F. Elevated CO_2 concentrations and stomatal density: observations from 17 plant species growing in a CO_2 spring in central Italy. Global Change Biol 1998; 4:17-22

Bolhàr-Nordenkampf H.R., Draxler G. "Functional Leaf Anatomy". In *Photosynthesis and Production in a Changing Environment.* D.O. Hall, J.M.O. Scurlock, H.R. Bolhàr-Nordenkampf, R.C. Leegood, S.P. Long, eds. London: Chapman and Hall, 1993

Bunce J.A. Stomatal conductance, photosynthesis and respiration of temperature deciduous tree seedlings grown outdoors at an elevated concentration of carbon dioxide. Plant Cell Environ 1992; 15:541-549

Bunce J.A. Responses of stomatal conductance to light, humidity and temperature in winter wheat and barley grown at three concentrations of carbon dioxide in the field. Global Change Biol 1999; 6:371-382

Bünning E. Die Entstehung von Mustern in der Entwicklung von Pflanzen. Encyclopedia of Plant Physiol 1965; 15:383-403

Cousson A. Analysis of the sensing and transducting processes implicated in the stomatal responses to carbon dioxide in *Commelina communis* L. Plant Cell Environ 2000; 23:487-495

Esau K. *Plant Anatomy.* Ed 2, John Wiley and Sons, Inc., 1965

Fahn A. *Plant Anatomy.* Ed 3, Pergamon Press, 1982

Gay A.P., Hauck B. Acclimation of *Lolium temulentum* to enhanced carbon dioxide concentration. J Exp Bot 1994; 45:1133-1141

Jarvis P.G., Mansfield T.A., Davies W.J. Stomatal behaviour, photosynthesis and transpiration under rising CO_2. Plant Cell Environ 1999; 22:639-648

Körner C Does global increase of CO_2 alter stomatal density? Flora 1988; 181:253-257

Körner C., Neumayer M., Menendez-Riedl S., Smeets-Scheel A. Functional morphology of mountain plants. Flora 1989; 182:353-383

Leuning R. A critical appraisal of a combined stomatal-photosynthesis model for C_3 plants: theoretical paper. Plant Cell Environ 1995; 18:339-355

Mauseth J.D. *Plant Anatomy*. The Benjamin/Cummings Publishing Company, Inc., 1988

Martin E.S., Donkin M.E., Stevens R.A. *Stomata*. London, UK: Edward Arnold Publishers Limited, 1983

Metcalfe C.R., Chalk L. *Anatomy of the Dicotyledons*. Vol 1. Oxford: Oxford University Press, 1950

Morison J.I.L. Plant growth and CO_2 history. Nature 1987; 327:560

Oren R., Sperry J.S., Katul G.G., Pataki D.E., Ewers B.E., Phillips N., Schäfer K.V.R. Survey and synthesis of intra- and interspecific variation in stomatal sensitivity to vapour pressure deficit. Plant Cell Environ 1999; 22:1515-1526

Radoglou K.M., Jarvis P.G. Effects of CO_2 enrichment on four poplar clones. I. Growth and leaf anatomy. Ann Bot 1990; 65:617-626

Radoglou K.M., Jarvis P.G. Effects of CO_2 enrichment and nutrient supply on growth, morphology and anatomy of *Phaseolus vulgaris* L. seedlings. Ann Bot 1992; 70:245-256

Raschke K. Stomatal action. Annu Rev Plant Physiol 1975; 26:309-340

Salisbury E.J. Leaf form and function. Nature 1949; 163:515-518

Thomas G.F., Harvey C.N. Leaf anatomy of four species grown under continuous CO_2 enrichment. Bot Gazette 1983; 144:303-309

Ticha I., Obermajer P., Snopek J. Stomata density and sizes in in vitro grown tobacco plantlets. Physiol Plantarum 1997; 29:101-107

Tognetti R., Longobucco A., Miglietta F., Raschi A. Transpiration and stomatal behaviour of *Quercus ilex* plants during the summer in a Mediterranean carbon dioxide spring. Plant Cell Environ 1998; 21:613-622

Vazzana C., Puliga S., Bochicchio A., Raschi A. Study of the stomatal density in an alfalfa (*Medicago sativa* L.) ecotype. Rivista di Agronomia 1988; 22:127-132

Willmer C.M. *Stomata*. Longman Group Limited, 1983

Woodward F.I. Stomatal numbers are sensitive to increasing in CO_2 from preindustrial levels. Nature 1987; 327:617-618

Woodward F.I., Bazzaz F.A. The responses of stomatal density to CO_2 partial pressure. J Exp Bot 1988; 39:1771-1781

Woodward F.I., Kelly C.K. The influence of CO_2 concentration on stomatal density. New Phytol 1995; 131:311-327

CHAPTER 18

HPLC TECHNIQUES - PHENOLICS

X. Carlos Souto[1], J. Carlos Bolaño[2], Luís González[2], and Xoan X. Santos[2]
[1]*Depto Producción Vexetal.. Universidade de Vigo. Spain*
[2]*Depto Bioloxía Vexetal e Ciencia do Solo. Universidade de Vigo. Spain*

GENERAL INTRODUCTION TO HPLC - HIGH PERFORMANCE LIQUID CHROMATOGRAPHY

It is estimated that approximately 60 % of all analyses world-wide can be attributed to chromatography and mainly to HPLC (high performance liquid chromatography) techniques (Kellner et al., 1998). Hundred years ago there were set up some kind of chromatography techniques (column chromatography) but in the past few years the field of chromatography has changed radically with the optimisation of adsorption chromatography, partition chromatography, affinity chromatography, gas chromatography and liquid chromatography coupled to mass spectrometers.

In general, chromatography, and HPLC in particular, is a method of separating a mixture of compounds into its components. Depending on the precision of the equipment and the response of the technique it enables to separate trace impurities or major concentrations from each other. From the information provided for that separation (time required for elution) we can identify each component. However final identification of an unknown chromatographic peak needs other analytical procedures, such as infrared spectroscopy (IR), nuclear magnetic resonance (NMR), or mass spectrometry (MS). Usually in plant ecophysiology, we need to know the variation in concentration of some known compounds, and quantitative analysis can be carried out by measuring the area of the chromatographic peaks. This makes HPLC techniques very useful to plant ecophysiologists.

M.J. Reigosa Roger, Handbook of Plant Ecophysiology Techniques, 251–282.

Liquid chromatography is important because most natural compounds are not sufficiently volatile for gas chromatography (Robinson, 1995). High-performance liquid chromatography uses high pressure to force solvent to pass through closed columns containing very fine particles that give high-resolution separations from liquid natural samples. The HPLC system in Figure 1 consists in a reservoir for the solvent delivery, a pump system, a sample injection valve, a column, a detector and a computer to control the system and display results.

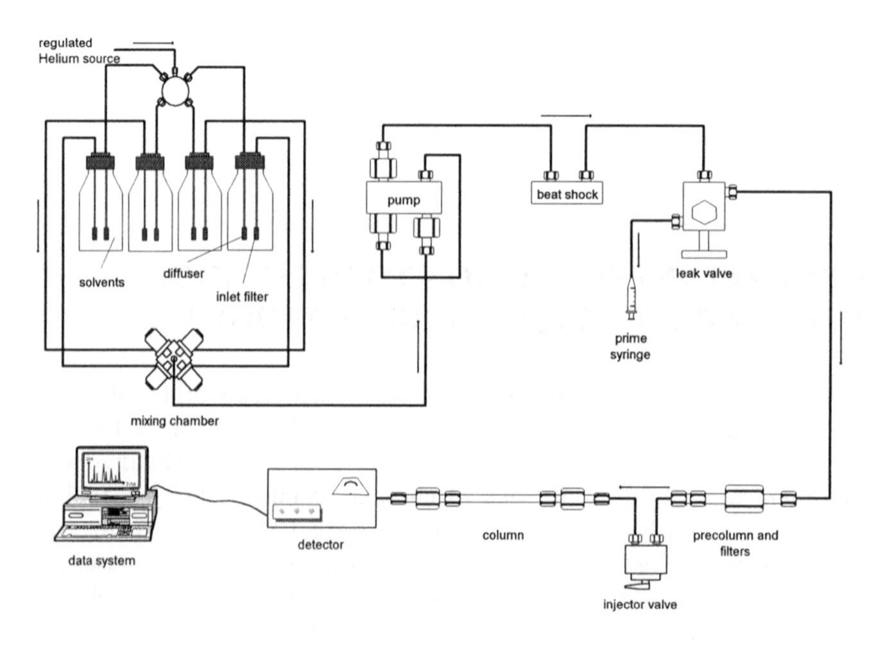

Figure 1. Structure of a HPLC unit with precolumn.

The principle of HPLC technique is based on the passage of the constituents to be separated between two non-miscible phases. For this, the sample is dissolved in the mobile phase, which can be composed by different liquids, and moved across a stationary phase, which is inside column. Due to the interactions of the constituents with the stationary phase they separate after sufficient running time and they appear at the end of the column at different times, where they can be detected externally by means of a detector properly attached to a registering device.

Solvents

The solvents used as mobile phase are stored in a glass reservoir or in stainless steel bottles. Dissolved gases, which can lead to the formation of bubbles and thus to interfere with the detector, as well as suspended matter, must be removed. The simplest way to clean the solvents is by suctioning them through a filter under vacuum. The dissolved gases can also be degassed by introducing a noble gas (usually helium) or by being processed in an ultrasound bath.

Frequently plant ecophysiologists use gradient elution more than isocratic methods. The composition of eluents is constantly changed according to a particular planning related to the physical and chemical characteristics of the compounds to separate. In liquid chromatography significant interactions appear between the substances and the mobile phase; therefore it is an important factor in the development of the method.

Pumping systems

The main requirement for pumps for HPLC is the reproducibility and the control of the flow with a relative error of less than 0.5%.

Nowadays preponderant reciprocating pumps can get the suppression of pulsation with great efficiency. High pressure gradients that are reached by these pumps provide more accurately different gradients. In the gradients up to four solvents are presented in a constant relationship. The gradient-making process is electronically controlled and programmable in very short levels of volume.

Injection systems

When the sample is introduced into the HPLC system, pressure should be kept. Modern injection systems allow it and stand for volumes in the range of 5 to 500 µL to be introduced by means of an injection system in the form of a sample loop. Sample solution is fed in through a needle inlet into the loop using a microlitre syringe.

Sample injection systems, which work automatically are preferred for prominent precision in sample introduction.

Column

Columns are packed preferentially on polished stainless steel and filled with particles containing the stationary phase. The columns are integrated in the system in the form of cartridges and the connections to the stainless steel tubes are made via conical metal sealing rings (fittings).

In order to reduce the use of solvents that must be very pure for HPLC purposes, minimised columns with very small inner diameters and short length are progressively being used. Short precolumns are often employed to protect the separation column or to get a preseparation.

Detectors

An ideal detector of any type is sensitive to low concentrations of every analyte, provides linear response, and does not expanded the eluted peaks.

There are two possibilities for detection in a HPLC system:

1. Indirect recognition of the analyte by the detection of changes in any of the characteristics of the mobile phase.
2. Direct detection by recognition of any characteristic of the analyte, such as its absorption in UV range or its fluorescence. This is the most frequently method used, and its application allows for over 70 % of detections (Kellner et al., 1998).

Spectroscopic Detectors.

- *Ultraviolet detector:* A liquid flow cell is employed to measure the light absorption of an eluent at the outlet of a column. The length of the cell varies between 2 and 10 mm and must be made of quartz for measurements in the UV range. Is the most common HPLC detector, because many solutes absorb ultraviolet light. Simple systems employ 254 nm emission of a mercury lamp but deuterium, xenon or tungsten lamps joined to a monochromator are more useful.

There are photodiode-array spectrometers that allow monitoring the full UV spectrum and the information is received as a 3-D graphic in which the absorbance is recorded as a function of the retention time and the wavelength.

- *Fluorescence detectors*. They have greater sensitivity than UV detectors. The excitation source is most frequently a mercury vapour lamp. Monochromators or a fluorescence spectrometer can select the excitation and emission wavelengths.

Refractive Index Detector.

The method for measuring the refractive index is completely unspecific, responding to almost every solute. The basis is the difference in the refractive indices between the pure eluent and the eluent containing the elements of the sample.

Evaporative Light-Scattering Detector.

It is sensitive to any analyte that is significantly less volatile than the mobile phase. Solvent evaporates from the droplets formed in the heated drift tube, leaving a fine mist of solid particles to enter the detection zone. The particles are detected by the light that they scatter from a diode laser to a photodiode.

The mass of the analyte is related to the evaporative light-scattering detector response.

Electrochemical Detectors.

They are not very common in plant ecophysiology uses but it is possible to find some investigations carry out with:

- *Conductometric detector*: It is necessary to use a flow conductometer.
- *Voltammetric detector*: It is necessary to use a dynamic or stationary electrode to record a current voltage characteristic.

Chromatogram

A chromatogram is a graph showing the detector response as a function of elution time (Figure 2).

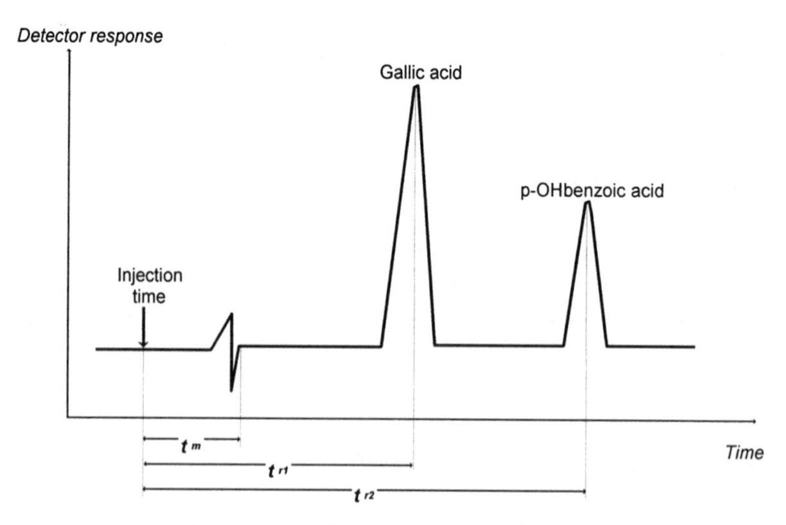

Figure 2. Schematic HPLC chromatogram for two phenolic compounds showing measurement of retention times.

The retention time (t_r) for each detected component that is detected is the time measured after injection of the mixture onto the column until that component reaches the detector. The associated quantity, retention volume (V_r) is the volume of mobile phase required to elute a particular solute from the column.

Mobile phase travels without restraint through the column, and t_m is the time that needs to be detected. The adjusted retention time for a solute is the additional time required for solute to travel the length of the column, minus the time required by unretained solvent.

$$t_r^{'} = t_r - t_m$$

The relative retention (α) or selectivity factor is the ratio of the adjusted retention times for any two components in a chromatogram,

$$\alpha = \frac{t'_{r2}}{t'_{r1}}$$

where:

$$t'_{r2} > t'_{r1} \text{ so } \alpha > 1$$

As a major relative retention, the separation between two components is greater.

For each peak in the chromatogram we can define the capacity factor, recently retention factor (k') as

$$k' = \frac{t_r - t_m}{t_m}$$

The longer the column retains a component, greater will be the capacity factor. To check the functioning of a particular column, it is a good practice to periodically measure the capacity factor of a standard.

Column efficiency:

The equation

$$\sigma = \sqrt{2Dt}$$

represents the standard deviation for a diffusive band spreading where D is the solute diffusion coefficient. If solute has travelled a distance x at the linear flow rate u_x (m/s), then the time it has been on the column is $t = x/u_x$. Therefore

$$\sigma^2 = 2Dt = 2D\frac{x}{u_x} = \left(\frac{2D}{u_x}\right)x = Hx$$

where the term

$$\left(\frac{2D}{u_x}\right)$$

is the plate height or the height equivalent to a theoretical plate (*H*)

$$H = \frac{\sigma^2}{x}$$

which is defined as the constant of proportionality between the variance of the band and the distance that it has travelled. In high performance liquid chromatography you should just regard plate height as a term that relates the width of a peak to the distance travelled through the column. Small plate height involves narrow peaks and better separation.

We can calculate the number of theoretical plates (*N*) in the entire column as a result of the division from height equivalent to the theoretical plate (*H*) and the length of the column (L), as:

$$N = \frac{L}{H}$$

Latter

$$N = \frac{Lx}{\sigma^2} = \frac{L^2}{\sigma^2} = 16\left(\frac{L}{w}\right)^2$$

because x = L, and $\sigma = w/4$ (Figure 3). If we express L and *w* (or σ) in units of time instead of length, we obtain the most useful expression for *N* by writing

$$N = 16\left(\frac{t_r}{w}\right)^2$$

Thus, the theoretical number of plates can be determined from a chromatogram by measuring the retention time and the base width of the peak. Better results are often obtained with a modified equation that uses the width at half-maximum, $w_{1/2}$, of the peak (see figure 3).

$$N = 5.54\left(\frac{t_r}{w_{1/2}}\right)^2$$

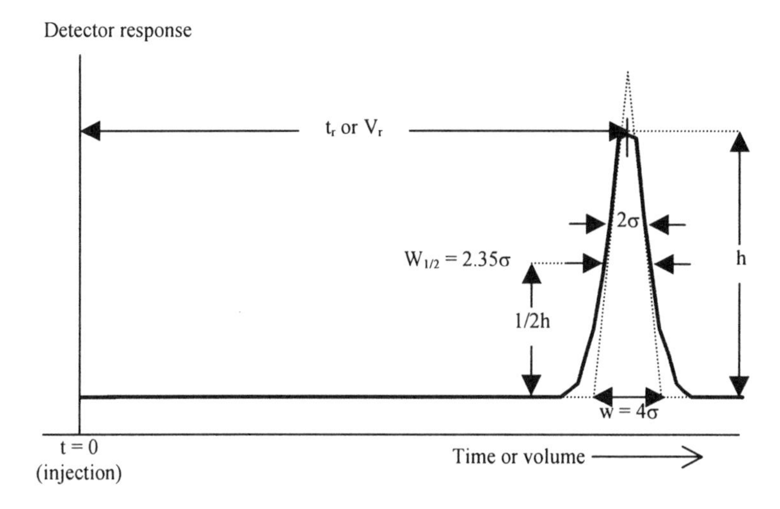

Figure 3. Modelled Gaussian chromatogram showing how w and $w_{1/2}$ are measured.

Peaks in a chromatogram from a mixture of natural compounds are frequently asymmetric due to tailing or overloading. To estimate the number of theoretical plates for the asymmetric peak (Figure 4) it is necessary to measure the asymmetry in a particular point of the peak height (usual at one-tenth of the maximum) and the number of plates would be:

$$N \approx 41.7\frac{\left(\frac{t_r}{w_{0.1}}\right)^2}{\left(A/B + 1.25\right)}$$

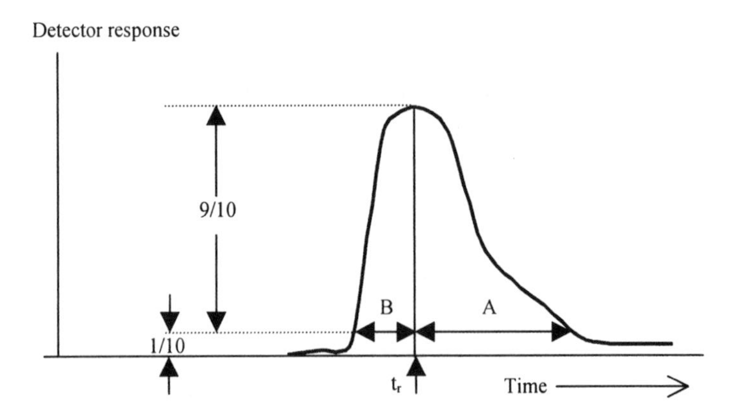

Figure 4. Asymmetric peak showing parameters used in this text to estimate the number of theoretical plates.

Separation efficiency

A common task in plant ecophysiology is to recognise the compounds that interact with plants from a crude extract or to know the different compounds that can change after a physiological interaction. Most of them are really similar in their physical and chemical properties and can be difficult to separate in a chromatogram. There are two factors that contribute to separate different compounds by chromatography. One is the difference in elution times between peaks: the further apart, the better their separation. The other factor is the peak broadening along the chromatogram, the width of a peak is in direct relation to the separation efficiency or column efficiency: the wider the peaks, the poorer their separation.

We use the chromatographic resolution (Rs) to characterise the ability of a chromatographic column to separate two analytes. Solute moving through a chromatographic column tends to spread into a Gaussian shape with standard deviation σ.

The more time takes the analyte in passing through the column wider becomes the band. Useful measurements are the width $w1/2$ measured at a height that equals to half of the peak height and the width w at the baseline between tangents drawn to the steepest parts of the peak. It is possible to show that $w1/2=2.35\sigma$ and $w=4\sigma$.

The resolution of two peaks, 1 and 2 with w1 and w2 approximately the same, from each other is defined as

$$R_s = \frac{t_{r2} - t_{r1}}{\frac{w_1 + w_2}{2}} = \frac{\Delta t_r}{w_{av}} = \frac{\Delta V_r}{w_{av}}$$

where Δt_r or ΔV_r is the separation between peaks (in units of time or volume) and w_{av} is the average width of the two peaks in corresponding units.

There is a relationship between α, k' and N (number of theoretical plates: based on the dependence of the resolution on α, k' separation can be optimised. The individual variables can be varied fairly independently. The selectivity factor (α) can be modified by choosing a different type of molecular interaction, that is by changing the stationary phase. The retention factor (k') can be changed by varying the composition of the mobile phase.

One of the advantages of HPLC is the possibility of making qualitative and quantitative analysis of dissolved organic compounds.

Qualitative analysis

The most widely method used for qualitative detection is based on the retention time of the compounds of a sample. If the conditions of the column are kept constant, the retention time of a particular compound will be constant. As a result, to identify two compounds it is necessary to check that the retention times of those compounds are different. This verification is not complete, since it is possible for two compounds to have identical or very similar retention times. Then, the identification is difficult if we do not have another factor for the recognition. If you are using a diode-array detector it is possible to record the spectra of the two compounds in the same peak and to know the presence of two compounds that you can identify by comparing the retention times and spectra with a library of standards.

Nowadays it is possible to identify directly unknown peaks by attaching a mass spectrograph at the end of the HPLC column. The different compounds emerging from the column are fed directly into the MS system.

Quantitative analysis

The area of a chromatographic peak is proportional to the number of molecules of the compound that reach the detector. We must use a calibration curve for the different compounds using standards. The response depends on the amount of sample that was injected.

PHYSIOLOGICAL ASPECTS

Phenolics have been considered as allelopathic substances in many cases (Inderjit and Dakshini, 1995). They are very extremely wide distributed, if not universal, in the vascular plant kingdom (Harborne, 1989). The term phenolic or polyphenol can be defined chemically as a substance which possesses an aromatic ring bearing a hydroxyl substituent, including functional derivatives as esters, methyl ethers, glycosides, etc. Most of them have two or more hydroxyl groups and are derived from one of the common dihydric or trihydric phenols (see examples in figure 5).

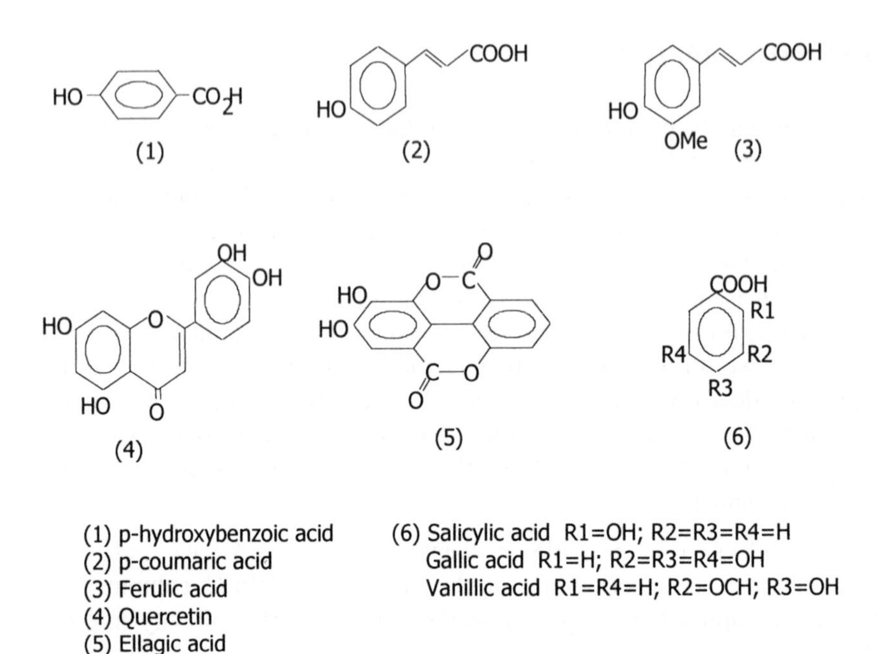

Figure 5. Examples of some phenolics.

Phenolics in plants arise from two main pathways, the shikimate pathway (Figure 6) which directly provides phenylpropanoids such as the hydroxycinnamic acids and coumarins, and the acetate pathway, which can produce simple phenols and also many quinones. Flavonoids, the largest group of phenolics, are derived from a combination of both pathways. Phenolic compounds are heterogeneously distributed in the plant kingdom: some types, such as flavonoids and hydroxycinnamic acids are widely distributed, while other types are present only in certain genera and species. Something similar can be told about the concentration of phenolics, which shows a great diversity among species.

Phenolic acids, which are part of the secondary metabolism of vascular plants, play an important role as precursors of lignins, and are also present in soils as part of humic acids. This kind of phenolics have been considered in many cases responsible of chemical interactions among plants, particularly hydroxybenzoic and cinnamic acids.

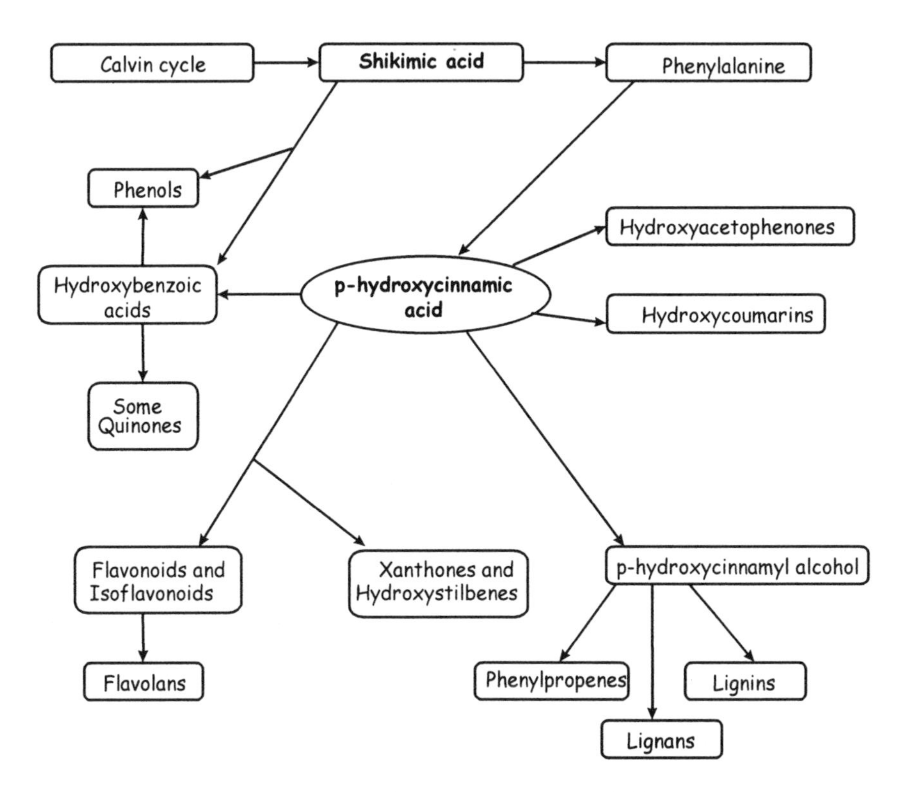

Figure 6. Biosynthetic origin of plant phenolics from shikimate and phenylalanine (Harborne, 1989).

Some hydroxybenzoic acids are commonly present in angiosperms, especially p-hydroxybenzoic, vanillic, siringic and protocatechuic acids. Other acids, such as gentisic, gallic and ellagic are also common, and the last two acids participate in the formation of more complex structures, the hydrolyzable gallotannins and ellagitannins. These benzoic acids are frequently linked to organic acids and sugars.

Most of hydroxycinnamic acids arise from combinations of the four basic acids ferulic, p-coumaric, sinapic and cafeic. Sometimes they occur as free molecules because of hydrolysis made by microbial activity, anaerobic conditions, etc.

Flavones, flavonols and their glycosides constitute a major group of phenolics, commonly called flavonoids. They are widely represented in the plant kingdom, but scarce in algae and fungi. Many functions in individual plants or plant groups have either been demonstrated or proposed. These include: protection of the plant from UV light, insects, fungi, viruses and bacteria; pollinator attractants; antioxidants; plant hormone controllers; stimulants of nodule production by *Rhizobium* bacteria; enzyme inhibitors; and allelopathic agents (Markham, 1989).

The diversity of structure and the uneven distribution of phenolic products between and within different plant families makes it almost certain that a single physiological role cannot be ascribed to these compounds (Table 1). In this sense, phenolic compounds have played a central role in the theories of plant-herbivore interactions. At first, they were proposed to have ecological activity as plant defences against pathogens and herbivores (Feeny, 1969). However their realised effects are much more diverse. Their effects may be positive as well as negative, and include feeding deterrence, feeding stimulation, digestion inhibition, digestion stimulation, toxicity, toxicity amelioration, disease resistance, signal inhibition, signal transduction, and nutrient cycle regulation (Appel, 1993). Furthermore, the type and magnitude of their effects vary with the situation and the organism. Appel (1993) suggested that phenolics have many modes of action, but that they require oxidation for most of their ecological activity; thus variation in phenolic activity often results from variation in oxidative conditions; physicochemical conditions of the environment control phenolic oxidation and thus generate variation in ecological activity.

A large number of plant phenolics have been proposed to be allelopathic by inhibiting the germination and/or growth of other plants (Table 2). Unfortunately, the importance of oxidation state has been investigated in few cases, for two allelopathic phenolics, juglone

(Rietveld, 1983) and sorgoleone (Einhellig and Souza, 1992). Produced in a reduced form in the plant, they are rapidly oxidised to the more stable and active quinone from when released as tissue leachates or root exudates.

Table 1. Factors affecting phenolics and their possible ecological roles.

	Some factors affecting Phenolic biosynthesis		Some possible ecological roles of phenolics
1	Light	1	Provision of mechanical support
	Ultra violet	2	Factors in plant/pathogen interactions
	Red/far red		Pre-infection factors
	Blue		Post-infection factors
2	Stress		(a) polymeric phenols
	Wounding		(b) phytoalexins
	Infection	3	Factors in plant pollination
	Low temperature		Flower colour
	Water logging		Honey guides
	Toxic chemicals		Flower scent
3	Plant growth regulators	4	Factors in plant/plant interactions
	Ethylene		Allelopathic agents
	Auxins	5	Factors in plant/animal interactions
4	Nutritional factors		Feeding attractants
	Nitrogen level		Feeding repellents
	Phosphorous level		Seed disposal by animals
	Boron deficiency		

Phenolics in plant litter are to a small extent water soluble, are released by rain water and can be detected in small amounts in throughfall and stem flow. However, their main origin is their release from decomposing leaf litter deposited on the soil surface and from decomposing fine roots (Kuiters, 1990). After reaching the soil, the effects of phenolics on the functioning of forest ecosystems are many and drastic. Soil phenolic substances directly affect bacteria, the development of mycelia and spore germination of saprotrophic and mycorrhizal fungi, and the germination and growth of higher plants (Hartley and Whitehead, 1985). The process of humification is also largely directed by the polyphenol content of litter. Thus, site quality characteristics such as mineralization rate, nitrification rate and soil productivity are highly influenced by phenolics, released from living, senescent and decomposing plant tissues.

Table 2. Some examples of phenolics with allelopathic activity.

Phenolic compound	Target species	Inhibitory effect
Emodin	*Phleum pratense*	Seedling growth
p-hydroxybenzaldehyde	*Vigna radiata*	Seedling growth
p-hydroxybenzoic acid	*Lactuca sativa*	Growth and germ.
p-coumaric acid	*Triticum aestivum*	Growth and germ.
Taxifolin-3-arabinoside	*Brassica juncea*	Growth and germ.
Hesperidin	*Raphanus sativus*	Growth and germ.
Ononin	*Brassica juncea*	Growth and germ.
Ferulic acid	*Trifolium repens*	Germination
Vanillic acid	*Dactylis glomerata*	Growth

Phenolics can affect membrane permeability of roots, and alter the uptake and flux of nutrients. They can be absorbed by the roots and influence physiological processes. For that, these compounds should be solved in the soil. Concentrations in the soil solution depend on soil structure, pH, contents in organic matter, vegetation type, etc. (Inderjit and Dakshini, 1995). The highest concentrations in temperate forests are generally found from June to September, when microclimate conditions are optimal.

Seed germination and seedling growth are probably the most widely studied physiological processes with respect to allelopathy. Many studies demonstrate that phenolic concentrations have to be higher than 1 mM to produce inhibitory effects on agronomic species. However, some of these studies suggested that the adverse effects of phenolics on germination could be, at least partially, due to an effect on water uptake by seeds.

Seedling growth seems to be more sensitive to phenolic compounds than germination, since most phenolics are inhibitory at millimolar concentrations, but are stimulatory at micromolar levels. The seedling stage seems to be a sensitive phase in the life cycle of plants with respect to phenolic compounds occurring in the litter and humus layer. Effects of litter type on herbaceous species composition may partly be mediated by interference of phenolics, released from decomposing litter at the early seedling stage.

An important thing that one must take into account is whether the concentrations of phenolic compounds in soils are in fact able to interfere on the growth of plants. Estimations of phenolics in soil solutions are generally low, below 0.1mM, so they are lower than concentrations required for inhibit plant growth or seed germination. However, under natural conditions it is possible that a mixture of phenolic compounds is the responsible of inhibitory effects on growth through synergic or additive effects.

Some stresses may increase secondary metabolites in plants (Levitt, 1980). Since stress has an inhibitory effect on growth, the excess of carbon provided by photosynthesis can be used in the synthesis of carbon-based compounds that accumulate in tissues. For example, water stress can stimulate the accumulation of phenolic compounds (Pedrol, 2000). These responses have been correlated with an increase of tolerance against stress (Einhellig, 1996) as an adaptive role that phenolics may play.

Besides, the occurrence of various stresses at the same time has an additive effect on the accumulation of phenolics in the plant. In evergreen species there are important temporal variations in the production of phenolics. They are probably due to complex interactions of internal factors, such as tissue age, and external factors both biotic and abiotic. These variations can determine changes in the competitive relationships with other organisms, having a subsequent impact on the functioning of the ecosystem (Nilsson et al., 1998). The most recent critical reviews on this topic focused on the need to study the phenolic compounds in plants considering the combined effects of all these factors, that is, from an ecophysiological point of view (Reigosa et al., 1999).

PHENOLICS ANALYSIS.

Once we know the important role of phenolics in plant ecophysiology, we will focus on the analysis of low molecular weight phenolics and flavonoids from plants by High Performance Liquid Chromatography (HPLC). This technique has been successfully employed in both quantitative and qualitative analysis of phenolics. The polar nature of these substances makes it possible to separate them using a reversed phase chromatographic column. Due to their chemical structure, phenolics absorb electromagnetic radiation in the low ultraviolet range (220 to 383 nm), so they can be analysed by a photodiode array detector.

Extraction of low molecular weight phenolics from plants

The method used to extract phenolics is very important since all results of the analyses are affected by the extraction step. Sample management does affect phenolic extractability, so all samples should be treated similarly. An efficient optimisation method should be employed during the technique development in order to deal with the optimal process conditions for each plant species.

In the classical studies of phenolics, it was common to separate the phenolic fraction of a plant extract either by precipitation with lead acetate or by extraction into alkali or carbonate, followed by acidification (Harborne, 1989). Alternatively, the powdered plant material might be extracted using several solvent systems in sequence, some to remove lipids and others such as ethyl acetate or ethanol to remove the phenolic fraction. However it is much more common to make a direct extract of plant material, using boiling methanol or ethanol with fresh or dried tissue.

The extraction of phenolics is commonly accomplished by repeated maceration of crushed or ground material with a small amount of HCl (0.1-1.0%) in methanol or ethanol at room temperature. The addition of water (10-50%) might be advisable in some cases to get complete extraction, depending on the plant material.

Several methods can be used for extracting phenolics. We have selected one of them because of its good extraction capacity and repeatability. The method was proposed by Macheix et al. (1990), and was optimised by Bolaño et al. (1997). The schema of the process can be seen in figure 7. As a result of our experience we start from 1 g dry weight of plant material, air dried because phenolics are not stable to drying in an oven at high temperature. The fresh plant tissue should be previously cut into small pieces with scissors.

Use a 100 ml Erlenmeyer flask to mix plant material with 50 ml of methanol/HCl (1000:1, v/v) and keep it in darkness for 12 h with hand shaking every three hours. Filter the mixture, with a filter paper and a funnel, saving the filtered in the refrigerator and repeat the process with the residual plant material for another 12 h by adding 50 ml of methanol/HCl. Filter the mixture again and store.

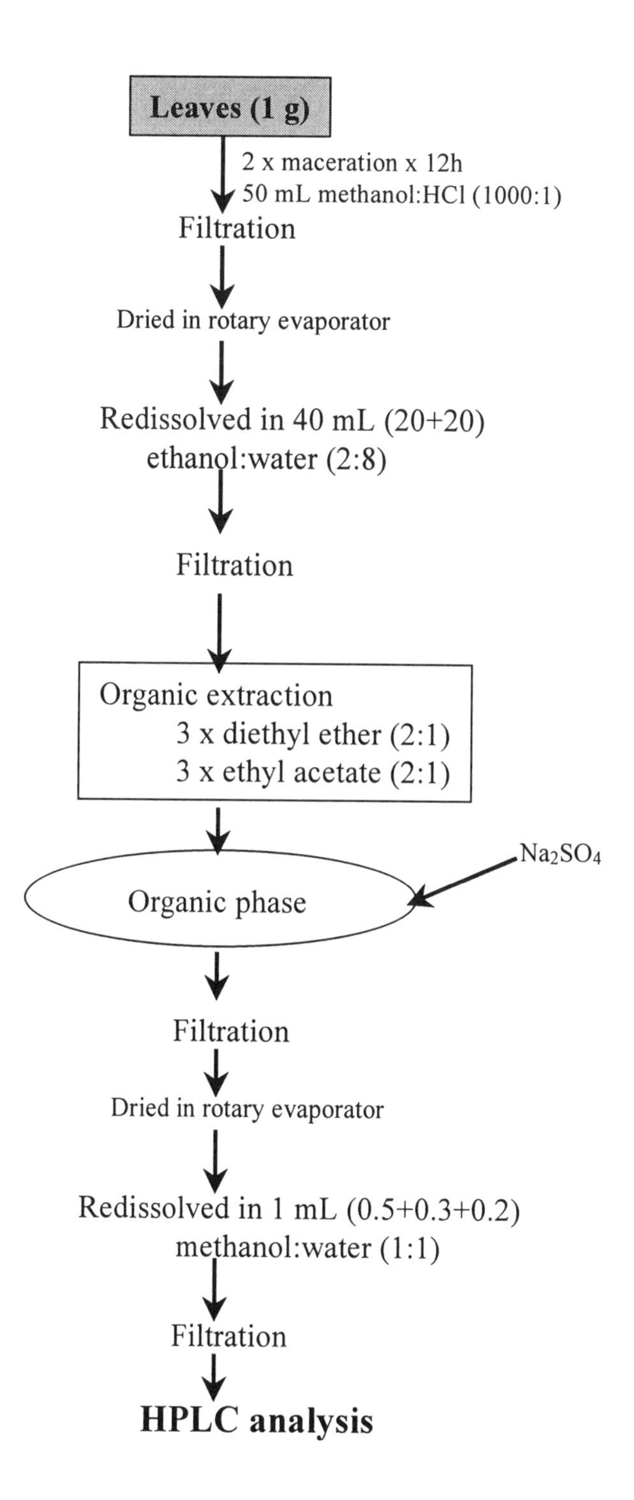

Figure 7. Extraction procedure for phenolic acids (Bolaño et al., 1997)

The combined methanolic extract (approximately 100 ml) is then completely dried with the help of a rotary evaporator under vacuum, taking into account that we will never exceed 60° C because phenolics are very sensitive to high temperature. The solid residue is brought to a final volume of 40 ml (20 + 20) of ethanol/water (2:8, v/v) and then filtered.

Afterwards, three sequential extractions with 20 ml of diethyl ether are performed. We use an extraction funnel and apply vigorous shaking for one minute each time, waiting until the complete separation in two phases: the aqueous one, in the lower part, and the organic one, in the upper part of the funnel, where there will be the ether extracted phenolics. This phase is removed and saved, collecting the three ethereal phases (approximately 60 ml) to an Erlenmeyer flask. Now complete another three sequential extractions with 20 ml of ethyl acetate on the aqueous phase, obtaining three new organic phases that will be collected and combined with the ethereal ones.

The total organic fraction obtained by this way is then dehydrated with anhydrous sodium sulphate for 30 minutes to remove the minimum water residues. Subsequently it is filtered to withdraw the sodium sulphate and evaporated to dryness in the rotary evaporator.

The final residue containing the phenolics is re-dissolved in 1 ml (0,5 + 0,3 + 0,2) of methanol/water (1:1, v/v), then filtered through a 0,45 μm pore size nylon membrane filter and saved in refrigerator at 4° C until its use for HPLC analysis.

HPLC analysis of phenolic acids

The photodiode array detector has led to great improvements in HPLC analysis of phenolic compounds, because not only the retention time but also the UV spectrum can be used for identification purposes. Quantification also becomes more exact due to the simultaneous recording of different wavelengths. Generally the most useful wavelengths to get a good resolution are 254, 280 and 320 nm for single phenolic compounds.

The most common chromatographic system used for the analysis of the phenolics present in the samples from plant material is built by a pump (which can be binary, ternary or quaternary), and a diode array detector with an attached computer. This computer is provided with a data storage system, and all the software needs to process the samples. It is also very common the use of an autoinjector, which makes the work much easier.

There are many different columns one can use for analysing phenolics. They are usually reversed phase columns, C-18 with 4 or 5 μm packing, and the length uses to be around 25-30 cm.

Solvents preparation

Many different solvents are described in the literature for analysing phenolics by means of HPLC. The most common solvents are water, methanol and acetonitrile, in different combinations and amounts depending on the phenolics and columns. It is also very common to add a small amount of a weak acid to one or more solvents. We are going to explain a routine procedure that has produced good results on allelopathy studies (Reigosa et al., 1999; González et al., 1995; Souto et al., 1995), the solvents (HPLC grade) are as follows: solvent A, water/acetic acid (98:2, v/v) and solvent B, water/methanol/acetic acid (68:30:2), v/v) filtered through 0,45 μ pore size membrane filter and degassed for 5 minutes in ultrasonic bath (Bransonic 1200). Gradient development and elution conditions are given in Table 3.

Table 3. Mobile phase gradient composition and flow-rate.

Time (min)	Flow-rate (ml/min)	A %	B %
0	0,8	100	0
59	0,8	20	80
69	0,8	20	80

Samples injection

Usually, a trial run is performed on a new sample, to determine the optimal volume to be injected on the HPLC system for the best determination of the phenolic content. An amount of 50 μL is routinely used for the initial run in leaf extract samples following this protocol.

Injections of phenolic acids and flavonoids standards dissolved in 50 % aqueous methanol at different concentrations were used to create a 'spectra library', useful to identify the peaks in the chromatogram.

Identification of phenols

The total running time for each chromatogram is very variable, and it depends on the length of the column and the solvents used. The routine we are describing, each run lasts about 69 minutes, with a flow rate of 0,8 ml/minute. Scanning a range of UV wavelengths from 220 to 400 nm usually performs the detection of phenolic peaks.

Each chromatograph has its own software, needed to identify and quantify the chemical compounds under study. However, most of them allow the user to monitor the chromatogram in different ways: for example in 3D (all wavelengths at the same time), or selecting one or more fixed wavelengths.

It is also useful to save each chromatogram as a file, which will be managed and analysed later. Many programs allow the user to export the data of the chromatogram as ASCII codes, which is very useful for transferring the data to other computer programs.

The detection of phenolic compounds is based on their UV absorbance in the mobile phase. The photodiode array detector allows saving the complete spectrum of a compound (one peak in the chromatogram). The computer registers its absorbance in a wide range of wavelengths, which is essential in an ulterior determination.

Different parameters in the UV spectrum from a peak will be measured with the photodiode array detector, in order to determine the identity of phenolics: retention time, position of maximum, absorbance ratios and the comparison of the spectrum shape of the unknown peak with those of the pure standards previously injected and analysed.

Peaks from chromatographic runs are then assigned or identified as specific phenolic acids. Examples of chromatograms of phenolic acids and some spectra can be seen in figures 8 and 9.

Quantification of phenolics.

To quantify the phenolics present in the extracts, each peak area is calculated from chromatograms at 280 nm, with the help of the proper data handling system, and compared with those from the standards injected at different known concentrations. A lineal regression gives a slope that allows calculation of the concentrations of each phenolic compound using a single equation.

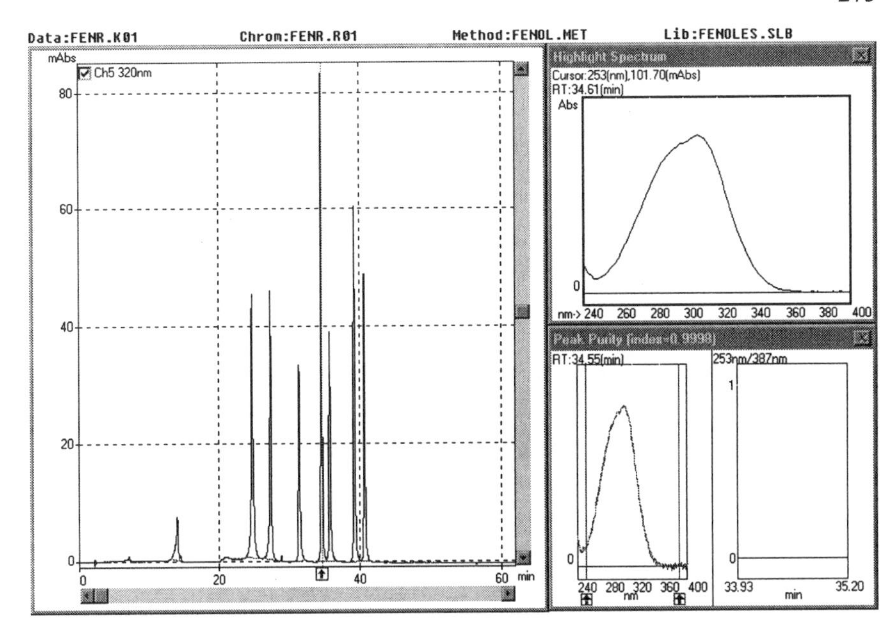

Figure 8. Example of a chromatogram. In the left window one chromatogram at 320 nm is shown. In the upper right window the spectrum of a selected peak of the chromatogram can be observed. The peak purity window is useful to clarify the purity of each peak.

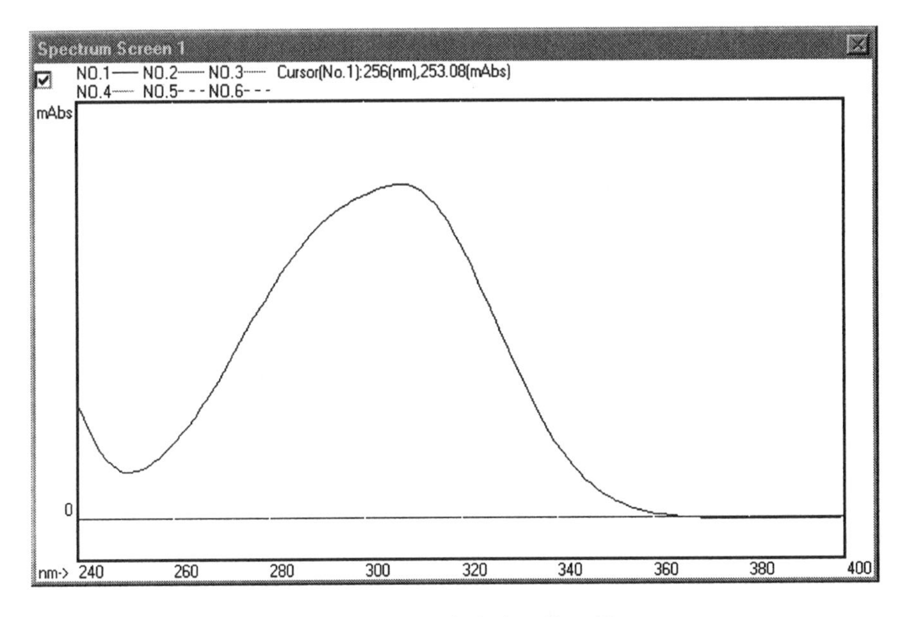

Figure 9. Spectrum of p-coumaric acid, a typical phenolic acid.

Peak Name	RT (min)	Area	Height	Base	Concentration
1	6,183	1699,7	42,46	0,054	1.92
2	19,542	1063,8	19,07	0,013	1.06
3	29,269	987,6	15,72	1,089	0.59
4	46,955	1428,7	21,27	0,567	1.34

(Example of data recovered from a data handling system)

These data should now be treated with a data sheet program to get the real concentration of each phenolic compound in the original tissue, expressed as mg/g of dry weight of plant material.

Extraction of flavonoids from plants

Flavonoids are compounds chemically related with phenolic acids, and many of their characteristics are similar or identical. As explained previously for phenolics, the method used to extract flavonoids is very important. However, it is also very easy to notice that most researchers working on flavonoid identification/quantification use the same or very similar methods.

It is estimated that about 2% of all carbon photosynthesised by plants is converted into flavonoids or closely related compounds. They constitute one of the largest groups of naturally occurring phenolics, and are virtually ubiquitous in green plants. And many of them have important physiological/ecological effects, including allelopathy (see previous section about low molecular weight phenolics).

Flavonoid aglycones posses the chemical properties of phenolics, that is, they are slightly acidic and will thus dissolve in alkali (Markham, 1982). Flavonoids possesing a number of unsubstituted hydroxyl groups, or a sugar, are polar compounds. They are generally moderately soluble in polar solvents such as ethanol, methanol, acetone, water, etc. The presence of an attached sugar (which is very commonly encountered) tends to render the flavonoid more water soluble and combinations of the above solvents with water are thus better solvents for glycosides. In contrast, less polar aglycones such as flavanones, methoxylated flavones and flavonols tend to be more soluble in solvents such as ether and chloroform.

As previously mentioned for phenolic acids, it is very common to make a direct extract of plant material. It is highly recommended to dry rapidly the fresh material in an oven (about 100°C) before storing, or ground to a fine powder with a mortar and a pestle for solvent extraction.

The extraction of flavonoids is commonly carried out in two steps (Markham, 1989), firstly with methanol:water (9:1) and secondly with methanol:water (1:1), although many times it is reduced to a unique extraction with methanol:water (80:20) or methanol:ethanol (1:1). It has almost universally been considered good practice to extract flavonoids with methanol:HCl (99:1) or methanol:weak acids e.g. acetic acid. However, several of these treatments (especially those with HCl) may result in pigment degradation.

In our case, we have selected a method that was used with success in extracting flavonoids and also phenolic acids (figure 10). The process starts macerating 10 g of plant material (dried and powdered) in 300 ml of methanol:water (80:20) for 24 hours at room temperature and constant shaking. Filtration to separate extract from plant material can be carried out rapidly using a Buchner funnel with filter paper.

The methanolic extract is concentrated with the help of a rotary evaporator under vacuum until all methanol is removed (approximately 60 ml).

The resultant aqueous extract can then be cleared of low polarity contaminants such as fats, terpenes, chlorophylls, xantophylls, etc. by extraction with petroleum ether using an extraction funnel (it is recommended to repeat this extraction three times). The aqueous extract is now performed as described for phenolic acids, that is, three sequential extractions with diethyl ether followed by three sequential extractions with ethyl acetate.

Some researchers analyse flavonoids by HPLC after clearing with petroleum ether: in that case, they evaporate the aqueous extract to dryness under vacuum in a rotary evaporator, and redissolve in methanol:ethanol. The next step is the analysis in the HPLC.

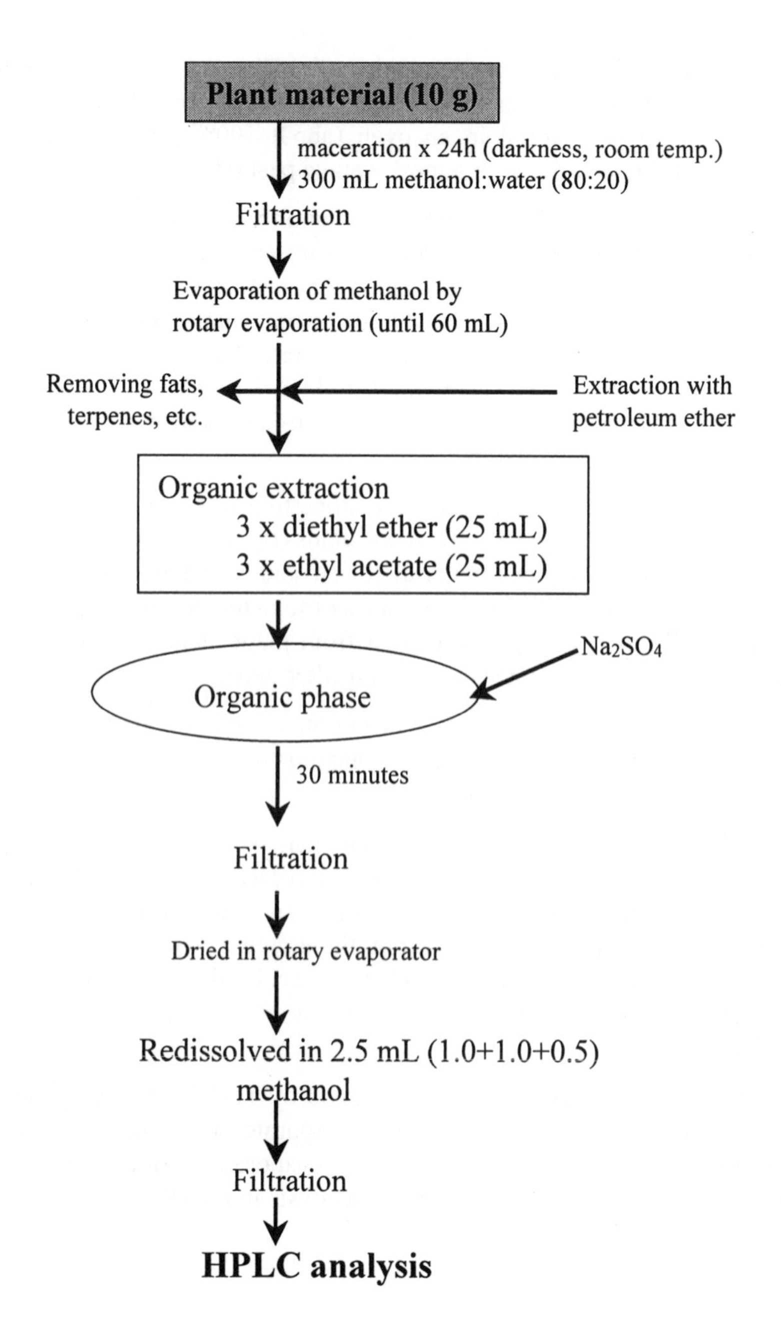

Figure 10. Extraction procedure for flavonoids.

In our method, the organic fraction obtained after extractions with dietyl ether and ethyl acetate is dehydrated with anhydrous sodium sulphate for 30 minutes to remove water residues. Subsequently it is filtered to withdraw the sodium sulphate and evaporated to dryness in the rotary evaporator.

The final residue containing the phenolics is redissolved in 2.5 ml (1.0 + 1.0 + 0.5) of methanol, then filtered through a 0,45 μm pore size nylon membrane filter and saved at -20°C until its use for HPLC analysis.

HPLC analysis of flavonoids.

As previously described for phenolic acids, the analysis of flavonoids is made using the useful technology of HPLC. The same assessments can be applied to flavonoid analysis, so if more information are required see section about Extraction of low molecular weight phenolics from plants.

Solvents preparation

A wide range of packing/solvent combinations have been reported in the literature for flavonoids. It is clear that for most applications reversed phase C-18 type columns (in which a hydrocarbon is bonded to the silica packing) are suitable.

Solvents such as water:methanol, water:methanol:acetic and water:acetonitrile (in varying proportions) have been used successfully to chromatograph flavonoids.

It is well known the problem of tailings of flavonoid peaks (which depends on the mobile and especially on the stationary phase, and disturbs the automatic integration), but this problem can be solved by adding a small amount of a weak acid (generally formic or acetic acids, or even orthophosphoric acid) into the solvents.

In our case, the solvents (HPLC grade) are as follows: solvent A, methanol:orthophosphoric acid (999:1) and solvent B, water:phosphoric acid (999:1) filtered through a 0,45 μm pore size nylon membrane filter and degassed by means of a vacuum filter degassifier. Gradient development and elution conditions are given in table 4.

Table 4. Mobile phase gradient composition and flow-rate for flavonoids.

Time (min.)	Flow rate (ml/min)	%A	%B
0	1	20	80
40	1	100	0
45	1	100	0
50	1	20	80
55	1	20	80

Samples injection

This section is identical to that of phenolic acids.

Identification of flavonoids

The flavonoid spectrum typically consists of two absorption maxima in the ranges 240-285 nm (band II) and 300-550 (band I). The precise position and relative intensities of these maxima give valuable information on the nature of the flavonoid and its oxygenation pattern. Characteristic features of these spectra are the low relative intensities of band I in dihydroflavones, dihydroflavonols, and isoflavones, and the long wavelength position of band I in the spectra of chalcones, aurones and anthocyanins. Some examples are presented in figures 11 and 12.

The detection of flavonoid peaks is usually performed by scanning a range of UV wavelengths from 220 to 400 nm (in our case the detection was effected at 325 nm, bandwidth from 220 to 400, optical bandwidth 4 nm). The total running time for each chromatogram is 55 minutes. The identification of each flavonoid is based on the comparison of its spectra and retention time with those of authentic standards (example in figure 13). The identification of the peaks was made with an automated library search system from the HPLC software. This application includes the peak purity (comparison of the spectra upslope, apex, downslope and correlation between the peak at different wavelengths) and compares the spectra of the peaks in the chromatogram with the spectra of the reference compounds stored in the library. This comparison also considers the retention time of the peaks in the chromatogram and in the library. In table 5 some retention times of flavonoids are presented.

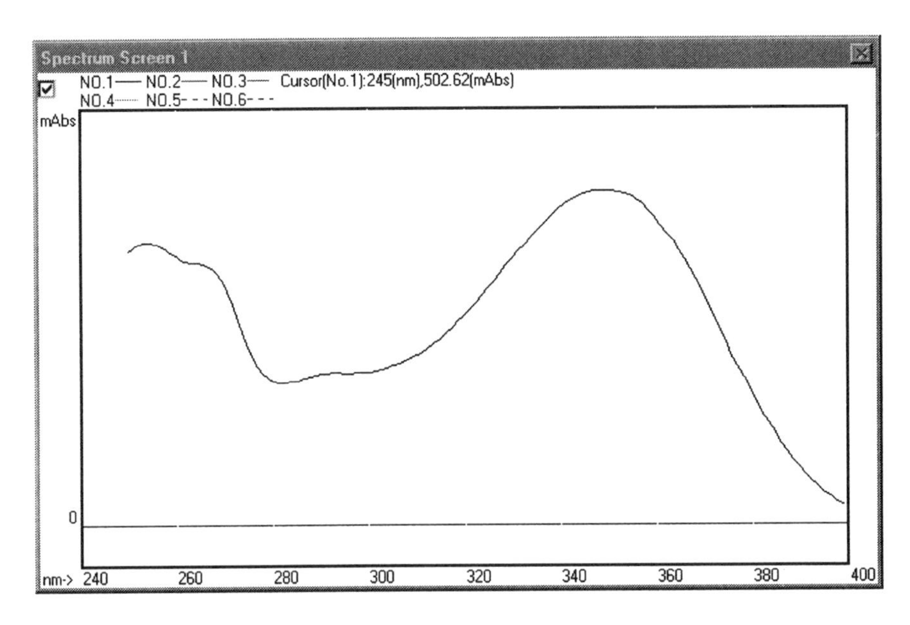

Figure 11. Spectra of a typical flavonoid, luteolin.

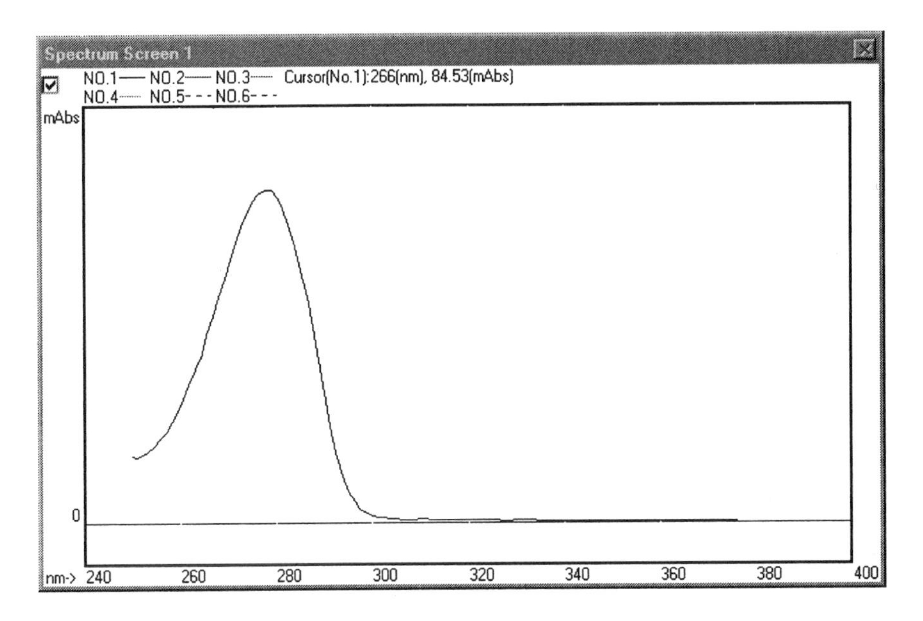

Figure 12. Spectra of a typical flavonoid, epicatechin.

Table 5. Retention time of different flavonoids using the described method

Flavonoid	Retention time (minutes)
Taxifolin	13.2
Eriodictyol	17.5
Ellagic acid	18.1
Quercetin	21.8
Luteolin	23.7
Kaempferol	25.1
Apigenin	26.7

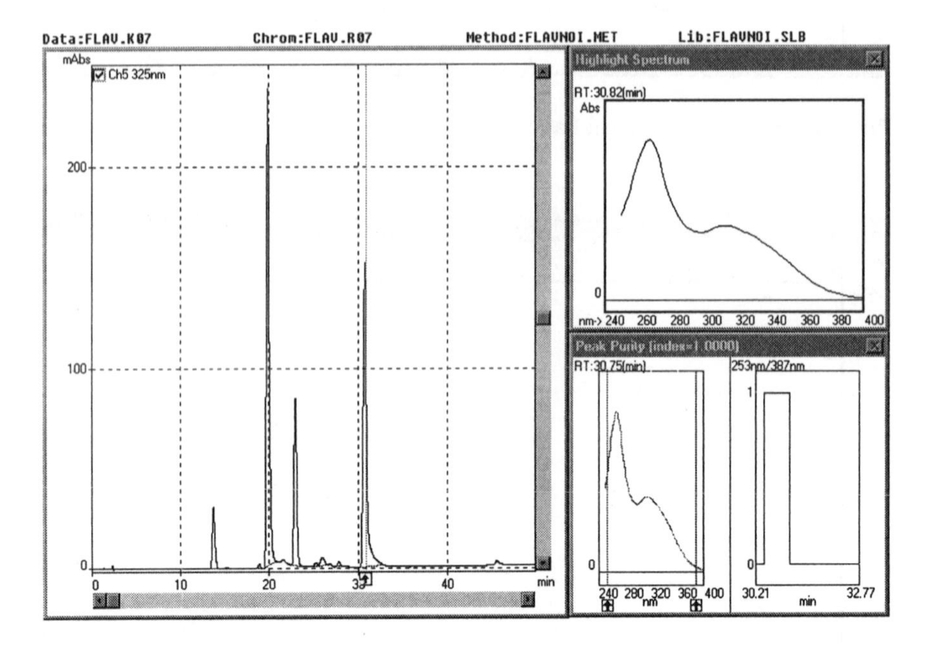

Figure 13. Example of a chromatogram. In the left window one chromatogram at 325 nm is shown. In the upper right window the spectrum of a selected peak of the chromatogram can be observed (the flavonoid chrysin). The peak purity window is useful to clarify the purity of each peak.

Quantification of flavonoids

Quantification of flavonoids follows the procedure explained for phenolic acids.

REFERENCES

Appel H.M. Phenolics in ecological interactions: the importance of oxidation. J Chem Ecol 1993; 19:1521-1552

Bolaño J.C., González L., Souto X.C. "Análisis de Compuestos Fenólicos de Bajo Peso Molecular Inducidos por Condiciones de Estrés, mediante HPLC." In *Manual de Técnicas en Ecofisiología Vegetal*. N. Pedrol, M.J. Reigosa, eds. Vigo, Spain: Gamesal, 1997

Einhellig F.A. Interactions involving allelopathy in cropping systems. Agron J 1996; 88:886-893

Einhellig F.A., Souza I.F. Phytotoxicity of sorgoleone found in grain sorghum root exudates. J Chem Ecol 1992; 18:1-11

Feeny P. Inhibitory effect of oak leaf tannins on the hydrolysis of proteins by trypsin. Phytochemistry 1969; 8:2119-2126

González L., Souto, X.C., Reigosa, M.J. Allelopathic effects of *Acacia melanoxylon* R.Br. phyllodes during their decomposition. For Ecol Manage 1995; 77:53-63

Harborne J.B. "General Procedures and Measurements of Total Phenolics." In *Methods in Plant Biochemistry, Vol. 1, Plant Phenolics*. P.M. Dey, J.B. Harborne, eds. London: Academic Press, 1989

Hartley R.D., Whitehead D.C. "Phenolic Acids in Soils and their Influence on Plant Growth and Soil Microbiological Processes." In *Soil Organic Matter and Biological Activity*. D. Vaughan, R.E. Malcolm, eds. Dordrecht, The Netherlands: Nijhoff/Junk, 1985

Inderjit, Dakshini K.M.M. On laboratory bioassays in allelopathy. Bot Rev 1995; 61:28-44

Kellner R., Mermet J.M., Otto M., Widmer H.M. *Analytical Chemistry*. Weinheim: Wiley-VCH, 1998

Kuiters A.T. Role of phenolic substances from decomposing forest litter in plant-soil interactions. Acta Bot Neerl 1990; 39:329-348

Levitt J. *Responses of Plants to Environmental Stresses*. New York, USA: Academic Press, 1980

Macheix J.J., Fleuriet A., Billot Y. *Fruit Phenolics*. Boca Ratón, Florida, USA: C.K.C. Press, 1990

Markham K.R. "Flavones, Flavonols and their Glycosides." In *Methods in Plant Biochemistry, Vol. 1, Plant Phenolics*. P.M. Dey, J.B. Harborne, eds. London: Academic Press, 1989

Markham K.R. *Techniques of Flavonoid Identification.* New York, USA: Academic Press, 1982

Nilsson M.C., Gallet C., Wallstedt A. Temporal variability of phenolics and batatasin.III in *Empetrum hermaphroditum* leaves over an eight-year period: interpretations of ecological function. Oikos 1998; 81:6-16

Pedrol N. *Caracterización Ecofisiológica de Gramíneas. Competencia y Estrés Hídrico.* Thesis dissertation. Vigo, Spain: University of Vigo, 2000

Reigosa M.J., Sánchez-Moreiras A.M., González L. Ecophysiological approach to allelopathy. Crit Rev Plant Sci 1999; 18:577-608

Reigosa M.J., Souto X.C., González L. Effect of phenolic compounds on the germination of six weeds species. Plant Growth Regul 1999; 28:83-88

Rietveld W.J. Allelopathic effects of juglone on germination and growth of several herbaceous and woody species. J Chem Ecol 1983; 9:295-308

Robinson J.W. *Undergraduate Instrumental Analysis.* New York, USA: Marcel Dekker, 1995

Souto X.C., González L., Reigosa M.J. Allelopathy in forest environment in Galicia, NW Spain. Allelopathy J 1995; 2:67-78

CHAPTER 19

PROTEIN CONTENT QUANTIFICATION BY BRADFORD METHOD

Nuria Pedrol Bonjoch and Pilar Ramos Tamayo
Depto Bioloxía Vexetal e Ciencia do Solo. Universidade de Vigo. Spain

THE BRADFORD METHOD

There are many procedures to quantify the protein content of an extract. A single protein solution would probably throw different results if measured with different methods, because they are based on different principles. Actually an absolute method does not exist; everyone has some advantages and disadvantages. The choice of an adequate method will depend on the nature of the proteins present in the sample, on the purity of the extracts, on the required sensibility and accuracy, and on the desired speed (Boyer, 1986). The most widely utilised methods for the analysis of protein contents are the assays of Biuret, Bradford, Kjendahl, Lowry, Smith, and Warburg-Christian. Nonetheless, many authors recommend the spectrophotometric assay of Bradford (1976) because of its multiple advantages if compared with other methods (Snyder and Desborough, 1978; Berges et al., 1993; and see Bio-Rad bulletin 1069EG, 1979). The most conspicuous *advantages* of Bradford method are: (i) the use of *a single reactive*, (ii) the *rapidity* of the reaction (just 5 min.), (iii) a high *stability* of the protein-dye complex, (iv) a *high reproducibility*, and (v) the occurrence of *minimal interferences.*

This simple, fast, and inexpensive method has *multiple applications* in experimental sciences. Besides being a well-suited method for repetitive determination of protein concentrations (e.g., in physiological, inmunological, cytological, clinical, and food quality *routine analysis*), the Bradford assay is often coupled to other techniques. It is a convenient

M.J. Reigosa Roger, Handbook of Plant Ecophysiology Techniques, 283–295.

means of monitoring *column chromatography* effluents and, due to the relative insensitivity to interferences, it is a valuable method for analysing fractions from *affinity*, *ion exchange*, *gel* and *adsorption chromatography*. This protein assay provides a useful means of monitoring sample loads for *gel electrophoresis* (see Chapter 20). The method also simplifies the assessment of protein concentration during enzyme purification of different tissues. Moreover, the Bradford assay is adaptable to semiautomated methods of protein quantification (Bio-Rad bulletin 1069EG, 1979).

Chemical basis

The Bradford method (1976) is based on the absorbance shift observed in an acidic solution of dye *Coomassie® Brilliant Blue G-250*. When added to a solution of protein, the dye binds to the protein resulting in a colour change from a reddish brown to blue. The dye has been assumed to bind to protein via electrostatic attraction of the dye's sulfonic groups to the protein. The bound points are primarily arginine residues, but the dye also binds to a lesser degree to histidine, lysine, tyrosine, tryptophan and phenylalanine (Compton and Jones, 1985). The peak absorbance of the acidic dye solution changes from 465 to 595 nm when binding to protein occurs. Therefore, measuring *absorbance of the protein-dye complex at 595 nm* allows an accurate quantification of the protein content of a sample.

The binding of the dye to protein is *a fast process*, being virtually completed in approximately 2 min. The protein-dye complex colour is *very stable*; complexes keep dispersed in the solution for around 1 h. This gives time enough to read a large number of samples, thus avoiding the 'critical time of measurement' of other methods (e.g., Lowry, 1951). After 1 h., complexes begin to aggregate, and the solution suffers a notorious decrease in colour (Bradford, 1976).

In order to evaluate the *accuracy* of the bioassay, Bradford (1976) assayed various proteins (bovine albumin, haemoglobin, cytochrome c, etc.) and compared the results with those obtained from Lowry assays of the same proteins. By calculation, she obtained that the dye binding assay is approximately four times more sensitive than the Lowry assay. Besides this, replicated Bradford assays of bovine serum albumin as a standard resulted in a *highly reproducible* response pattern. Statistical analysis gives a standard deviation around 1 % of mean value for the assay. Moreover, there is a *extreme sensitivity* in the assay: a final assay volume containing solely 5 μg protein · ml^{-1} will give an absorbance change of about 0.27 optical density (OD) units.

Some helpful hints

When selecting a *protein standard*, the best choice would be a purified preparation of interest, but it is not always possible. Commercial bovine seroalbumin is generally used as standard because it gives a representative colour response from which relative concentration values can be obtained. Nonetheless, when determining protein concentrations of immunoglobulins, commercial suppliers recommend the use of IgG standards (Pierce Chemical Co. Bulletin, 1990).

The extinction coefficient of a dye-albumin complex solution is constant over a 10-fold concentration range; thus, Beer's Law may be applied for accurate *quantification* of protein by selecting an appropriate ratio of dye volume to sample concentration. Over a broader range of protein concentrations, the dye-binding method gives an accurate, but not entirely linear response (Bio-Rad bulletin 1069EG, 1979). The source of the nonlinearity is in the reagent itself since there is an overlap in the spectrum of the two different colour forms of the dye. The background value of the reagent is continually decreasing as more dye is bound to protein. Nonetheless, there is no difficulty in obtain more accurate results if the assay is run with a set of standards and unknowns measured against the response curve of the standards, instead of calculated by Beer's Law (Bradford, 1976).

The protein-free reagent '*blank*' (sample buffer plus Bradford dye reagent) will result in a reddish-brown solution with optical density (*vs.* H_2O or sample buffer) characteristically about 0.45 OD units. The relatively high colour yield of the blank is normal, and does not interfere with the linearity, reproducibility, or sensitivity of the assay (Bio-Rad Bulletin 1069EG, 1979).

As a general precaution, the use of strongly alkaline buffering agents besides large quantities of detergents (greater than about 0.1 %) must be avoided, because they cause some *interference* in the assay system. Bradford (1976) observed a lack of effect by magnesium chloride, potassium chloride, sodium chloride, ethanol, and ammonium sulphate. Tris, acetic acid, 2-mercaptoethanol, sucrose, glycerol and EDTA, as well as trace quantities of Triton X-100, sodium dodecyl sulfate (SDS) and commercial glassware detergents, have solely small effects which can be easily eliminated by running the proper buffer control with the assay. Nevertheless, the presence of large quantities of the detergents produces abnormalities too great to overcome. Reagents that are compatible with the Bradford standard procedure are enumerated in Table 1.

Table 1. List of reagents that have been demonstrated to cause little or no interference in the Bradford protein assay, i.e., they do not affect the development of dye colour (From Bio-Rad Bulletin 1069EG, 1979 and Pierce Chemical Co. Bulletin, 1990).

Reagent	Conc.	Reagent	Conc.
Acetate	0.6 M	KSCN	3.0 M
Acetone	-	Malic acid	0.2 M
Adenosine	1 mM	$MgCl_2$	1.0 M
Amino acids	-	Mercaptoethanol	1.0 M
Ammonium sulphate	1.0 M	MES	0.7 M
Ampholytes	0.5 %	Methanol	-
Acid pH	-	MOPS	0.2 M
ATP	1 mM	NaCl	5 M
Aside	0.5 %	NAD	1 mM
Barbital	-	NaOH	0.25 M
BES	2.5 M	NaSCN	3 M
Boric acid	-	NP 40	0.25 %
Brij 35	0.5 %	Peptones	-
Brij 56	0.5 %	Phenol	5 %
Brij 58	0.25 %	Phosphate	1.0 M
Cacodylate-Tris	0.1 M	PIPES	0.5 M
CDTA	0.05 M	Polyadenylic acid	1 mM
CHAPS	1.0 %	Polypeptides	-
CHAPSO	1.0 %	Pyrophosphate	0.2 M
Citrate	0.05 M	rRNA	$0.25 mg \cdot ml^{-1}$
Deoxycholate	0.25 %	tRNA	$0.4 mg \cdot ml^{-1}$
Dithiothreitol	1 M	total RNA	$0.3 mg \cdot ml^{-1}$
DNA	$1 mg \cdot ml^{-1}$	SDS	0.05 %
EDTA	0.1 M	Sodium phosphate	-
EGTA	0.05 M	Streptomycin sulphate	20 %
Ethanol	-	Sucrose	10 %
Eagle's MEM	-	Tricine	-
Earle's salt solution	-	Triton X-100	0.5 %
Formic acid	1.0 M	Triton X-114	1.0 %
Fructose	-	Triton X-405	0.1 %
Glucose	-	Tyrosine	1 mM
Glutathione	-	Thymidine	1 mM
Glycerol	99 %	Tris	2.0 M
Glycine	0.1 M	Tween 20	0.5 %
Guanidine-HCl	-	Tween 80	0.25 %
Hank's salt solution	-	Urea	8 M
HEPES buffer	0.1 M	Vitamins	-
KCl	1.0 M		

The Bradford protein assay is *temperature dependent*, therefore wide fluctuations in temperature should be avoided (Pierce Chemical Co. Bulletin, 1990).

Finally, it was observed a tendency of the protein-dye complex in solution to bind to quartz cuvettes, although the error in assay readings due to cuvette blueness is negligible (Bradford, 1976). Therefore, removal of the residual blue colour between each sample reading is unnecessary. After the assay, the residual coloration on quartz cuvettes can be easily removed with methanol, acetone or HCl 0.1 M. This difficulty may be eliminated by using, if possible, either glass or plastic cuvettes.

PLANT PROTEIN CONTENTS ARE AFFECTED BY STRESS

When a plant is subjected to any biotic or abiotic stress factor, the first observed response is a decrease in its normal metabolic activities, with a consequent reduction of growth. In this 'alarm phase', *protein synthesis* is one of the most negatively affected anabolic processes (Bohnert and Jensen, 1996) together with photosynthesis, transport of metabolites, and uptake and translocation of ions (Hsiao, 1973; Larcher, 1995; Lichtenthaler, 1996).

Moreover, in those plants that possess only low or no stress tolerance mechanisms, acute damage and *senescence* will occur rapidly. Besides loss of chlorophyll, RubisCO and other chloroplast proteins are hydrolysed and exported via phloem, followed by hydrolysis of mitochondrial proteins and vascular tissues (Pell and Dann, 1991; Gan and Amasino, 1997). Therefore, low protein concentrations should be interpreted as a clear symptom of stress damage in plants.

However, in response to unfavourable conditions most plants will activate their stress coping mechanisms such as acclimation of metabolic fluxes, activation of repair processes, and long-term metabolic and morphological adaptations, which conform the named *general adaptation syndrome* (Lichtenthaler, 1996). Such mechanisms include *de novo* synthesis of proteins with specific adaptive functions (i.e., stress proteins, see Chapter 20), osmotic adjustment (e.g., proline accumulation, see Chapter 22) and antioxidative defence (e.g., polyamines accumulation, see Chapter 21), among others.

Further mechanisms include metabolic responses such as re-adjustment of the metabolic fluxes and changes in biomass partitioning (Lambers et al., 1998). Whatever the adaptive response, all of them carry *modifications in the levels and/or activities of many enzymes* of multiple metabolic pathways, which would be reflected in the total protein levels of plants.

Besides the overexpression of specific stress-related genes, many adaptive responses are time-dependent, so that they are solely shown up in long-term acclimation to stress (Conroy et al., 1988; Amthor and McCree, 1990; Bray, 1993). Pankovic et al. (1999) proved that *de novo* synthesis of RubisCO (see Chapter 23) made the major contribution to increases of total leaf soluble protein induced by water deficit in tolerant sunflower plants. RubisCO accumulation was observed even in full-expanded mature leaves and was discussed as a symptom of *photosynthetic acclimation* to stress. The progressive acclimation of seedlings to long-term stress must be considered as an integrated plant response (Chapin, 1991; Shangguan et al., 1999). In such cases, the accumulation of soluble proteins is usually accompanied by the maintenance of turgor through the synthesis of compatible solutes (e.g., proline); both responses together contribute to a high water and nitrogen use efficiency at the level of whole plant (e.g., Ashraf and Yasmin 1995; Pedrol et al., 2000).

As a result of growth inhibition caused by stress the photosynthate is usually diverted to *storage of non-structural biomass*, including organic compounds such as proline and proteins (Amthor and McCree, 1990; Setter, 1990). The accumulation of leaf soluble proteins may represent a *reserve of nitrogen* to be used during recovery once stress has ceased (Millard 1988, Stitt and Schulze 1994). Storage usually includes *recycling of nutrients*, so that C and N can be remobilise from one tissue and subsequently used for the growth or maintenance of another. Thus, an increase of soluble proteins in young leaves can occur as a consequence of N remobilization from senescent tissues, in which shares of metabolically active proteins (mainly RubisCO) are implied. Recycling of N can also occur as a consequence of root turnover (Chapin, 1980; Millard, 1988).

Phenotypic plasticity regarding biomass partitioning contributes notably to success in changing environments where biotic and abiotic stress factors are continually interacting (Austin, 1990; Dunson and Travis, 1991; Aerts, 1999). Nitrogen storage confers multiple ecophysiological advantages to plants, especially if considered in the context of interspecific relationships (Aerts, 1999; Lemaire and Millard, 1999). Nitrogen storage will allow growth to occur when the external availability of N is low, and will enable more rapid recovery from catastrophic events such as defoliation. The efficient internal cycling of C and N within a plant through the processes of storage and remobilization minimises the loss of resources from the individual (e.g., through leaf litter) which otherwise would be available to competitors. Furthermore, a rapid turnover of nutrients together with a high morphological plasticity may also be important determinants of the competitive ability for light interception (Wedin, 1994; Aerts, 1999).

Under the perspective of ecological strategies, fast-growing species have generally higher leaf nitrogen concentrations (i.e., protein contents) and higher levels of nitrogen use efficiency than the slow-growing ones. Such characteristics determine the great competitive abilities of fast-growing species in productive habitats (Lambers and Poorter, 1992; Aerts, 1999).

Whatever of the discussed features will affect the general metabolism of nitrogen, including polyamine and proline metabolisms (see Chapters 21 and 22). The study of stress-induced changes in protein contents, together with other ecophysiological parameters and stress metabolites would throw much light in the integration of the whole plant responses to multiple stress factors (Chapin et al., 1987; Chapin, 1991).

ANALYSING PROTEIN CONTENT IN PLANT TISSUES BY BRADFORD METHOD

Now we will describe the basic steps for the analysis of protein contents in different plant tissues, by means of the Bradford assay. The procedure here proposed (summarised in Figure 1) is adequate to determine the protein concentrations of young to mature leaves, cotyledons and/or fruit tissue, among others.

Fresh *plant material* must be processed immediately after being harvested, kept in refrigerator at 2 °C for 5-10 days, or kept in freezer at -20 °C until extraction. For field or greenhouse bioassays, fresh masses of the different tissues should be better collected when day-time temperatures does not exceed 30 °C, considering proteins are thermolabile; samples must kept at 0 to 4 °C (preferably on ice) during collection. Particularly in comparative studies concerning stress, protein contents are better to be measured in young actively growing leaves, or fully expanded leaves that have been developed during stress exposure.

EXTRACTION of proteins
Plant tissue (100 to 300 mg *fresh mass*)
+ 0.050 g PVPP
+ 1 ml 0.05 M Tris buffer (Arulsekar and Parfitt 1986)
Powder (with liquid N) and homogenise in cool mortar
Keep tubes on ice
Centrifugation 14,000 to 19,000 g_n 4 °C 20 min.
take **Supernatant**
Protein dye-binding REACTION 0.1 ml + 3 ml Bradford reactive
Mix avoiding foam formation
Wait 5 min.
Reading of ABSORBANCE at 595 nm
determination and quantification of
PROTEIN CONTENT

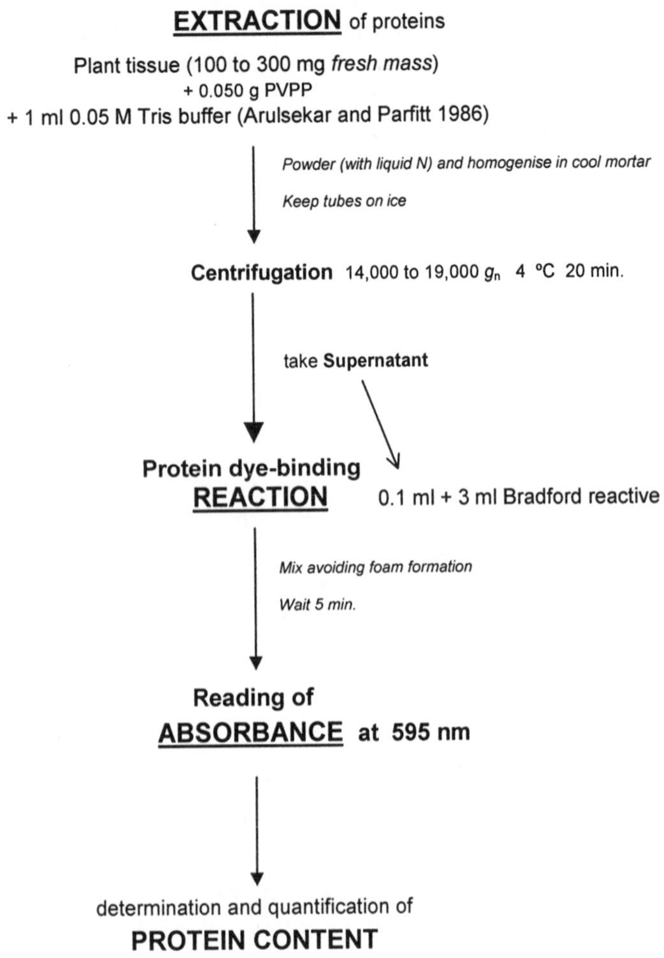

Figure 1. Procedure for the analysis of protein contents by Bradford method (Bradford, 1976). Extraction procedure as described by Pedrol (1997).

Preparation of reagents

Extraction solution

An appropriate solution or buffer to extract proteins from plat tissues can be selected by consulting the pertinent literature, considering reagent composition is compatible to Bradford assay (see Table 1). Here we propose the *extraction buffer* formulated by Arulsekar and Parfitt (1986), which provides an optimal extraction of soluble proteins from different plant tissues. It consists in a 0.05 M Tris buffer (pH 8.0) with the following composition in distilled water:

Reactive:	**Concentration:**
Tris base	0.05 M
ascorbic acid (Na salt or free acid)	0.1 % (w/V)
cysteine hydrochloride	0.1 % (w/V)
polyethylene glycol (M_r 3500 to 4000)	1 % (w/V)
citric acid (monohydrate)	0.15 % (w/V)
2-mercaptoethanol	0.008 % (V/V)

This buffer can be used in a maximum period of 1 wk. after preparation.

Standard solutions

A *stock solution* of commercial bovine serum albumin is prepared at the concentration of 2 g · L^{-1} (2000 μg · ml^{-1}) in the same buffer as the protein samples whose concentration is to be determined. Aliquots of the stock solution can be stored at -20 °C and used whenever are needed.

The series of protein standard *working solutions* should cover the range of concentration between 50 and 1500 μg · ml^{-1}, in which the sample concentrations would be most often represented. Convenient standard data points (prepared from successive dilutions of the stock in the buffer) are 50, (100), 125, 250, (400), 500, 1000, (1500) and 2000 μg · ml^{-1}.

Bradford protein reagent

The Bradford protein reagent is prepared at the following concentrations of its different components in distilled water (Bradford, 1976):

Reactive:	**Final concentration:**
Coomassie® Brilliant Blue G-250	0.01 % (w/V)
Ethanol 95 %	4.7 % (w/V)
Phosphoric acid 85 % (w/V)	8.5 % (w/V)

As an example, to prepare 1 L of protein reagent, Coomassie® Brilliant Blue G-250 (100 mg) is dissolved in 50 ml 95 % ethanol. To this solution 100 ml 85 % (w/V) phosphoric acid is added. The resulting solution is then diluted to a final volume of 1 L. The Bradford reagent has a relatively long duration. Always mix reagent solution prior to use by repeatedly inverting reagent bottle. Do not shake! Reagent solution will foam if shaken vigorously.

Extraction procedure

A trial-and-error approach will be necessary to determine the ratio at which extraction must be carried out. The required amount of plant material will depend on the water content of each tissue and species. Different extraction procedures usually include centrifugation or filtration to remove particulate matter in the sample. Our proposed extraction procedure (Pedrol, 1997; Pedrol et al., 2000) is carried out as follows (see Fig. 1):

Samples of plant material are powdered with liquid nitrogen, using precooled mortars and pestles. Proteins are extracted by homogenising in the described cold 0.05 M Tris buffer (Arulsekar and Parfitt, 1986), at a ratio of 100 to 300 mg fresh mass · ml^{-1} buffer. A small quantity (e.g., 0.05 g) of the antioxidant polyvinyl polypyrollidone (PVPP) is added to each sample during the homogenisation procedure. Homogenates are transferred to cold centrifuge tubes (e.g., 2 ml Eppendorf tubes) and centrifuged at 14,000 to 19,000 g_n for 20 min. at 4 °C.

Sample tubes must be kept on ice and closed until centrifugation, in order to prevent oxidation. After centrifugation, clear supernatants can be used immediately for the protein assay, or frozen at -20 °C and used later.

Protein-dye binding reaction

0.1 ml of each sample supernatant and working standard solution are transferred to assay tubes (e.g., 16 × 100 mm test tubes). A blank containing 0.1 ml extraction buffer must be also prepared. Then, 3 ml of the Bradford reagent (containing the dye) are added to each tube. The tubes are softly vortexed or mixed by gentle inversion to prevent the generation of foam in the sample, which would reduce significantly the colour yields (Bio-Rad Bulletin 1069EG, 1979).

Absorbance measurement and quantification

The absorbance at 595 nm (*vs.* the blank) is measured spectrophotometrically after 5 min and before 1 h., taking into account the stability of the protein-dye complex.

The net absorbances at 595 nm for each albumin standard are subjected to regression analysis (known concentration *vs.* absorbance). The protein concentration for each unknown sample (expressed in $\mu g \cdot ml^{-1}$) is calculated by the resulting equation. Final protein concentrations of samples in $\mu g \cdot g^{-1}$ (fresh mass) are determined taking into account the different dilutions made during the extraction procedure.

Fresh masses will vary with different growing conditions (e.g., stressed *vs.* unstressed), thus protein contents are recommended to be expressed in relation to dry mass, previous calculation of dry mass/fresh mass ratio for each sample.

Finally, in some cases, we will observe that the protein concentration of the sample is too dilute for the described standard assay, being in the range of 1 to 25 $\mu g \cdot ml^{-1}$ (e.g., phytoplankton samples, Berges et al., 1993). Then, a *micro-assay procedure* can be performed as follows: The protein standard series would cover the range of concentration, being convenient standard points 1, 5, 10, 15 and 25 $\mu g \cdot ml^{-1}$. Take 0.5 to 1 ml (instead of 0.1 ml) of dilute standard or unknown protein sample, and add 1 ml (instead of 3 ml) of Bradford reagent. Further steps are equal to standard procedure.

REFERENCES

Aerts R. Interespecific competition in natural plant communities: mechanisms, trade-offs and plant-soil feedbacks. J Exp Bot 1999; 50:29-37

Amthor J.S., McCree K.J. "Carbon Balance of Stressed Plants: A Conceptual Model for Integrating Research Results." In *Stress Responses in Plants: Adaptation and Acclimation Mechanisms*. R.G. Alscher, J.R. Cumming, eds. New York: Wiley-Liss Inc. Publishers, 1990

Arulsekar S., Parfitt D.E. Isozyme analysis procedures for stone fruits, almond, grape, walnut, pistachio, and fig. Hortscience 1986; 21:928-933

Ashraf M., Yasmin N. Responses of 4 arid zone grass species from varying habitats to drought stress. Biol Plantarum 1995; 37:567-575

Austin M.P. "Community Theory and Competition in Vegetation." In *Perspectives on Plant Competition.* J.B. Grace, D. Tilman, eds. San Diego: Academic Press, 1990

Berges J.A., Fisher A.E., Harrison P.J. A comparison of Lowry, Bradford and Smith protein assays using different protein standards and protein isolated from the marine diatom *Thalassiosira pseudonana.* Mar Biol 1993; 115:187-193

Bohnert H.J., Jensen R.G. Metabolic engineering for increased salt tolerance - the next step. Aust J Plant Physiol 1996; 23:661-667

Boyer R.F. *Modern Experimental Biochemistry.* Massachusetts: Addison-Wesley Publishing Co., 1986

Bradford M.M. A rapid sensitive method for the quantification of microgram quantities of protein utilising the principle of protein-Dye Binding. Anal Biochem 1976; 72:248-254

Bray E.A. Molecular responses to water deficit. Plant Physiol 1993; 103:1035-1040

Chapin III F.S. The mineral nutrition of wild plants. Annu Rev Ecol Syst 1980; 11:233-260

Chapin III F.S., Bloom A.J., Field C.B., Waring R.H. Plant responses to multiple environmental factors. BioScience 1987; 37:49-57

Chapin III F.S. Integrated responses of plants to stress. BioScience 1991; 41:29-36

Compton S.J., Jones C.G. Mechanisms of dye response and interference in the Bradford protein assay. Anal Biochem 1985; 151:369-374

Conroy J.P., Virgona J.M., Smillie R.M., Barlow E.W. Influence of drought acclimation and CO_2 enrichment on osmotic adjustment and chlorophyll *a* fluorescence of sunflower during drought. Plant Physiol 1988; 86:1108-1115

Dunson W.A., Travis J. The role of abiotic factors in community organization. Am Nat 1991; 138:1067-1091

Gan S., Amasino R.M. Making sense of senescence. Plant Physiol 1997; 113:313-319

Hsiao T.C. Plant responses to water stress. Annu Rev Plant Physiol 1973; 24:519-570

Lambers H., Poorter H. Inherent variation in growth rate between higher plants: a search for physiological causes and ecological consequences. Adv Ecol Res 1992; 23:187-261

Lambers H., Chapin III F.S., Pons T.L. *Plant Physiological Ecology.* New York: Springer-Verlang, 1998

Larcher W. *Physiological Plant Ecology.* Berlin: Springer-Verlag, 1995

Lemaire G., Millard P. An ecophysiological approach to modelling resource fluxes in competing plants. J Exp Bot 1999; 50:15-28

Lichtenthaler H.K. Vegetation stress: An introduction to the stress concept in plants. J Plant Physiol 1996; 148:4-14

Lowry O.H., Rosebrough N.J., Farr A.L., Randall R.J. Protein measurement with Folin-phenol reagent. J Biol Chem 1951; 193:265-275

Millard P. The accumulation and storage of nitrogen by herbaceous plants. Plant Cell Environ 1988; 11:1-8

Pankovic D., Sakac Z., Kevresan S., Plesnicar M. Acclimation to long-term water deficit in the leaves of two sunflower hybrids: photosynthesis, electron transport and carbon metabolism. J Exp Bot 1999; 50:127-138

Pedrol N. "Analítica de proteínas totales por colorimetría." In *Manual de Técnicas en Ecofisiología Vegetal.* N. Pedrol, M.J. Reigosa, eds. Vigo, Spain: Gamesal, 1997

Pedrol N., Ramos P., Reigosa M.J. Phenotypic plasticity and acclimation to water deficits in velvet-grass: a long-term greenhouse experiment. Changes in leaf morphology, photosynthesis and stress-induced metabolites. J Plant Physiol 2000; 157:383-393

Pell E.J., Dann M.S. "Multiple Stress-induced Foliar Senescence and Implications for Whole-plant Longevity." In *Response of Plants to Multiple Stresses.* H.A. Mooney, W.E. Winner, E.J. Pell, eds. San Diego: Academic Press, 1991

Setter T.L. "Transport/Harvest Index: Photosyntate Partitioning in Stressed Plants." In *Stress Responses in Plants: Adaptation and Acclimation Mechanisms.* R.G. Alscher, J.R. Cumming, eds. New York: Wiley-Liss Inc., 1990

Shangguan Z., Shao M., Dyckmans J. Interaction of osmotic adjustment and photosynthesis in winter wheat under soil drought. J Plant Physiol 1999; 154:753-758

Snyder J.C., Desborough S.L. Rapid estimation of potato tuber total protein content with Coomassie brillant blue G-250. Theor Appl Genet 1978; 52:135-139

Stitt M., Schulze D. Does Rubisco control the rate of photosynthesis and plant growth? An exercise in molecular ecophysiology. Plant Cell Environ 1994; 17:465-487

Wedin D.A. "Species, Nitrogen and Grassland Dynamics: the Constraints of Stuff." In *Linking Species and Ecosystems.* C. Jones, J.H. Lawton, eds. New York: Chapman and Hall, 1994

CHAPTER 20

TWO-DIMENSIONAL ELECTROPHORESIS. STRESS PROTEINS

Adela M. Sánchez-Moreiras and Nuria Pedrol Bonjoch
Depto Bioloxía Vexetal e Ciencia do Solo. Universidade de Vigo. Spain

INTRODUCTION

Proteome analysis, as its name indicates, is the 'analysis of the entire PROTEin complement expressed by a genOME, or by a cell or tissue' (Wilkins, 1996). So, proteomics is the science that correlates proteins with their genes. The most important characteristic of this analysis, the necessity to resolve a plurality of complex protein extracted from a biological sample, only can be done with high resolution by the two-dimensional electrophoresis, 2D-electrophoresis (Görg et al., 1988a,b). This technique allows the separation of hundreds of proteins in only two steps with single sorting parameters, which are the isoelectric point (pI) and the molecular weight of proteins.

The demand of an efficient method for protein separation induced to scientists to search for a technique that should permit, in a mixture of protein complexes, the perfect sorting of proteins with the analysis of a single protein in a single conformation. Along these years, several techniques were applied for it, like first electrophoresis, quantification of enzymatic activity by colorimetric techniques, inmunological methods, etc. All of these techniques are limited in the number and conformation of measurable proteins, and normally only a range or a type of protein can be determined by these methods.

In 1975 two published works of O'Farrell and Klose opened the door to 2D-electrophoresis. Since the beginning this technique appeared to be a

M.J. Reigosa Roger, Handbook of Plant Ecophysiology Techniques, 297–333.

good biochemical method for protein separation, but it was in the last few years when its application became more and more important. The reasons for this reappearance were:

- the development of broad genome and protein databases,
- the appearance of a simpler but also more accurate image analysis method,
- the development of new advanced characterisation techniques like the peptide mass fingerprinting (mass spectrometric quantification of the masses of some fragments from an unknown protein can give us the exact protein identification; Yates, 1998), or like the N-terminal EDMAN sequencing, and finally,
- the improving of 2D-electrophoresis with a high resolution and reproducibility of the samples.

At this moment, 2D-electrophoresis is usually used for the attainment of different objectives with a broad field of application. 2D-electrophoresis is an excellent technique for applying in the proteome analysis, mutagenesis test, and cell differentiation analysis, cancer research, detection of disease markers and the development of new medicines, for example (Berkelman and Stenstedt, 1998).

This plurality of applications is caused by the fact that 2D-electrophoresis is the best first step for identification and sequenciation of single proteins in a complex protein mixture, being a link between different sciences. The contribution of these different sciences is the principal reason for a so fast and plural growth of this technique in the last decade.

A new old protein analysis technique

The technique 2D-electrophoresis is based in a double separation of the proteins by using two consecutive electrophoresis (see Fig. 1).

The first of them separates the proteins according with their isoelectric point (pI), it is an isoelectric focusing; and the second electrophoresis separates the proteins according with their molecular weight. Both electrophoresis require a fast and automated procedure for a good resolution and reproducibility of the samples.

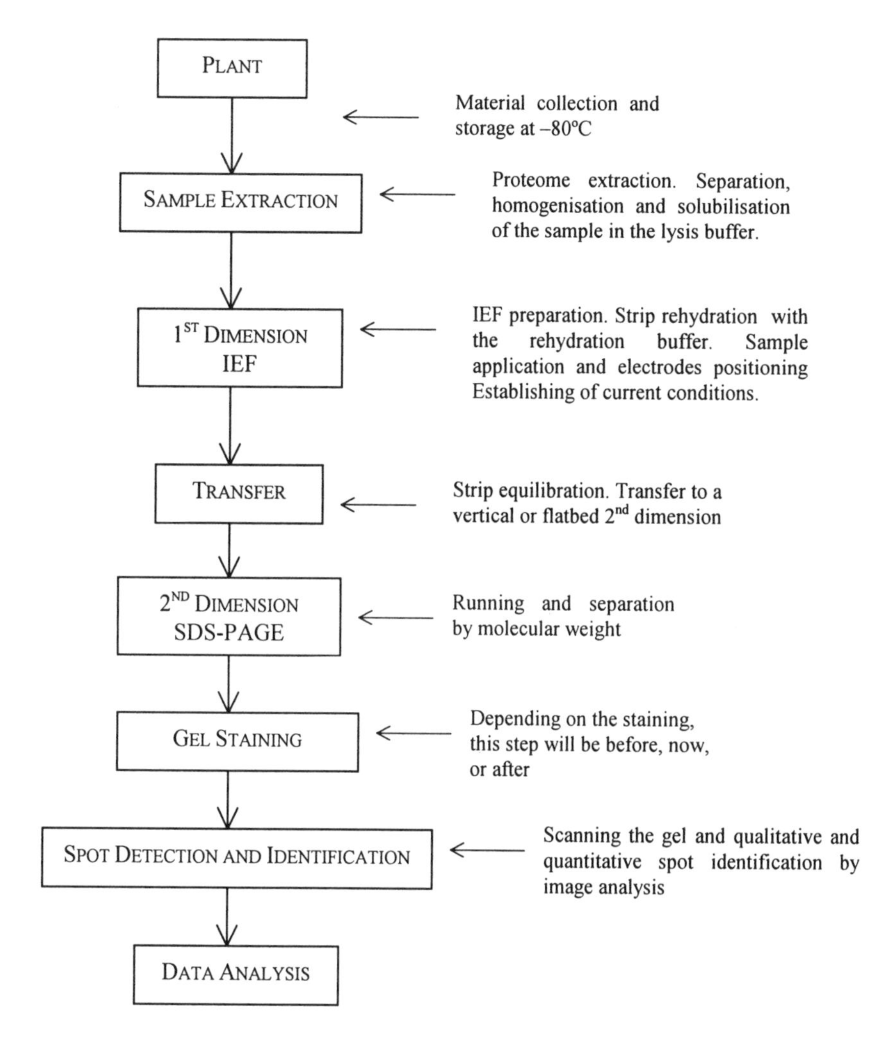

Figure 1. Scheme of a two-dimensional electrophoresis technique

The first step, sample preparation, is one of the most critical parts in this technique, because the type and amount of proteins appeared in the final solution depends highly on the selection of the optimal preparation procedure of the biological sample. So, it is necessary to be careful during the extraction, solubilisation and denaturation processes to avoid later problems with the loss of important proteins. Because of the different origin of the samples and the complex protein mixtures present in them, there is not a unique procedure for sample preparation, and so it is necessary to optimise the preparation of the samples in our conditions (Quadroni and James, 1999).

Different treatments, more or less vigorous, can be used for protein extraction. Osmotic lysis, detergent lysis, sonication, mechanical homogenisation, or others, are used in sample preparation, depending on the origin of the sample (microorganisms with cell walls, tissue cultures, plant membranes, fungal cells, blood cells, etc.) and the goal of the research (Ames et al., 1976; Theillet et al., 1982; Dignam, 1990; Geigenheimer, 1990; Portig et al., 1996). We must be sure that the race for extracting more and more proteins does not limit the diversity and origin of the final protein mixture.

In the same way, purification methods as precipitation with TCA and/or acetone, ammonium sulphate precipitation, etc. (Görg et al., 1988; Halloway and Arundel, 1988; Flengsrud and Kobro, 1989) or methods used for the removal of contaminants from the sample (dialysis, filtration, ultracentrifugation, precipitation, etc) can help or difficult the final result of the analysis. For example, the presence of ammonium sulphate or TCA sometimes avoids the solubilisation of the proteins and so, if the objective is to separate as many proteins as possible, we can suffer a loss of information with a too strong purification.

In general, a solubilisation buffer must contain (Berkelman and Stenstedt, 1998):

- a chaotropic agent to denature proteins and expose their hydrophobic residues. It allows proteins to unfold presenting an only pI.
- a non-ionic detergent to solubilise the proteins. This non-ionic surfactant solubilises the hydrophobic residues exposed after the denaturation by the chaotropic agents.
- IPG Buffer (Immobilised pH Gradient) to avoid protein carbamilation. It gives more conductivity to the sample and allows nucleic acids for precipitation.
- a reductant (charged or not) to ensure complete reducing conditions along the IEF.
- a tracking dye to monitor the evolution of the electrophoresis.

One of the objectives to use denatured conditions in sample solubilisation (use of urea and thiourea in the solubilisation buffer) is to obtain the protein in an only conformation avoiding different pKs for a same protein. On the other hand, denaturation is used to avoid also aggregates of protein to protein, to maintain solubilisation of hydrophobic proteins along the total electrophoresis process and to assure that teoric and denatured pIs are the same. It is important to know that each protein appears in the gel with only one spot in the last part of the technique (image analysis of the results).

First dimension

The first dimension is the isoelectric focusing (IEF). Here the proteins are separated according to their isoelectric point (pI). Isoelectric point is that pH at which the net charge of the protein is zero. In a sample, we will have a mixture of positively and negatively charged proteins, but if we expose them to an electric field in presence of a pH gradient these proteins will move in the gradient to the pH value where the net charge is zero. So, a negatively charged protein will migrate to the anode while a positively charged protein will go toward the cathode. Thanks to this we will get the proteins separated along the gradient pH gel.

IEF must be performed in gels with a pH gradient of high stability, including acidic and basic extremes without losing proteins present in these extremes as occurred with the first pH gradients (Westermeier, 1997). Thus, immobilised pH gradients were recently introduced in 2D-electrophoresis (Bjellqvist et al., 1982; Görg et al., 1988). Until nowadays tube gels of pH gradient with ampholites for IEF were used in the first dimension, but these vertical gel roads revealed several troubles in the resolution, reproducibility and separation of the most acidic and basic proteins. The long time needed for the separation produced the loss of proteins from these extremes. Likewise, the result of IEF was too dependent of handling, which reduced specially the reproducibility.

The new immobilised pH gradient (IPG) 'is created by covalently incorporating a gradient of acidic and basic buffering groups into a polyacrylamide gel at the time it is cast' (Berkelman and Stenstedt, 1998). The solutions containing these basic and acidic buffering groups will determine the pH range and so, also the resolution of the technique.

Today, these IPG gels are available in several sizes and with several pH gradients, normally cut and dried to simplify the handling. This availability allows the focusing of proteins present in a specific pH range. So, if we start the sample separation with a 3-10 pH gradient, and we detect an interesting protein around a pH value of 6, we can then use a IPG gel strip with a 4-7 pH gradient. If we restrict every time more and more this gradient, we can obtain the most exact focusing of our protein (see Fig. 2).

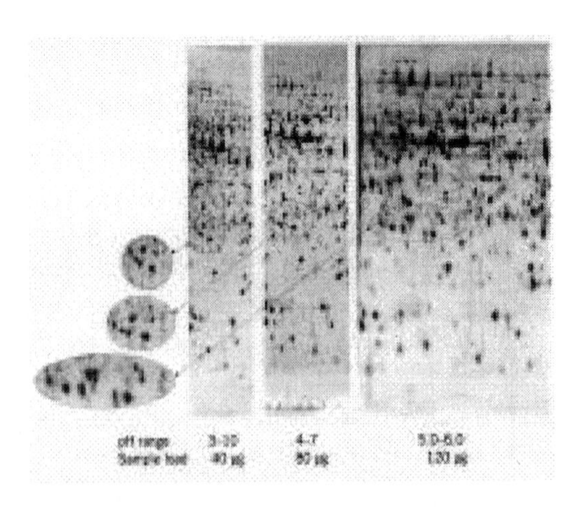

Figure 2. Using the different available pH gradients (3-10; 4-7; 5-6) for focusing a specific protein appeared in a pH value nearly to 5. (By courtesy of Amersham Pharmacia Biotech).

These gels can be stored during a long time. When needed, IPG gel strips are rehydrated in a rehydration solution and placed in a horizontal or flatbed electrophoresis unit to start the sample separation. Usually, a rehydration solution contains similar components to the solubilisation buffer, and so in this solution we can also find chaotropic agents, non-ionic detergents, reducing agents, etc.

The sample can be applied to the IPG gel with the rehydration solution or in the rehydrated IPG gel during the first steps of the IEF. The amount of sample depends on the amount of the protein/s of interest in our sample and also on the staining technique that will be used for the characterisation of the spots once finished the second dimension. Once applied the sample and fixed all conditions for running the first dimension IEF, the separation of the proteins according with their pI starts.

Transfer

Now, after IEF, the IPG strips contain the proteins separated along the gel, but for their characterisation it is necessary to start a second dimension. In this second dimension, SDS-polyacrylamide gel electrophoresis (SDS-PAGE), a protein separation occurs according to the molecular weight of each protein. Here, the electrical charge of the protein is not important because we add an anionic detergent to the sample (sodium dodecyl sulphate, SDS), which denatures proteins masking them real charge. So, SDS forms with the proteins anionic complexes with constant net negative charge. These complexes will migrate faster or slower in the gel according to the number of SDS molecules bound to the protein and so according also to the molecular weight of it.

IPG gel strips will be provided with the SDS saturated conditions in an equilibration step made after the first dimension (IEF) and just before the second dimension (SDS-PAGE). Before starting the equilibration step, we must select which SDS-PAGE system will be used, because at this moment we can opt for a vertical or flatbed second-dimension system.

The equilibration solution must contain the following components needed for a good separation and resolution in the SDS-PAGE:

- *Equilibration buffer*, to ensure a good pH range along the second dimension.
- *Chaotropic agent,* together with an alcohol prevents electroendosmosis giving more viscosity to the buffer and improving the transfer of proteins to the 2nd dimension gel.
- *Anionic detergent*, to denature proteins and create negatively charged complexes for the 2nd dimension.
- *Reductant,* to preserve the total reduced state.
- *Dye,* to follow the development of the technique.

Second dimension

The equilibrated IPG strip will be positioned on the selected SDS-PAGE system and it will be run in specific current conditions. The resolution of the separation according with the molecular weight depends also on the used polyacrylamide matrix. Therefore, we should keep this in mind when a polyacrylamide percentage is selected.

When the second dimension has finished the proteins are totally separated in the gel and it is the moment to identify them.

Detection methods

Several detection methods can be applied for the visualisation of the results. There are methods for applying before IEF, between IEF and SDS-PAGE, after SDS-PAGE or after blotting (Rabilloud, 2000). The most used methods are those applied after the SDS-PAGE, because of their flexibility.

There are also different staining types as radioactive, fluorescence, metal ion reduction, etc., with more or less sensitivity. For example, autoradiography and fluorography are the most sensitive detection methods but both demand complicated and expensive conditions for its development. On the other hand, silver staining is the most sensitive nonradioactive method but also one of the most complex, with a lot of complications by handling or impurities in the gel or water. By contrary, coomassie staining reduces these complications, but it is 50-fold less sensitive than silver staining. Therefore, we must approach and personalise the staining type to our sample.

Once that the staining process has finished, we can obtain up to 15,000 spots. A good characterisation is reached when around 4,000-7,000 spots (depending on the sample) are identified.

Image analysis

The software used for the evaluation of the results must be able to detect, correct and compare spots in a same gel or in different gels. With the software we pretend to quantify the protein spots as exact as possible.

FROM PLANT GENOME TO PLANT PROTEOME

The entire genomic DNA sequence of *Arabidopsis* is available since December, 2000 [see Plant Physiology 124 (4), Special Issue on *Arabidopsis* genome]. The reward for plant sciences can hardly be overlooked: having full genomic data will allow us to understand the orchestration of multiple genes in elaborating programs of homeostasis, development, defence, and disease (Smith, 2000). But, as it has been the case for a number of prokaryotic and eukaryotic species whose genomic DNA sequences have been completed, many 'open reading frames' in the *Arabidopsis* genome will encode *proteins of unknown functions* (Li and Assman, 2000).

The first post-genomic approach is expression profiling, also known as TRANSCRIPTOMICS. Transcriptomics involves identifying the mRNAs expressed in the genome at a given time. In words of Abbott (1999) 'this gives a snapshot of the genome's plans for protein synthesis under the cellular conditions at a moment'. As she describes, messenger RNAs can be 'fished out' of a cellular soup onto *microarrays*. These are chips onto that are stuck thousands of complementary DNAs, which specifically bind the mRNA transcribed from the gene. Finally, small amounts of mRNA can be amplified by PCR (*polymerase chain reaction*) techniques. Therefore, transcriptomics can provide important clues about expression patterns and thus the functions of gene products (Abbott, 1999; Somerville and Somerville, 1999).

However, for a substantial number of proteins, there may be only a loose correlation between mRNA and protein levels because the rates of degradation of individual mRNAs and proteins differ (e.g., Gygi et al., 1999). Moreover, the functions of proteins depend notably on *post-translational modification* and *interaction with other proteins*, so that one mRNA can give rise to more than one protein. Post-translational modification of proteins is important for biological processes, particularly in the *propagation of cellular signals* (i.e., the stress signal, see below) where, for example, phosphorylation can trigger either activation or inactivation of a signalling cascade (see Abbott, 1999; Li and Assmann, 2000). Thus, gene transcription, even when defined on a genome wide scale, 'does not offer the complete molecular picture of the organism that genomics as a science had promised' (Smith, 2000). Measuring proteins directly would undoubtedly give us a 'more accurate picture of a cell's biology' (Abbott, 1999).

The interface between protein biochemistry and molecular biology has become known as PROTEOMICS. As said before, it includes efficient approaches for the following items:

- identifying proteins
- determining expression in different tissues and under *different conditions* (e.g., stressing *vs.* unstressing conditions)
- identifying post-translational modification of proteins in response to different stimuli (e.g., stress factors)
- characterising protein interactions

Proteomics will therefore be critical for understanding plant biological processes in the post-genome era (Li and Assmann, 2000; Smith, 2000).

As said before, besides the *separation* of complex protein mixtures by *2-D Electrophoresis* (e.g., Görg et al., 1988b), proteome analysis comprise the *characterisation* of the separated polypeptides and, finally, *database search.*

The most commonly used technology for characterisation is *mass spectrometry*, being nowadays an essential tool for post-electrophoresis analysis of proteins (Li and Assmann, 2000). Typically, proteins separated by 2-D gel electrophoresis are first digested *in situ* with trypsin. When a complete genome sequence is available (as the completely sequenced line of *Arabidopsis*), mass spectrometric quantification of the masses of a few tryptic fragments from an unknown protein, followed by the use of algorithms to compare the observed peptide masses against those predicted for the theoretical tryptic fragments of all expressed sequences, will often be sufficient for exact protein identification (Yates 1998). This process is known as peptide mass mapping or *peptide mass fingerprinting*.

But, what to do when a complete genome sequence is not available and we want to identify stress-products from any plant species? Then, amino acid sequencing by *tandem mass spectrometric sequencing* is often used. Amino acid sequence information provided by mass spectrometric analysis can allow homology searching and cloning or database identification of the corresponding gene (Shevchenko et al., 1997). Microanalytical protein characterisation with multidimensional liquid chromatography/mass spectrometry (MALDI ToF: matrix-assisted laser desorption/ionisation time-of-flight) will become a widely used method for determination of peptides and proteins (see PharmaBiotech report, 2001).

Functional proteomics technologies include *yeast two-hybrid system* for studying protein-protein interactions. This approach uses the yeast (*Saccharomices cereviseae*) genome as a matrix for studying interactions between two proteins from a given species. As Abbott (1999) explains, 'hybrids are made of two proteins whose potential interaction is to be elucidated: one test protein is attached to a yeast DNA-binding protein, and a second to the transactivation domain (or gene activator) of a yeast transcription factor. If both proteins interact, a 'reporter' gene (typically a gene that turns yeast blue) is activated by the transactivation factor that has been pulled into the gene's vicinity on the yeast genome by the DNA-binding protein'.

Proteome projects are inherently large scale, and generate massive amounts of experimental data. The use of *bioinformatics* is essential for analysing all the data generated from both genomics and proteomics. The largest expansion of proteomics in future will be in *protein biochip technologies*, and, undoubtedly, in informatics (PharmaBiotech report, 2001).

Proteomics and stress-proteins in plants

Plants are *constantly subject to adverse environmental conditions* such as drought, flooding, extreme temperatures, excessive salts, heavy metals, high-intensity irradiation, allelochemicals, and infection by pathogenic agents, among others. Because of their immobility, plants have to make necessary metabolic and structural adjustments to cope with the stress conditions. Stress-induced changes in plant metabolism and development can often be attributed to *altered patterns of gene expression*. In response to stress, some genes are expressed more intensively, whereas others are repressed. The *protein products of stress-induced genes* often accumulate in response to unfavourable conditions. The functions of these proteins, named **stress proteins**, and the mechanisms that regulate their expression are currently a central topic of research in stress physiology (see Bray et al., 2000 as a review).

A large number of the stress-responsive genes have been identified and isolated by the differential screening of cDNA libraries and protein identification of electrophotograms (see Ingram and Bartels, 1996;

Grover, 2000, for water-stress induced genes and products). The precise identity of the isolated genes has been looked into through search for homology of the corresponding nucleotides/amino acids. From these analysis, most genes have appeared to be novel as no corresponding sequences are found in the databases. Clearly, more efforts are needed to find out the functional role of the genes which are up regulated in response to stress (Grover, 2000). Furthermore, the intricate signalling pathways that are assumed to participate in alterations of plant gene expression in response to stress are also yet to be elucidated (see below).

Under the exposed framework, one of the most amazing **applications of proteomics in Plant Physiology** will be undoubtedly the *identification of novel stress-proteins in plants and their functions*, as well as the *elucidation of regulatory pathways leading to adaptative responses*. Many researchers are already working on this enormous challenge (e.g., Abromeit et al., 1992; Usuda and Shimogawara, 1995; Chang et al., 2000; Hasegawa et al., 2000).

STRESS PROTEINS IN THE FRAME OF PLANT ECOPHYSIOLOGY

An ecophysiological classification of stress proteins

Under an ecophysiological point of view, most of the stress proteins have a role either in (1) helping the plants survive, or (2) minimising the effectiveness of the stress agent (Ho and Sachs, 1989):

- In ***helping plants survive under stress conditions***, the stress proteins perform the following functions: (i) maintenance of the basic metabolism in the stressed cell, e.g. the induction of ADH (alcohol dehydrogenase) and some glycolysis enzymes in anaerobic stressed cells; (ii) protection of cellular components from being damaged by the stressful condition, e.g. the association of heat shock proteins (HSPs, see below) with enzymes or organelles resulting in tolerance to thermodenaturation; (iii) removal of damaged cellular components, e.g. the tagging of denatured proteins by ubiquitin (see below) for proteolysis in heat-stresses cells. Although the stressed cells are usually not metabolically active, the induction of stress proteins can keep the cells from being killed and the cells can recover once the stress condition is relieved.

- In ***minimising the effectiveness of the stress agents***, the stress proteins take up a more active set of functions: (i) the physical blockage of entry of stress agents, such as the induction of cell wall hydroxy-proline-rich glycoproteins, and lignin synthesis by fungal elicitors; (ii) the sequestration of stressful agents, as in the induction of phytochelatin to chelate heavy metal ions; (iii) the impairment of the biological stress agents, such as the induction of proteinase inhibitors and enzymes capable of hydrolysing fungal cell walls, and the synthesis of phytoalexins.

Under a more strictly physiological perspective, Grover (2000) and Hasegawa et al. (2000), focussing mainly on water stress (Bray, 1993, 1997; Ingram and Bartels, 1996; Shinozaki and Yamaguchi-Shinozaki, 1997), classify water stress-related genes regarding their physiological roles in stress tolerance:

- ***Stress adaptation effectors*** or *metabolic proteins* (see Fig. 3), such as water channel proteins and transmembrane ion-transport proteins (involved in the movement of water and ions through membranes); enzymes required for the biosynthesis of various osmoprotectants; proteins which may protect macromolecules and membranes, including LEA proteins (*Late embryogenesis abundant*, i.e., polypeptides that are synthesised late in embryogenesis -just prior to seed desiccation-, and in seedlings in response to dehydration stress), osmotin, chaperones, and proteases for protein turnover; and the detoxification enzymes such as glutathione S-transferases, catalases, superoxide dismutases (see below) and ascorbate peroxidases.
- ***Regulatory molecules*** or gene products with regulatory functions in the *stress signal transduction*, including various protein kinases and transcription factors, which regulate the stress adaptation effectors enumerated above.

Stress proteins in the actual theory of stress in plants

Current knowledge about stress-related proteins firmly supports the actual theories of stress in plants: *GAS* and *co-stress* (Larcher, 1995;

Lichthenthaler, 1996; Prasad, 1997). As Leshem et al., (1998) summarised, 'converging data indicate the possible existence of a *general adaptation syndrome* (GAS) in which different types of stress induce identical coping mechanisms. Consequently, this implies a *co-stress* response whereby one type of stress resistance may impart co-resistance to others.'

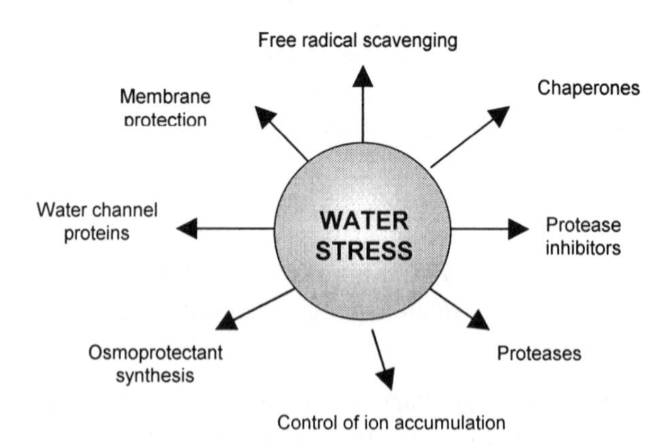

Figure 3. Some *stress adaptation effectors* induced by water stress. These stress-related proteins have been reported to play an adaptative role in plants subjected to drought, salinity, chilling, osmotic stress, or other stresses involving cellular dehydration. After Bray (1993).

It has been shown that all of the following groups of factors: ABA and/or jasmonic acid derivatives, free radical scavengers and antioxidants, osmoregulation, HSPs such as ubiquitin and chaperone complex (see below), and often ethylene, are significantly involved in coping mechanisms when plants are challenged by the following environmental stress categories: high light intensity, ozone and SO_2, heat, hypoxia and anoxia, flooding, chilling, freezing, drought and/or salinity, herbicide exposure, heavy metal toxicity, deficiency of essential elements, wounding, and pathogen infection (Leshem and Kuiper, 1996; Prasad, 1997; Heiser and Elstner, 1998; Tomashow, 1999). It appears that an *inherent multiple stress resistance mechanism* is developmentally advantageous, and may be pleiotropically encoded (i.e. controlled by few genes) by evolutionary selection (Leshem et al., 1998). Possibly, during evolution direct responses of plants to each environmental change were gradually replaced by *environmental signal perception/transduction pathways*, which enable plants to cope with stress in a 'cheaper' and more efficient way, thus increasing the plant fitness (Kuiper, 1998).

Stress signalling

The origin and development of signal perception/transduction pathways in green plants is a basic question to understand the functioning of plants in nature, where environmental conditions are continually changing (Kuiper, 1998). Recently it has been proved the existence of either positive or negative interactions among different stress factors in which concerns to gene expression (e.g., Xiong et al., 1999). Signalling is nowadays matter of active research in many plant biology laboratories.

Lichtenthaler (1998) summarises stress signalling as follows (see Fig. 4): 'All biotic and abiotic stressors, natural and anthropogenic, represent external signals. There are many different forms of signal perception and transduction in the plant and its organs (leaves, root, stem, flowers), which will lead (i) to direct metabolic responses, and (ii) to the ***activation of gene expression***: enzyme formation, synthesis of stress proteins, stress metabolites, stress hormones, etc. The latter further modify the plants' metabolic responses under stress; i.e., there are fluent transients and feedback controls between gene expression on the one hand and metabolic responses on the other hand'.

Little is known about *how plants recognise stresses* at the cellular level. The best clues come from yeast and bacterial proteins that initiate signal transduction in response to abiotic stresses such as low osmotic potential. Plants probably contain similar proteins, but comparable functions have not yet been demonstrated (Bray et al., 2000).

Yeast mutants are extremely useful tools for rapidly identifying *genes encoding proteins functioning in a particular pathway or with a specific biochemical activity*. However, the study of protein localisation and function in plants cannot rely solely on the results obtained with yeast DNA complementation, because 'plant cells are not just green yeast' (Bassham and Raikhel, 2000). After Bassham and Raikhel, information obtained with yeast needs to be confirmed as much as possible in the plant because, due to subcellular compartmentation, the localisation of a protein in yeast does not necessarily correspond with the localisation of that protein in its native plant species.

Moreover, the activity of the stress proteins in plants may be modulated by their interactions with other proteins that are absent in yeast, and the *expression of the genes may be regulated by the environment and/or the stage of development*, which may contribute to their specific functions (Bassham and Raikhel, 2000). There is already evidence of signalling cascades in plants that are not known to exist in the unicellular eukaryote (Hasegawa et al., 2000).

The use of proteomics will in next future throw much light on these questions.

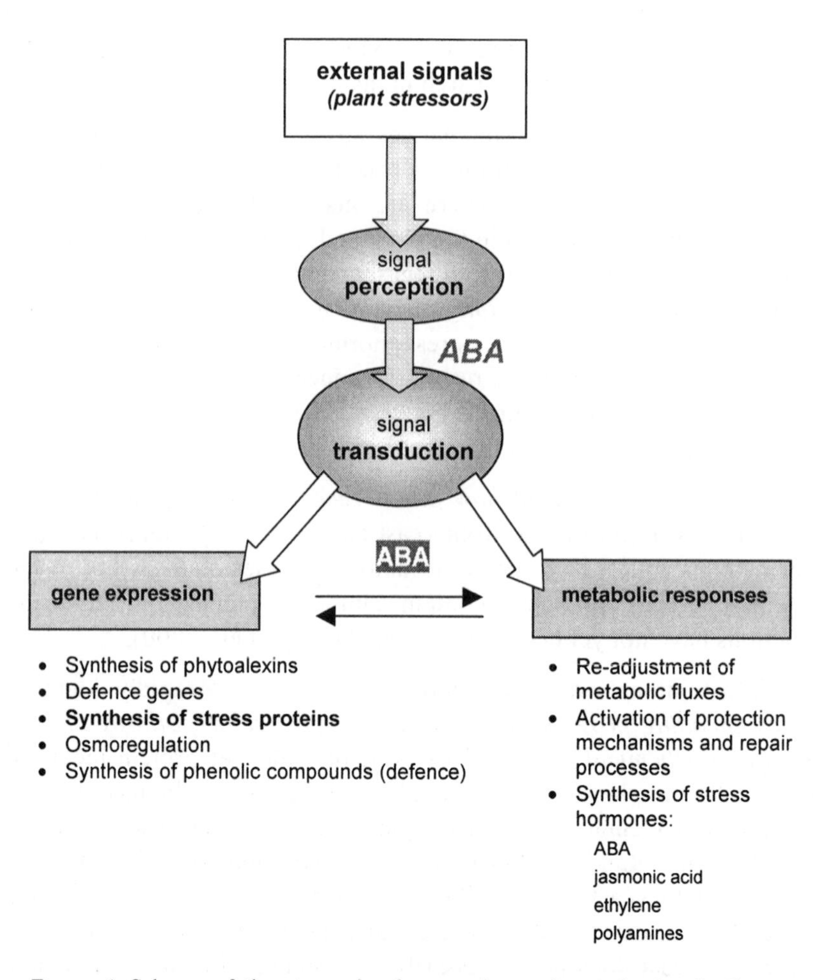

Figure 4. Scheme of the stress signal perception and transduction leading to metabolic responses and gene expression (usually mediated by abscisic acid – ABA-) as well as stress-induced plant responses. Drawn after Lichtenthaler (1998).

☆ ABA plays a crucial role in stress-induced gene expression

Considerable evidence indicates that the regulation of plant stress responses involves hormones and plant growth regulators, especially abscisic acid (ABA; see Chapter 26), jasmonic acid, ethylene, polyamines, and secondary messengers such as Ca^{2+} (Bray et al., 2000; Tiburcio et al., 2001).

The accumulation of ABA as an early response to stress, leading to stomatal closure, is a general feature in plants (Hsiao 1973); in fact, many (but not all) of the genes induced by different stress factors are also induced by exogenous application of ABA (Shinozaki and Yamaguchi-Shinozaki, 1997). Abscisic acid (ABA) plays undoubtedly a central role in stress manifestations by *converting multiple environmental stress signals in gene expression* (Bray, 1993; Chandler and Roberston, 1994; Leung and Giraudat, 1998). But, how does it work?

Abscisic acid transmits the environmental adverse stimuli from the contact zone (e.g., roots in a dry soil or leaves attacked by pathogens) to the whole plant, and then acclimation mechanisms are induced (see Fig. 1) Intermediate points of signal transduction mediated by ABA (reversible protein phosphorylation, modification of cytosolic pH and calcium levels) besides cascade signals at the cellular level are relatively well known (see Leung and Giraudat, 1998). But the nature of ABA receptor/receptors has not yet been elucidated. The most recent works evidence an initial signal consisting on a rise in the levels of cytoplasmic Ca^{2+}, which initiates the liberation of an ABA precursor. A rise of pH due to the decompensation of K^+-H^+ fluxes induces ABA liberation to xylem stream, which promotes stomatal closure and the initiation of reparation processes, including stress-proteins (see Netting, 2000).

The gene expression associated with stresses involving cellular dehydration (drought, salinity, osmotic stress, cold, freezing, etc.) are primarily controlled at the transcriptional level (Ingram and Bartels, 1996, Shinozaki and Yamaguchi-Shinozaki, 1997; Zhu, 2000). The up-regulation of water stress-responsive genes involves a concomitant increase in the levels of the corresponding transcripts. It has been further noted that most of the water stress-inducible genes are also upregulated by exogenous application of ABA (Shinozaki and Yamaguchi-Shinozaki, 1997). To explain this, it was proposed that water stress triggers the production of ABA, which in turn induces expression of various water stress-induced genes.

Involvement of ABA in plant co-stress manifestations (Fig. 5) is nowadays a priority area of research (Prasad and Rengel, 1998).

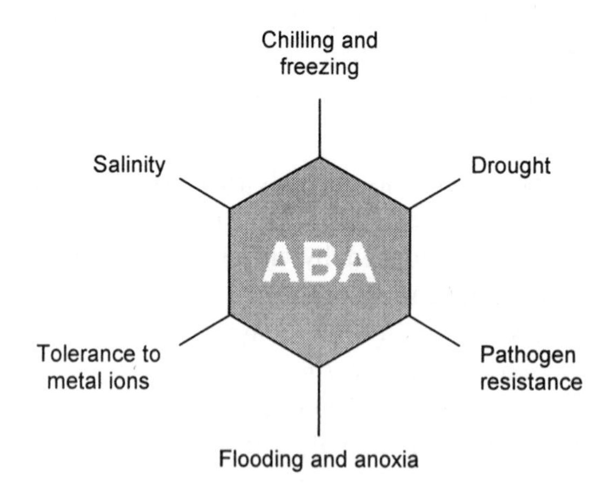

Figure 5. ABA plays a key role in many co-stress manifestations, by converting different stressful environmental signals to gene expression. After Prasad and Rengel (1998).

However, as several water stress-responsive genes are not triggered in response to ABA, existence of both *ABA-dependent and ABA-independent signal transduction pathways* for expression of specific genes has been implied (Bray, 1997; Shinozaki and Yamaguchi-Shinozaki, 1997; Bray et al., 2000).

It have been shown genes that are specifically induced by water deficit or salinity, but no by ABA neither cold nor heat (e.g., Iuchi et al., 1996, Joshee et al., 1998). Increased levels of both *ozone* and *UV-B* radiation, the typical combination of stresses for high-altitude sites, showed an obvious interaction at the level of gene expression on Scots pine seedlings, by affecting the mRNA transcript levels of cinnamyl alcohol dehydrogenase and stilbene synthase, both enzymes involved in the synthesis of protective phenolic compounds (Zinser et al., 2000). Thus, despite the evidences of 'co-stress' at the level of gene expression, specially for those agents causing cellular dehydration, other recent experiments indicate that *plants can sense the differences between different primary potential causes of water stress*. Water stress triggers the co-ordinated induction of mRNAs involved in different aspects of the adaptative stress response of the plants. In this sense, varied stressors, like drought or salinity, cause different transcription levels of several genes; thus, *the molecular stress response of plants must be triggered by multiple signals* (e.g., Tabaei-Aghdaei et al., 2000; Jin et al., 2000).

Moreover, synthesis pathways that are induced by ABA under lab conditions (where studied until now) probably are not successfully induced in field conditions (Bray, 1993; Thompson et al., 1997) or under multiple stress (e.g., Xiong et al., 1999).

Some overall roles of stress proteins in acclimation to stress

Osmoregulation and osmoprotection

Osmoregulation through the synthesis of low molecular weight organic compounds in the cytoplasm is a major manifestation of co-stress. Different signals having cellular dehydration as common feature (chilling, drought, salinity, osmotic stress) induce the synthesis of compatible solutes such as betaines and related compounds (Rhodes and Hanson, 1993; McNeil et al., 1999), polyols and sugars (Stoop et al., 1996) and aminoacids like proline (see Chapter 22). Besides their osmoregulatory function, they have osmoprotective roles such as protein and membrane stabilisation, and free radical scavenging. Many authors have reported the overexpression of several genes that codify enzymes of the biosynthetic routes of different osmolytes, besides other proteins involved in transport mechanisms that occur during osmotic adjustment (Bray, 1993; Taylor, 1996; McNeil et al., 1999; and see Chapter 22). The knowledge of these genes has opened new doors to genetic engineering, regarding osmolytes synthesis could be installed in crop species with low osmoregulation capacity, thus making them tolerant to drought, salinity and other stresses (see Grover, 2000; Sakamoto and Murata, 2000).

Some stress-proteins prevent oxidative damage

Another common factor of co-stress manifestations is the *prevention of oxidative damage*. The *oxidative stress* results from deleterious effects of reduced oxygen species such as superoxide and hydrogen peroxide. These oxygen species can form hydroxyl radicals (OH·). Hydroxyl radicals cause lipid peroxidation, protein denaturation, DNA mutation, photosynthesis inhibition, etc. Aerobic organisms have evolved a defence mechanism against oxidative stress produced under physiological conditions (e.g., during photosynthetic electron transport, β-oxidation of fatty acids, etc.), based on a rapid elimination of superoxide and hydrogen peroxide, before they produce free radicals. The central role in the plant antioxidative mechanism is played by *superoxide dismutase* (SOD), an enzyme that converts superoxide into hydrogen peroxide, which is then broken down by either *catalase* (which efficiently scavenges H_2O_2 to release water and dioxygen in peroxisomes and mitochondria) or various *peroxidases* (ascorbate peroxidase, glutathione reductase, guaiacol peroxidase, and dehydroascorbate reductase) (see Allen, 1995; Sen Raychaudhuri and Wang Deng, 2000).

Free radical scavenging is associated with every type of plant stress mentioned here (including allelochemicals), apparently without exception, and this indicates a clear-cut general adaptation function of this mode of stress coping (Leshem et al., 1998; Prasad and Rengel, 1998). Therefore, *expression of stress-responsive genes* results altered: genes encoding catalase have been shown to be induced by chilling, UV-B, ozone, and SO_2. SOD has been reported to have greater activities in plants showing resistance to salt stress, waterlogging, ozone stress, and chilling. SOD genes are differentially regulated and respond to a variety of stress conditions, such as herbicide application, flooding, drought, and chilling. The efficiency of SOD is stress tolerance is enhanced due to the *compartmentation of different isozymes* (Sen Raychaudhuri and Wang Deng, 2000).

Heat shock proteins (HSPs)

A sudden elevation in temperature triggers a stress response, found in all organisms, that brings about a global transition in gene expression. Typically, the expression of most genes is either shut down or greatly attenuated, and a specific group of genes, called heat shock (HS) genes, is rapidly induced to high levels (Schlesinger *et al.*, 1982). Some of the genes coding for heat-shock proteins in plants are homologous with those from animals; in fact, heat shock proteins were first discovered in *Drosophyla*.

Proteins encoded by HS genes enable plant cell survive the harmful effects of heat *by two general strategies* (Lee and Vierling, 2000): (1) one group of heat shock proteins acts as molecular *chaperones* that counteract protein denaturation and aggregation, and (2) other, including *ubiquitin* and certain *proteases*, target non-native proteins for degradation. The HS response is transient in nature, usually peaking 1 to 2 h. after onset, providing protection from acute episodes of thermal stress. Other new factors distinct to HSPs, some *chloroplast-encoded proteins* that regulate the synthesis of some nucleus-encoded proteins, have been shown to enhance the thermal stability of the oxygen-evolving machinery (Tanaka *et al.* 2000).

Heat shock proteins are commonly designated by the abbreviation HSP followed by the molecular weight in kDa. Most major *classes* of known HSPs are present in plants and include the small HSPs (ranging in molecular weigh from 15 to 28 kD), HSP60, HSP70, HSP90 and HSP100 (see Vierling 1991 as a review). As the case in other organisms, refolding of non-native proteins in plants is thought to occur in *HSP complexes* ('refolding machines') (see Heldt, 1997). HSPs facilitate growth and survival of plants not only over the course of transient extremes of temperature but also under conditions of *severe heat stress* whereby lethal temperature can be tolerated for short periods. Protection from severe heat stress usually requires a *preconditioning* by prior exposure to moderate HS conditions. This phenomenon, known as *induced* or *acquired thermotolerance,* has long attracted the attention of researchers with the prospect that plants may be engineered to better tolerate heat stress (see Gurley, 2000)

Although thermotolerance can be improved by manipulating HS factors, the approach may potentially lead to unwanted consequences because many of the HSPs induced may have roles in normal growth and development. In fact, HSPs are produced either constitutively (at the whole cell level, or in the mitochondria and the chloroplast) or under cell cycle or developmental control in some cells (Vierling, 1991). Recently, HSP101 was identified in *Arabidopsis* mutants as an essential HS factor to survive heat stress which can be over- and underexpressed without affecting normal growth and development (Queitsch *et al.*, 2000). Moreover, the constitutive expression of HSP101 can nearly eliminate the need to precondition plants to survive heat stress, thus having important implications for efforts to improve stress tolerance in plants (Gurley, 2000).

Under the point of view of co-stress, *most, if not all, environmental stress-coping factors involve HSPs*. Apart from heat shock, HSPs are also synthesised and accumulated in plants in response to arsenine, ethanol, heavy metals, UV-B radiation, water stress, excess NaCl, chilling, abscisic acid, wounding, anoxic conditions, and pathogen infection (Singla *et al*., 1997; Király, 1998). Especially important is the *ubiquitin function* since in the final outcome all stresses cause disfunction of proteins; thus, an *overall GAS role* may well be assigned to HSPs. The mode of HSP-mediated stress coping is pleiotropic, and involves mRNA protection, prevention of enzyme-especially photosynthesising-denaturation and/or their stress-induced aggregation and post-stress ubiquitin and chaperonin-aided repair (Leshem *et al*., 1998).

Other stress proteins, such as antifreezing proteins, some transmembrane transport proteins, jasmonate specific proteins, and pathogenesis related proteins (i.e., osmotin) have more restricted functions, improving plant resistance to freezing, salt stress, disease, and pathogen infection, respectively (see Prasad and Rengel, 1998; Thomasow, 1999; Bray et al., 2000; as reviews).

Further considerations

Pareek *et al*. (1997) stated that for *relating gene expression to stress tolerance*, parameters such as (i) analysis of the contrasting types, (ii) mutants, (iii) ABA-inducibility and (iv) analysis in relation to natural or induced tolerance at the level of whole plant, can prove useful. The last decade has seen tremendous successes of genetic analysis using the *Arabidopsis* model system (see Ausubel, 2000), which allowed shed light on until now unknown tolerance mechanisms operating in plants.

Mutational analysis is specially suited for making inroads to study complex systems (i.e., salt tolerance) because each component can be specifically mutated to reveal its effect on the entire system (see Grover, 2000; Hasegawa *et al*., 2000; Zhu, 2000).

Let us see an 'eloquent' *example of the complete functional characterisation of an stress-related protein* by combining mass spectrometry, two-dimensional gel electrophoresis, and mutational analysis (Li and Assmann, 2000):

> Li and Assmann identified a 48-kD ABA-activated and Ca^{2+}-independent protein kinase (AAPK) from bean, a low abundance protein that can only be detected in guard cells. AAPK is activated upon treatment of intact guard cells with ABA, but is not activated *in vitro*. Due to the very low amount of protein starting material from a

limited quantity of purified guard cells, purification of AAPK requires the use of 2-D gel electrophoresis (see Li and Assmann, 2000 for specific separation protocol).

Once separated, proteins were subjected to an in-gel autophosphorylation assay. A 48-kD Ca^{2+}-independent ^{32}P-labeled spot was detected from guard cells treated with ABA, but not from control guard cells lacking the ABA treatment, thus demonstrating that the spot corresponded to the target protein AAPK. Furthermore, silver staining of the 2-D gels showed that the AAPK spot was separate from other protein spots.

About 1 pmol of AAPK excised from 2-D gels (six units) was digested with trypsine (a residue-specific protease) and subjected to tandem mass spectrometry for amino acid sequence analysis. Two peptide sequences were obtained, both of which highly conserved in a previously described family of protein kinases whose transcription is induced by ABA (PKABA1).

The AAPK peptide sequence information obtained by mass spectrometry allowed to clone the *AAPK* cDNA (through *reverse transcription*). Expression in guard cells of GFP-tagged AAPK (through *insertional mutagenesis* of the sequence tag to identify) ultimately allowed to determine the function of this kinase: it mediates ABA-induced stomatal closure via activation of guard cell anion channels.

The advent of transgenic technology has enabled scientist to combine genetics with Plant Physiology (Grover, 2000). Now, in the proteome era, **we have the chance to combine proteomics with Plant Ecophysiology!** Let's begin trying 2D-Electrophoresis...

THE 2D-ELECTROPHORESIS PROCEDURE

Within all of the possibilities previously mentioned for the development of the 2D-electrophoresis, the procedure described below (see Fig. 6) was selected to be one of the most sensitive and reproducibility methods. It avoids losses of proteins by handling and reduces the costs compared to other techniques.

Plant proteome extraction (Görg et al., 1988)

An amount of 3-4 g fresh weight of one-month-old lettuce leaves are crushed with liquid nitrogen in a cooled mortar (in ice), and the powder is resuspended in approximately 6 mL buffer of 10 % TCA in acetone with 0.07 % β-mercaptoethanol (v/v). Proteins are allowed to precipitate for 45 min at –18 °C (Damerval et al., 1986).

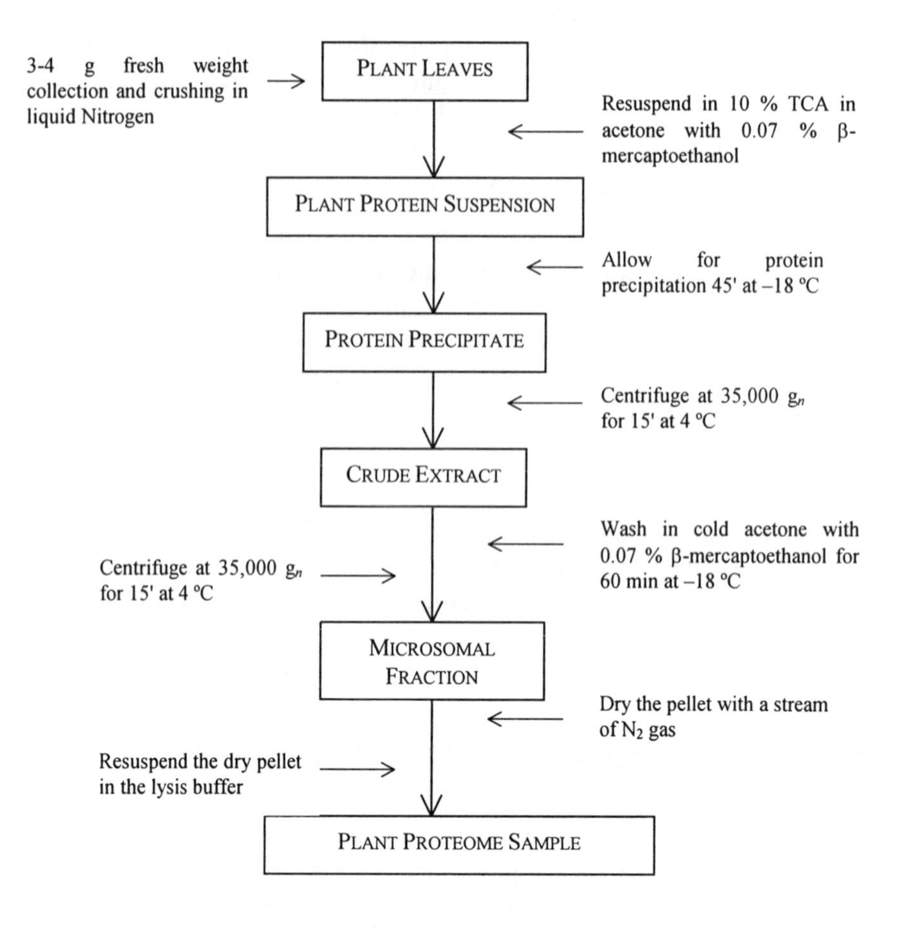

Figure 6. Scheme of the plant proteome extraction procedure

After a centrifugation at 35,000 g_n for 15 min at 4 °C, the supernatant is discarded and the pellet is washed with cold acetone containing 0.07 % β-mercaptoethanol for 60 min at –18 °C.

After a second centrifugation at 35,000 g_n for 15 min at 4 °C, the supernatant is discarded again and the pellet is dried under a stream of Nitrogen gas. Then, 25 mg of the dry pellet are resuspended in the lysis buffer (Rabilloud, 1998) in a proportion of 50 μL buffer per mg of pellet.

The mixture is centrifuged at 42,000 g_n for 10 min at 15 °C, for removing nucleic acids from the sample. The supernatant can be stored at –80 °C for later analyses.

Total protein determination is conducted according to Bradford's method (1976) (see Chapter 19).

Horizontal first dimension

Preparing first dimension

To prepare the first dimension, one of the first important steps is the rehydration of the IPG gel strips. Here we have selected the IPG gel strips of 18 cm large and with a pH range of 3-10 Linear (Amersham Pharmacia Biotech), because our objective is to see the total proteome of the sample.

For each IPG gel strip of this size, 350 μL rehydration buffer (Herbert, 1999) are added in the medium of the reservoir slot of the Reswelling Tray (see Fig. 8a).

Once that the IPG gel strips are placed in the buffer, 2 mL oil are added to avoid urea crystallisation. IPG gel strips are allowed for rehydration at room temperature a minimum of 10 h.

IEF is performed with the Multiphor II connected to the EPS 3501 XL Power Supply from Amersham Pharmacia Biotech (see Fig. 7). For cooling at 20 °C the MultiTemp III Thermostatic Circulator (Amersham Pharmacia Biotech) is used.

Running the first dimension (IEF) (Berkelman and Stenstedt, 1998)

After washing with water the IPG gel strips, and removing the excess of water with a filter paper, the strips are placed in the cooling plate of

Multiphor II, according with the position of the electrodes, and all strips are aligned in it (see Fig. 8b).

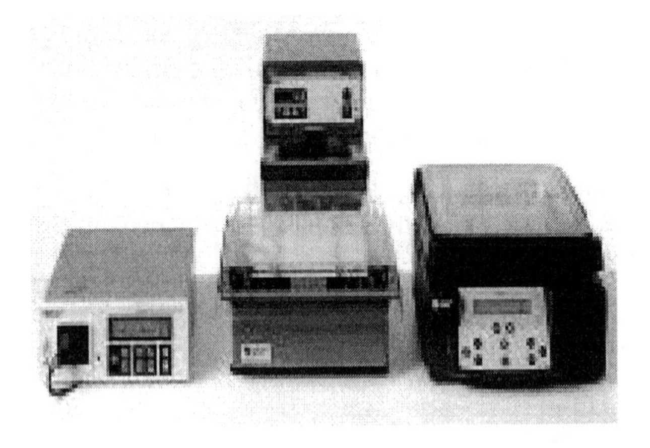

Figure 7. From right to left are presented the EPS 3501 XL Power Supply, the Multiphor II and the MultiTemp III Thermostatic Circulator. (By courtesy of Amersham Pharmacia Biotech)

Once positioned the electrodes (see Fig. 8c), the sample cup is placed over the IPG gel strips and the sample is added (see Fig 5d). When strips with a pH range of 3-10 are used, the sample must be added near the anode, and if the strips are of 18 cm large 15-30 μg proteins (20-50 μL) are needed.

Electric conditions for IEF are set in the power supply. For IPG Strips of 18 cm and pH: 3-10L (Berkelman and Stenstedt, 1998; Görg et al., 2000):

Temperature: 20 °C
Current max.: 0.05 mA per strip
Voltage max.: 3500 V
Power max.: 0.2 W per strip

Voltage	Time	Voltage/time	Total
500 V	1 min	1 V/h	1 V/h total
500 V	2 h	1000 V/h	1000 V/h total
3500 V	1:30 h	3000 V/h	4000 V/h total
3500 V	6 h	21000 V/h	25000 V/h total

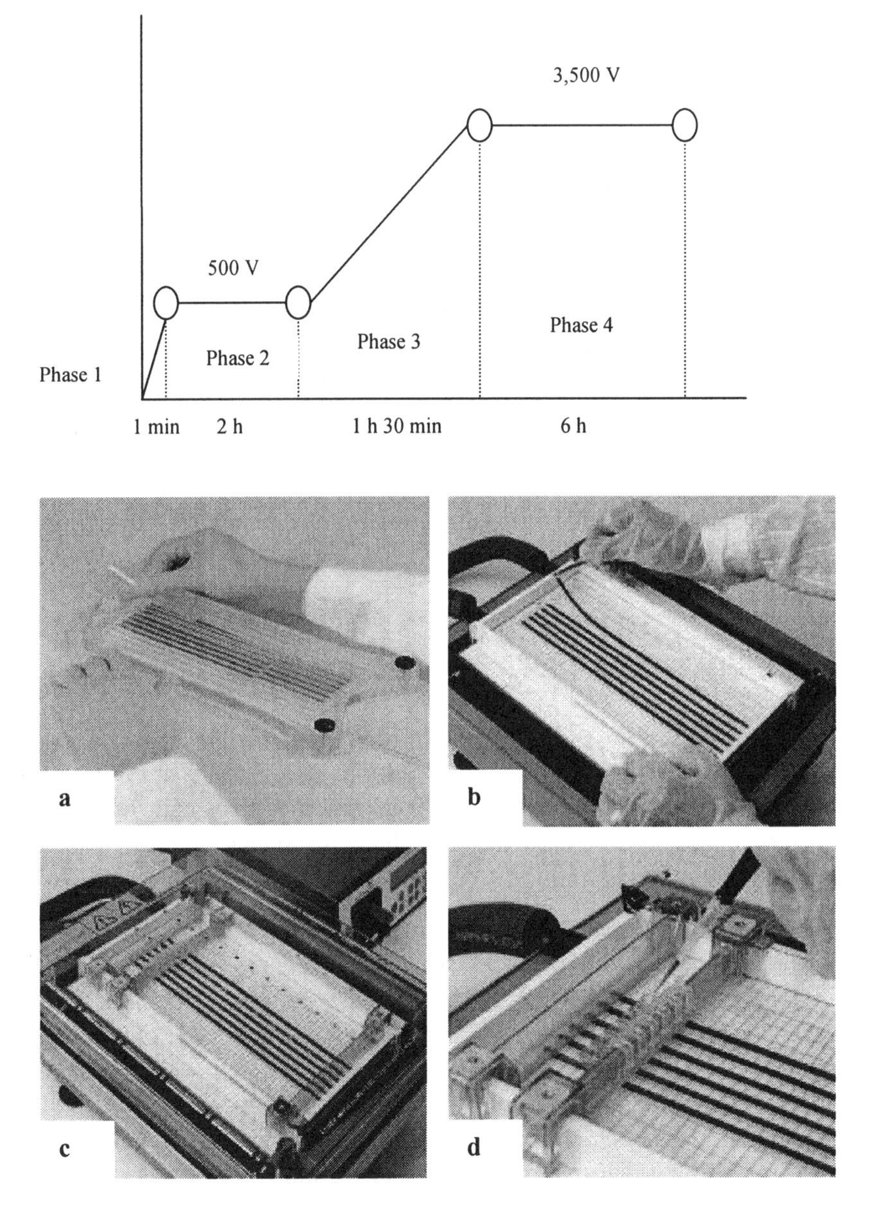

Figure 8. **a-** after IPG gel strip rehydration, the strip is removed from the Reswelling Tray and washed with a little water
b- The IPG gel strips are parallel placed in the cooling plate of the Multiphor II, that is the first dimension system used
c- The electrodes are carefully positioned over the IPG gel strips
d- The sample is applied by mean a sample cup and the separation can start
(By courtesy of Amersham Pharmacia Biotech)

Once finished IEF, the IPG strips can be stored in plastic films at –80 °C.

Transfer to second dimension

To transfer the IPG gel strips to the second dimension of the 2D-electrophoresis, it is necessary to make two equilibration steps. For making the first equilibration, 10 mL equilibration buffer with 100 mg DTT are added to each IPG strip in a tube from 25 mL allowing to equilibrate for 15 min. After this time, the first buffer is discarded. Then, a second equilibration is made for 15 min in the same tubes with 400 mg iodoacetamide in 10 mL equilibration buffer. Iodoacetamide is used to remove the DTT residual in the sample and to avoid streaking on silver staining (Görg et al., 1987).

Horizontal second dimension

SDS-PAGE is run horizontally also in the Multiphor II. The temperature is set in this case at 15 °C. For Multiphor II, Amersham Pharmacia Biotech has available prepared gels from different sizes and for different polyacrylamide matrix. So, we only have to select our gel and it is directly prepared for running.

Once that the gel is orientated in the plate and the dry buffers are positioned over it, the IPG gel strip is placed on the SDS gel. The cathodic buffer (-) and the IPG strip must be placed parallel with a distance of 2-3 mm. If we are going to run more than one IPG strip in the same SDS gel, the IPG strips must be placed with a distance of 1 cm. The molecular weight marker proteins are applied also at this time (for silver staining it is necessary an approximate amount of 10-50 ng for each component). Now, the current values are established. For gels with a gradient of 8-18 %:

Step	V	mA	W	h:min
1	600	20	30	0:30
2	600	50	30	1:10

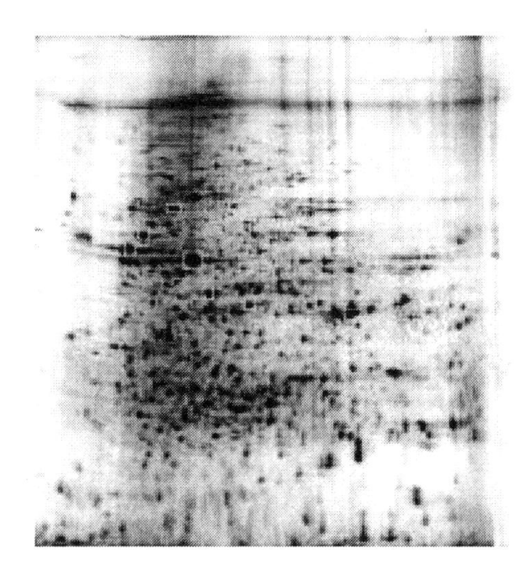

Figure 9. Proteins stained by silver method. We can identify several spots in an only gel. Every spot represents every protein.

Silver staining

The Hoefer Automated Gel Stainer (Amersham Pharmacia Biotech Europe GmbH) is used for silver staining. The protocols 2 and 3 are prepared for staining proteins by silver method (see Fig. 9).

But if we want to analyse this gel later by mass chromatography, it is necessary to make some little changes because some products, for example glutardialdehyde, can alter the identification of our proteins by this technique.

When the silver staining is finished the gel is dried and stored in a plastic film for the next visualisation and interpretation of the results.

Image analysis

The ImageMaster 2D Elite Software and 2D Database Software together with the Sharp JX-330 Scanner from Amersham Pharmacia Biotech, are used for the evaluation of the spots appeared after the silver staining. This equipment allows capturing the spots, correcting them as much as possible, quantifying the stops and comparing them with the other protein spots appeared in this gel or in others.

PREPARATION OF BUFFERS

Lysis buffer

	Final Concentration	*Amount*
Urea[1]	5 M	12 g
Thiourea	2 M	6.1 g
CHAPS	4 %	1,6 g
SB 3-10	2 %	0,8g
Spermine	20 mM	0,16 g
IPG buffer[2]	1 %	0,4 mL
TCEP-HCl	1.5 mM	
Bromophenol blue	traces	≅4 mg
Distilled water		to 40 mL

Rehydration buffer

	Final Concentration	*Amount*
Urea[1]	5 M	7.51 g
Thiourea	2 M	3.81 g
CHAPS	2 %	0.5 g
SB 3-10	2 %	0.5 g
Spermine	20 mM	0.1 g
IPG buffer[2] (50 μL/2.5 mL)	1-2 %	0.5 mL
TCEP-HCl	1.5 mM	
Bromophenol Blue	traces	≅2 mg
Distilled water		to 25 mL

[1] Heat the solution for a best urea dilution, but never over 30 °C!
[2] Use 1 % IPG buffer when horizontal gels are used for 2nd dimension and 2 % when vertical gels are used

Functions of the compounds into the Lysis and Rehydration buffers:

UREA and THIOUREA: are the two-selected chaotropic agents. Both together denature proteins exposing their hydrophobic residues. But thiourea increases the solubility of proteins and in particular of integral membrane preparations
CHAPS (3-[(3-cholamidopropyl) dimethylammonio]-1-propanesulfonate): is a non-ionic surfactant for the solubilisation of the hydrophobic residues exposed after the denaturation by the chaotropics. It removes also lipids from the sample.
SB 3-10 (*N*-decyl-*N*, *N*-dimethyl-3-ammonio-1-propane sulfonate): another sulfobetaine more efficient than CHAPS but also with less solubility.
SPERMINE: allows the precipitation of polyanions (DNA, polysaccharides, etc.)
IPG BUFFER (Immobilised pH Gradient): avoids protein carbamilation. Gives more conductivity to the sample and allows nucleic acids for precipitation.
TCEP-HCl (Tris-2-carboxyethyl-phosphine hydrochloride): is an uncharged reducing agent, to ensure reducing conditions during IEF.
BROMOPHENOL BLUE: the dye used to monitor the evolution of electrophoresis.

Equilibration buffer

	Final Concentration	*Amount*
1.5 M Tris-Cl, pH 8.8	50 mM	6.7 mL
Urea	6 M	72.07 g
Glycerol (87 % v/v)	30 % (v/v)	69 mL
SDS	2 % (w/v)	4 g
Bromophenol Blue	traces	few grains
Distilled Water		to 200 mL

Tris-Cl solution:

Tris base	27.23 g
Distilled Water	100 mL
HCl	adjust to pH 8.8
Distilled Water	to 150 mL

Filter the solution with a 0.45 μm filter and store at 4 °C

Functions, into the Equilibration buffer, of the selected compounds:

TRIS-CL: equilibrates pH for the 2nd dimension.

UREA and GLYCEROL: both together prevent electroendosmosis and improving the transference of proteins to the 2nd dimension gel.
Sodium Dodecyl Sulfate (SDS): denatures proteins and creates negatively charged complexes for the 2nd dimension.
DTT (Dithiotreitol): preserves the total reduced state.
IODOACETAMIDE: alkylates the thiol groups of proteins and preserves their reoxidation removing the all residual DTT.

REFERENCES

Abbot A. A post-genomic challenge: learning to read patterns of protein synthesis. Nature, 1999; 402:715-720

Abromeit M., Askman P.A., Sarnighausen E., Dörffling K. Accumulation of high-molecular-weigh proteins in response to cold hardening and abscisic acid treatment in two winter wheat varieties with different frost tolerance. J Plant Physiol, 1992; 140:617-622

Allen R.D. Dissection of oxidative stress tolerance using transgenic plants. Plant Physiol, 1995; 107:1049-1054

Ames G.F.L., Nikaido K. Two-dimensional gel electrophoresis of membrane proteins. Biochemistry 1976; 15:616-623

Ausubel F.M. Arabidopsis Genome. A Milestone in Plant Biology. Plant Physiol, 2000; 124:1451-1454

Bassham D.C., Raikhel N.V. Plant cells are not just green yeast. Plant Physiol, 2000; 122:999-1001

Berkelman T., Stenstedt T. 2-D Electrophoresis. Using Immobilized pH Gradients: Principles and Methods. Amersham Pharmacia Biotech, USA, 1998.

Bjellqvist B., Ek K., Righetti P.G., Gianazza E., Görg A., Westermeier R., Postel W. Isoelectric focusing in immobilized pH gradients: Principle, methodology, and some applications. J Biochem Biophys Methods 1982; 6:317-339

Bradford M.M. A rapid and sensitive method for the quantification of microgram quantities of protein utilizing the principle of protein-dye binding. Anal Biochem 1976; 72:248-254

Bray E, Bailey-Serres J, Weretilnyk E. Responses to abiotic stress. In Buchanan BB, Guissem W, Jones RL, eds. Biochemistry and Molecular Biology of Plants, 1158-1203. ASPP, Rockville, Maryland, USA, 2000

Bray E.A. Molecular responses to water deficit. Plant Phisiol 1993; 103:1035-1040

Bray E.A. Plant responses to water deficit. Trends in Plant Science 1997; 2:48-54

Chandler P.M., Robertson M. Gene expression regulated by abscisic acid and its relation to stress tolerance. Annual Review of Plant Physiol and Plant Molecular Biology 1994; 45:113-141

Chang W.W.P., Huang L., Shen M., Webster C., Burlingame A.L., Roberts K.M. Patterns of protein synthesis and tolerance of anoxia in root tips of maize seedlings acclimated to a low-oxygen environment, and identification of proteins by mass spectrometry. Plant Physiol 2000; 122:295-317

Damerval C., de Vienne D., Zivy M., Thiellement H. Technical improvements in two-dimensional electrophoresis increase the level of genetic variation detected in wheat-seedling proteins. Electrophoresis 1986; 7:52-54

Dignam J.D. Preparation of extracts from higher eukaryotes. Methods Enzymol 1990; 182:194-203

Flengsrud R., Kobro G. A method for two-dimensional electrophoresis of proteins from green plant tissues. Anal Biochem 1989; 177:33-36

Geigenheimer P. Preparation of extracts from plants. Methods Enzymol 1990; 182:174-193

Görg A., Obermeier C., Boguth G., Harder A., Scheibe B., Wildgruber R., Weiss W. The current state of two-dimensional electrophoresis with immobilized pH gradients. Electrophoresis 2000; 21:1037-1053

Görg A., Postel W., Domscheit A., Günther S. Two-dimensional electrophoresis with immobilized pH gradients of leaf proteins from barley (Hordeum vulgare): Method, reproducibility and genetic aspects. Electrophoresis 1988b; 9:681-692

Görg A., Postel W., Günther S. The current state of two-dimensional electrophoresis with immobilized pH gradients. Electrophoresis 1988a; 9:531-546

Görg A., Postel W., Weser J., Günther S., Strahler J.R., Hanash S.M., Somerlot L. Elimination of point streaking on silver stained two-dimensional gels by addition of iodoacetamide to equilibration buffer. Electrophoresis 1987; 8:122-124

Groverr A. Water stress responsive proteins/genes in crop plants. In Yunus M., Pathre U., Mohanty P. eds. Probing photosynthesis. Mechanisms, regulation and adaptation, pp398-409. Taylor and Francis, London, UK, 2000

Gurley W.M. HSP101: A key component for the acquisition of thermotolerance in plants. Plant Cell 2000; 12:457-460

Gygi S.P., Rochon Y., Franza B.R., Aebersold R. Correlation between protein and mRNA abundance in yeast. Mol Cell Biol 1999; 19:1720-1730

Halloway P., Arundel P. High-resolution two-dimensional electrophoresis of plant proteins. Anal Biochem 1988; 172:8-15

Hasegawa P.M., Bressan R.A., Zhu J-K, Bohnert H.J. Plant cellular and molecular responses to high salinity. Annual Review of Plant Physiol and Plant Molecular Biology 2000; 51: 463-499

Heiser I., Elstner E.F. The biochemistry of plant stress and disease. Oxigen activation as a basic principle. In Csermely P. ed. Stress of Life: From Molecules to Man. Annals of The New York Academy of Sciences. Vol. 851, pp224-232. The New York Academy of Sciences. New York, New York, EE.UU., 1998

Heldt H.W. Plant biochemistry and molecular biology 1st Ed. Oxford University Press, Oxford, UK. 522 pp, 1997

Herbert B. Advances in protein solubilisation for two-dimensional electrophoresis. Electrophoresis 1999; 20:660-663

Ho T-HD, Sachs M.M. Environmental control of gene expression and stress proteins in plants. In Jones H.G., Flowers T.J., Jones M.B. eds. Plants under Stress. Biochemistry, Physiology and Ecology and Their Application to Plant Improvement. Cambridge University Press, Cambridge, UK, 1989

Hsiao T.C. Plant responses to water stress. Annual Review of Plant Physiol 1973; 24:519-570

Ingram J., Bartels D. The molecular basis of dehidration tolerance in plants. Annual Review of Plant Physiol and Plant Molecular Biology 1996; 47: 377-403

Iuchi S., Yamaguchi-Shinozaki K., Urao T., Shinozaki K. Novel drought-inducible genes in highly drought-tolerant cowpea: Cloning of cDNAs and analysis of the expression of corresponding genes. Plant and Cell Physiology 1996; R37:1073-1082

Jin S., Chen C.C.S., Plant A.L. Regulation by ABA of osmotic-stress-induced changes in protein synthesis in tomato roots. Plant Cell and Environment 2000; 23: 51-60

Joshee N., Kisaka H., Kitagawa Y. Isolation and characterization of water stress-specific genomic gene, pswi 18, from rice. Plant and Cell Physiology 1998; 39:64-72

Király Z. Plant Infection-Biotic Stress. In Csermely P., ed. Stress of Life: From Molecules to Man. Annals of The New York Academy of Sciences. Vol. 851, pp233-240. The New York Academy of Sciences. New York, New York, EE.UU., 1998

Klose J. Protein mapping by combined isoelectric focusing and electrophoresis of mouse tissues. A novel approach to testing for induced point mutation in mammals. Humangenetik 1975; 26:231-243

Kuiper P.J.C. Adaptation mechanisms of green plants to environmental stress. The role of plant sterols and the phosphatidyl linolenoyl cascade in the functioning of plants and the response of plants to global climate change. In Csermely P. ed. Stress of Life: From Molecules to Man. Annals of The New York Academy of Sciences. Vol 851, pp209-215. The New York Academy of Sciences. New York, Ney York, EE.UU., 1998

Larcher W. Physiological Plant Ecology. III ed. Springer- Verlag, Berlin, Alemania. 506 pp, 1995

Lee G.J., Vierling E. A small heat shoch protein cooperates with heat shock protein 70 systems to reactivate a heat-denatured protein. Plant Physiol 2000; 122:189-198

Leshem Y.Y., Kuiper P.J.C. Is there a GAS (general adaptation syndrome) response to various types of environmental stress? Biologia Plantarum 1996; 38:1-18

Leshem Y.Y., Kuiper P.J.C., Erdei L., Lurie S., Perl-Treves R. Do Selye's mammalian 'GAS' concept and 'co-stress' response exist in plants? In Csermely P. ed. Stress of Life: From Molecules to Man. Annals of The New York Academy of Sciences. Vol 851, pp199-208. The New York Academy of Sciences. New York, New York, EE.UU., 1998

Leung J., Giraudat J. Abscisic acid signal transduction. Annual Review of Plant Physiol and Plant Molecular Biology 1998; 49:199-222

Li J., Assmann S.M. Mass spectrometry. An essential tool in proteome analysis. Plant Physiol 2000; 123: 807-809

Lichtenthaler H.K. The stress concept in plants: an introduction. In Csermely P. ed. Stress of Life: From Molecules to Man. Annals of The New York Academy of Sciences. Vol 851, pp187-198. The New York Academy of Sciences. New York, New York, EE.UU., 1998

Lichthenthaler H.K. Vegetation stress: An introduction to the stress concept in plants. Journal of Plan Physiol 1996; 148:4-14

McNeil S.D., Nuccio M.L., Hanson A.D. Betaines and related osmoprotectants. Targets for metabolic engineering of stress resistance. Plant Physiol 1999; 120:945-949

Netting A.G. pH, abscisic acid and the integration of metabolism in plants under stresses and nonstresses conditions: cellular responses to stress and their implications for plant water relations. Journal of Experimental Botany 2000; 51:147-158

O'Farrell P.H. High resolution two-dimensional electrophoresis of proteins. J Biol Chem 1975; 250:4007-4021

Pareek A., Singla S.L., Grover A. Salt responsive genes/proteins in crop plants. In Jaiswal P.K., Singh R.B., Gulati A. eds. Strategies for improbing salt tolerance in higher plants, 365-391. Oxford and IBH, New Delhi, India, 1997

Peltier J.B., Frison G., Kalume D.E., Roepstorff P., Nilsson F., Adamska I., van Wijk K.J. Proteomics of the chloroplast: Systematic identification and targeting analysis of lumenal and peripheral thylakoid proteins. Plant Cell 2000; 12:319-341

PharmaBiotech. Proteomics – Technologies and commercial opportunities. 3rd Ed. 280 pp, 2001

Plant Physiology 124 (4) Special Issue: Arabidopsis genome: A Milestone in Plant Biology

Portig I., Pankuweit S., Lottspeich F., Maisch B. Identification of stress proteins in endothelial cells. Electrophoresis 1996; 17:803-808

Prasad M.N.V., Rengel Z. Plant acclimation and adaptation to natural and antropogenic stress. In Csermely P. ed. Stress of Life: From Molecules to Man. Annals of The New York Academy of Sciences. Vol 851, pp216-223. The New York Academy of Sciences. New York, New York, EE.UU., 1998

Prasad M.N.V. Ed. Plant Ecophysiology. John Wiley and Sons. Inc. New York, NY, EE.UU, 1997

Quadroni M., James P. Proteomics and automation. Electrophoresis 1999; 20:664-677

Queitsch C., Hong S.W., Vierling E., Lindquist S. Heat shock protein 101 plays a crucial role in thermotolerance in arabidopsis. Plant Cell 2000; 12:479-492

Rabilloud T. Detecting proteins separated by 2-D gel electrophoresis. Anal Chem 2000; 1:48-55

Rabilloud T. Use of thiourea to increase the solubility of membrane proteins in two-dimensional electrophoresis. Electrophoresis 1998; 19:758-760

Rhodes D., Hanson A.D. Quaternary ammonium and tertiary sulphonium compounds in higher plants. Annual Review of Plant Phisiol and Plant Molecular Biology 1993; 44:357-384

Sakamoto A., Murata N. Genetic engineering of glycinebetaine synthesis in plants: current status and implications for enhancement of stress tolerance. Special Issue Journal of Experimental Botany 2000; 51:81-88

Schlesinger M.J., Ashburner M., Tissiéres A. eds. Heat Shok, from Bacteria to Man. Cold Spring Harbor Laboratory Press, Cold Spring Harbor, 1982

Sen Raychaudhuri S., Wang Deng X. The role of superoxide dismutase in combating oxidative stress in higher plants. The Botanical Review 2000; 66:89-98

Shevchenko A., Wilm M., Mann M. Peptide sequencing by mass spectrometry for homology searches and cloning of genes. J Protein Chem 1997; 16:481-490

Shinozaki K., Yamaguchi-Shinozaki K. Gene expression and signal transduction in water-stress response. Plant Physiol 1997; 115:327-334

Singla S.L., Pareek A., Grover A. 'High Temperature.' In *Plant Ecophysiology*, M.N.V.Prasad, ed. New York, USA: John Wiley and Sons. Inc., 1997.

Smith H.B. Proteomics: Broad strokes of expressionism? Plant Cell 2000; 12:303-304

Somerville C., Somerville S. Plant functional genomics. Science 1999; 285:380-383

Stoop J.M.H., Williamson J.D., Pharr D.M. Mannitol metabolism in plants: a method for coping with stress. Trends in Plant Sciences 1996; 1:139-144

Tabaei-Aghdaei S.R., Harrison P., Pearce R.S. Expression of dehydration-stress-related genes in the crowns of wheatgrass species [Lophopyrum elongatum (Host) A. Love and *Agropyron desertorum* (Fisch. ex Link.) Schult.] having contrasting acclimation to salt, cold and drought. Plant Cell and Environment 2000; 23:561-571

Tanaka Y., Nishiyama Y., Murata N. Acclimation of the photosynthetic machinery to high temperature in Chlamydomonas reinhardtii requires synthesis de novo of proteins encoded by the nuclear and chloroplast genomes. Plant Physiol 2000; 124:441-449

Taylor C.B. Proline and water deficit. Ups, downs, ins and outs. The Plant Cell 1996; 8:1221-1224

Theillet C., Delpeyroux F., Fiszman M., Reigner P., Esnault R. Influence of the excision shock on the protein metabolism of Vicia faba L. meristematic root cells. Planta 1982; 155:478-485

Thompson S., Wilkinson S., Bacon M.A. Davies W.J. Multiple signals and mechanisms that regulate leaf growth and stomatal behaviour during water deficit. Physiologia Plantarum 1997; 100:303-313

Tiburcio A.F., Altabella T., Masgrau M. Polyamines. In: Plant Hormone Research. K. Palme and J. Schell eds. Springer-Verlag, 2001. (In press).

Thomashow M.F. Plant cold acclimation: Freezing tolerance genes and regulatory mechanisms. Annual Review of Plant Physiol and Plant Molecular Biology 1999; 50:571-599

Usuda H., Shimogawara K. Phosphate deficiency in maize. VI. Changes in the two-dimensional electrophoretic patterns of soluble proteins from second leaf blades associated with induced senescence. Plant Cell Physiol 1995; 36:1149-1155

Vierling E. Roles of heat shock proteins in plants. Annu. Rev. Plant Physiol. Plant Mol. Biol. 1991; 42:579-620

Westermeier R. Electrophoresis in Practice. VCH-Verlag, ed., Weinheim, Germany, 1997

Wilkins M.R., Pasquali C., Appel R.D., Ou K., Golaz O., Sánchez J.C., Yan J.X., Gooley A.A., Hughes G., Humphrey-Smith I., Williams K.L., Hochstrasser D.F. From proteins to proteomes: Large-scale protein identification by two-dimensional electrophoresis and amino acid analysis. Biotechnol 1996; 14:61-65

Xiong L., Ishitani M., Zhu J.K. Interaction of osmotic stress, temperature and abscisic acid in the regulation of gene expression in *Arabidopsis*. Plant Physiol 1999; 119:205-211

Yates J.R. Mass Spectrometry and the age of the proteome. J Mass Spectrom 1998; 33:1-19

Zhu J-K. Genetic analysis of plant salt tolerance using Arabidopsis. Plant Physiol 2000; 124: 941-948

Zinser C., Jungblut T., Heller W., Seidlitz H.K., Schnitzler J-P., Ernst D., Sandermann H. Jr. The effect of ozone in Scots pine *Pinus sylvestris* L.: gene expression, biochemical changes and interactions with UV-B radiation. Plant Cell Envir, 2000; 23: 975-982.

CHAPTER 21

POLYAMINES DETERMINATION BY TLC AND HPLC

Nuria Pedrol[1] and Antonio F. Tiburcio[2]
[1]*Depto Bioloxía Vexetal e Ciencia do Solo. Universidade de Vigo. Spain*
[2]*Unit. Fisiologia Vegetal. Facultat de Farmàcia, Universitat de Barcelona. Spain*

THIN-LAYER CHROMATOGRAPHY (TLC)

What is TLC?

Thin-layer chromatography (TLC) is a chromatographic technique that is useful for separating organic compounds, a very frequent problem in the laboratory of Plant Physiology. Together with paper chromatography comprise 'planar' or 'flat-bed' chromatography. TLC is much simpler and less automated than High Performance Liquid Chromatography (HPLC, Chapter 18) with regards to methodology and equipment requirements; a suitable closed vessel containing solvent and a coated plate (the thin layer) are all that are required to carry out separations.

TLC is widely used in experimental Biology to analyse lipids, amino acids and vitamins. Its usefulness has been also proven in pharmaceutical, food and environmental analysis of multiple organic compounds (drugs, pesticides, herbicides, etc.). Because of its simplicity and rapidity, it is often used in analytical Chemistry to monitor the progress of organic reactions, and to check the purity of products (Sherma, 1990; Fried and Sherma, 1999; Hahn-Deinstrop, 2000).

M.J. Reigosa Roger, Handbook of Plant Ecophysiology Techniques, 335–363.

The TLC procedure

Despite its simplicity TLC shows certain analogies with HPLC (see Chapter 18); in fact, TLC is just another kind of liquid chromatography. Sherma (1990) summarises the basic TLC procedure as follows: The sample (either liquid or dissolved in a volatile solvent) containing a mixture of components is deposited as a spot near one end of the *stationary phase*, a thin layer immobilised on a plate. Then the sample is dried; and the end of the stationary phase with the initial zone is placed into a *mobile phase*, usually a mixture of pure solvents, inside a closed chamber. The components of the mixture migrate at different rates during movement of the mobile phase through the stationary phase (by capillary action), which is termed the *development of the chromatogram*. When the mobile phase has moved an appropriate distance, the stationary phase is removed from the chamber, the mobile phase is rapidly dried, and the separated *zones* or spots are detected by application of a suitable visualisation reagent.

Why do solutes separate?

Differential migration is the result of varying degrees of affinity of the mixture components for the solvents and the thin layer, depending on the nature of both phases and solutes. The interactions determining chromatographic retention and selectivity include hydrogen bonding, charge transfer, ion-ion, ion-dipole, and van der Waals interactions (Sherma, 1990).

Thin-layers and solvents

- The *thin layer* consists in an adsorbent material made of cellulose, polyamide, silica gel, alumina or a mixture of these components, usually installed on a glass, plastic or aluminium foil support. There are a great variety of commercial precoated layers for TLC which can be used as received, without preparation, leading to an enormous timesaving (see Fried and Sherma 1999). Silica gel adsorbent is by far the most frequently used layer material for HPTLC (High Performance TLC) plates. Compared to TLC, HP plates are more efficient, leading to tighter zones, better resolution, and more sensitive detection, due to their physical and chemical properties (narrower pore, smaller particle diameter, etc).

- The *solvents* for TLC are chosen in relation to the nature of the layer and mixture to be separated. The strength (polarity) of the mobile phase influences the distances moved by the solutes, while the chemical classification of the solvent components determines the interactions and selectivity of the system. Appropriate solvents are most often selected by consulting the pertinent literature, followed by a trial-and-error approach to adequate the mobile phase to a particular layer and/or local conditions (see Cimpoiu and Hodisan, 1997; as a review). The versatility and flexibility of TLC relative to HPLC is enhanced by the greater choice of solvents available for preparing TLC mobile phases. The choice of solvents for HPLC is limited regarding the exigent nature of the method: a close system operated under high pressure, with on-line detection, besides a column that is continually reused (see Chapter 18). Solvent components with high vapour pressure (e.g., ethyl ether) or UV absorbance (e.g., benzene) or those that might degrade the column (e.g., NaOH) are difficult to use in HPLC but are readily applicable to TLC.

Sample extraction and derivatisation

Sample collection, preservation, and purification are problems common to TLC and all other chromatographic methods. For complex samples, the TLC development will usually not completely resolve the analyte (the substance to be determined) from interferences unless a prior purification is carried out. This is often done by *selective extraction* (including powdering, extracting solutions, buffers, centrifugation, etc.) and, in some cases, *purification* by Column chromatography is required. Usually substances are converted, prior to TLC, to a derivative compound (coloured, fluorescent, radioactive, etc.) that is more suitable for separation, detection, and/or quantification than the parent compound. Such process is named *derivatisation*. Frequently, prechromatography derivatisation confers extra advantages for TLC, because the derivative results less volatile, more stable, and/or easier to separate, extract and clean up than its parent (Sherma, 1990).

Detection and identification of compounds

When the compounds of interest are naturally coloured or fluorescent or absorb UV light, *detection* or visualisation of separated zones is very simple. Despite the compounds have not easy-detectable natural properties, we also can perform simple detection when they have been previously derivatised, as will be the case for polyamines (see below). Otherwise, post-TLC application of a location or visualisation reagent by spraying or dipping is usually required to produce colour or fluorescence for most compounds. This procedure is named 'postchromatography derivatisation'. Depending on the compound properties (natural or acquired), separated zones or 'spots' are detected on thin layers by their colour, fluorescence, fluorescence under UV-light, quenching of fluorescence on a phosphor–containing layer or UV-absorbing. Additional methods are based on radiochemical detection, including autoradiography, liquid scintillation counting, direct *in situ* scanning, and biological properties (e.g., enzyme inhibition and bioautography techniques). Fluorogenic reagents and enzyme inhibition allow detection and confirmation at *low parts-per-billion concentrations* in many samples, at which level most chromogenic reagents would not be effective (Sherma, 1990).

Qualitative *identification* of the compounds separated by TLC (which migrated differently regarding their affinity for the phases, see above) is based initially on their characteristic R_f values compared to the known R_f values of the standards, being

$$R_f = \text{distance moved by the solute / distance moved by mobile-phase front}$$

Nevertheless, R_f values are generally not exactly reproducible from lab to lab, or even in different runs in the same laboratory, so they should be considered mainly as guides to relative migration distances and sequences. Factors causing R_f values to vary include: sample preparation methods, dimensions of the chamber, type of layer, composition of mobile phase, equilibration conditions, temperature, etc. (Sherma, 1990; Hahn-Deinstrop, 2000). Thus, we are better to apply internal standard references together with samples in each plate, in order to compare them directly. Standards will be of known concentration whether further analyte quantification is desired.

Criteria for identification of an analyte by TLC or HPTLC were published in 1989 (*Journal of the Association of Official Analytical*

Chemists 72:487-490). Among these criteria, the R_f value of the analyte should agree within ±3 % compared to the standard material under the same conditions; their visual appearance should be indistinguishable from each other; and the centre of the spot nearest to that of the analyte should be separated from it by at least half the sum of the spot diameters.

Quantification

A rude method used for semiquantitative analysis by TLC consists in comparing visually the size and intensity of the standards (containing known weights of analyte) and sample zones. Other more accurate methods widely used are 'zone elution' and 'scanning densitometry'.

The *zone elution* method comprises drying the layer (to remove the mobile phase); locating the separated analyte zones; scraping the portion of layer containing the analyte, collecting the sorbent and elution of the analyte from the sorbent; and measurement against standards by an independent microanalytical method, most usually visible/UV absorption or fluorescence spectrometry (Sherma, 1990).

This procedure is relatively time-consuming and must be carried out with many precautions, in order to minimise handling errors. On the other hand, *scanning densitometry* is the preferred technique for quantitative TLC. Substances separated by TLC or HPTLC are quantified by *in situ* measurement of absorbed visible or UV light, or emitted fluorescence upon excitation with UV light.

Scanning densitometers (TLC scanners) are manufactured by different companies, having a common work basis: The measurement of the signal diminution (absorbance) or increase (fluorescence) between the zone and a blank area of the layer provides the signal for quantitative measurement. The relative standard deviation of scanning densitometry can be maintained below 2 %, making it a reliable quantitative tool (see Katz, 1987).

Finally, video systems together with *image analysis* are nowadays being widely utilised for accurate qualitative and quantitative evaluation of TLC (see Hahn-Deinstrop, 2000).

Advanced TLC

TLC can be *automated* using forced solvent flow, running the plate in a vacuum-capable chamber to dry the plate. The ability to program the solvent delivery makes it convenient to do multiple developments in which the solvent flows for a short period of time; the TLC plate is dried, and the process is repeated. This method refocuses the spots to achieve higher resolution than in a single run (e.g., Poole and Poole, 1994). Robotics and automation either speed up the operation of TLC or allow for more samples to be processed at one time; nonetheless, completely automated systems are not yet available (Fried and Sherma, 1999; Hahn-Deinstrop, 2000).

Two-dimensional TLC uses the TLC method twice to separate spots that are unresolved by only one solvent. After running a sample in one solvent, the TLC plate is removed, dried, rotated 90°, and run in another solvent. Any of the spots from the first run that contain mixtures can now be separated. The finished chromatogram is a two-dimensional array of spots (Sherma, 1990; Hahn-Deinstrop, 2000).

Great advance in analytes identification has been achieved by *coupling TLC with other instrumental techniques*, such as infra red spectrometry and Raman spectrometry (see Fried and Sherma, 1999). For further characterisations, TLC can be also combined with Nuclear magnetic resonance or Mass spectrometry. The spot containing the unknown analyte must be scraped from the plate and eluted prior to Mass spectrometry.

POLYAMINES IN PLANTS

What are polyamines?

Polyamines are positively charged aliphatic nitrogen-containing compounds of low molecular weight that are widely distributed in living organisms. The major forms are *putrescine*, *spermidine*, and *spermine* (Fig. 1).

Polyamines occur in nature not only as free molecular bases but also as conjugates. In plants, the hydroxycinnamic acid amides are the most common polyamine conjugates, but polyamines also occur linked to proteins or as part of macrocyclic alkaloids. Due to their affinity to acidic components (phospholipids, nucleic acids), polyamines are also bound to DNA, RNA, cell walls and membranes (Tiburcio et al., 1990, 1997; Martin-Tanguy, 1997). The exact localisation of polyamines is still controversial, but it seems that they are mainly localised in the cytoplasm and in some specific compartments like chloroplast and mitochondria (see Bagni, 1991; Rajam 1997).

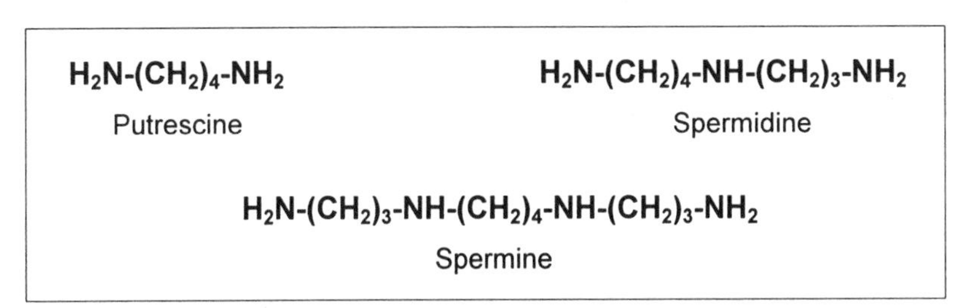

Figure 1. The most common polyamines in eukariote cells: putrescine, spermidine and spermine.

The role of polyamines in the regulation of plant growth and development has attracted considerable attention over the last decade (see Galston, 1983; Smith, 1985; Evans and Malmberg, 1989; Tiburcio et al., 1990; Slocum and Flores, 1991). In spite of this, there is still no consensus on their physiological significance since almost all of the evidence has been correlational. Recently, a new impetus in polyamine research has come from several experimental advances which include the cloning of most of the genes coding for polyamine biosynthetic enzymes in different plant species and the generation of mutants and transgenic plants modified with respect to their polyamine metabolism (see Galston et al., 1997; Kumar et al., 1997; Tiburcio et al., 1997; Walden et al., 1997; Martin-Tanguy, 1997; Malmberg et al., 1998; Tiburcio et al., 2001).

Biosynthesis and catabolism

Arginine decarboxylase (ADC) is a key enzyme in one of the two pathways leading to putrescine in plants. It converts arginine to agmatine, which is an intermediate in the production of putrescine (see Fig. 2). An alternative pathway converts directly ornithine to putrescine, in a reaction catalysed by ornithine decarboxylase (ODC) that is predominant in animals and fungi (Tiburcio et al., 1997). Spermidine and spermine are synthesised by addition of an aminopropyl group donated by decarboxylated S-adenosylmethionine (SAM), which is catalysed by the enzymes spermidine synthetase and spermine synthetase (Tiburcio et al., 1997).

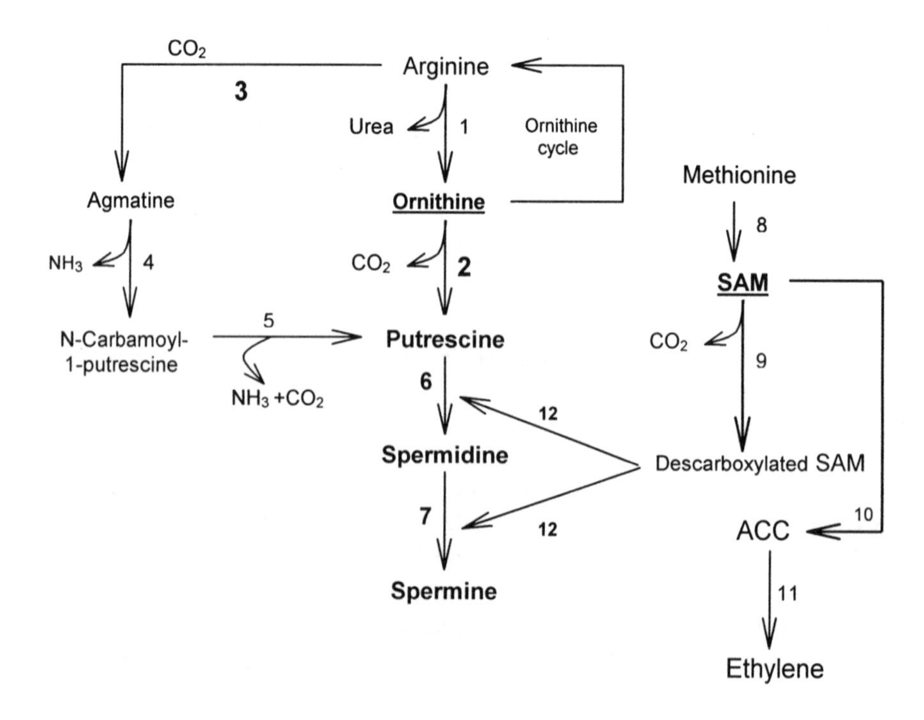

Figure 2. Biosynthesis of the most common di- and polyamines in plants. (1) arginase; (2) ODC=ornithine decarboxylase; (3) ADC=arginine decarboxylase; (4) agmatine iminohydrolase; (5) N-carbamoylputrescine amidohydrolase; (6) spermidine synthetase; (7) Spermine synthetase; (8) S-adenosylmethionine (SAM) synthetase; (9) SAM decarboxylase; (10) 1-aminocyclopropane-1-carboxylic acid (ACC) synthetase; (11) ACC oxidase; (12) propylamine group $[(CH_2)_3\text{-}NH_2]$. Aminoacid ornithine is underlined as being common precursor for proline synthesis (see Chapter 22); and SAM, as being common precursor for ethylene synthesis. (Drawn after Rajam, 1997).

For many years the attention to polyamine catabolic processes in plants (see Fig. 3) has been mainly focussed on the physiological and biochemical aspects of amine oxidases (Federico and Angelini, 1991). These enzymes include the copper proteins, diamine oxidases (DAO), defined on the basis of their higher substrate specificity towards diamines, and the flavoproteins, polyamine oxidase (PAO), which oxidise spermidine and spermine at their secondary amino groups (Federico and Angelini, 1991). DAO reaction products from putrescine are pyrroline, hydrogen peroxide and ammonia, while PAO yields pyrroline and 1-(3-aminopropyl)-pyrroline, respectively, from spermidine and spermine, along with *diaminopropane* and hydrogen peroxide (Fig. 3). Diaminopropane can be converted into alanine, whereas pyrroline can be further catabolised to aminobutyric acid (GABA) in a reaction catalysed by pyrroline dehydrogenase. GABA is subsequently transaminated and oxidised to succinic acid, which is incorporated into the Krebs cycle. Thus, this pathway ensures the recycling of carbon and nitrogen from putrescine and spermidine (Tiburcio et al., 1990).

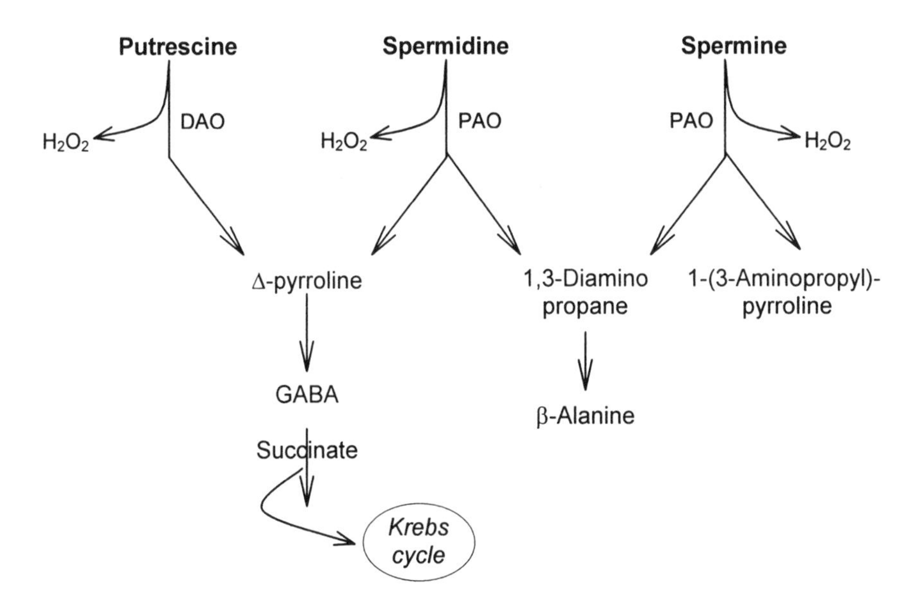

Figure 3. Catabolism of the most common di- and polyamines in plants. DAO, diamine oxidase; GABA, γ-aminobutiric acid; PAO, poliamine oxidase. Drawn after Tiburcio (1990) and Rajam (1997).

Polyamines in an ecophysiological context

The role of polyamines on the physiology of plants has been extensively studied during the last 20 years (see Flores, 1990; Tiburcio et al., 1993, 1997; Galston et al., 1997; Rajam, 1997; Malmberg et al., 1998). In spite of this, studies are still in progress to understand how polyamines do exactly work. In this section, the present knowledge concerning the mechanisms of polyamine action on plant senescence and plant responses to environmental stresses will be reviewed.

Plant growth regulators

The polyamines putrescine, spermidine and spermine regulate protein, RNA and DNA biosynthesis playing a role in the initiation and control of cell cycle (Serafini-Fracassini, 1991) and affecting different aspects of plant growth and development (Smith, 1985; Evans and Malmberg, 1989; Slocum and Flores, 1991). The interaction of polyamines with most plant hormones suggests that these compounds act as true plant growth regulators. It is assumed that binding to a specific receptor, still non-characterised, initiates the signal transduction cascade leading to differential expression of specific genes. This hypothesis is reinforced by the effect of polyamines on the expression of several genes and interactions with signal transduction pathways, both in animals and plants (Tiburcio et al., 2001).

Genetic analysis of polyamine-response mutants will allow in the future unravelling the mechanisms of polyamine signal perception and transduction.

Senescence prevention

The biosynthetic pathways of polyamines and ethylene (the senescence-inducer plant hormone) depend on the common precursor SAM (see Fig. 2). Polyamines can inhibit the conversion of ACC to ethylene through the inhibition of ACC-synthetase, together with the scavenging of active oxygen species that are required for the reaction; therefore, the senescence onset is avoided or, at least, delayed (Drolet et al., 1986; Kushad and Dumbroff, 1991; Pell and Dann, 1991).

The antisenescence effects of polyamines were firstly observed in freshly isolated oat protoplasts (Altman et al., 1977). More recently, a progressive loss of chlorophyll during dark-incubation of oat leaves was correlated with a decrease in the levels of the thylakoid proteins D1, D2 and cytochrome f, and the stromal protein Rubisco large subunit (Besford et al., 1993). Treatment with spermidine or spermine prevented the loss of chlorophyll, stabilised the molecular composition of the thylakoid membranes and delayed senescence (Besford et al., 1993). Furthermore, ultrastructural observations showed that the integrity of the thylakoid membrane system is better preserved with spermine (Tiburcio et al., 1994).

In view of the reported antisenescence properties of polyamines and of their effects on cell membranes, the influence of polyamines on lipid peroxidation during dark-incubation of oat leaves has been examined (Borrell et al., 1997). The results suggested that inhibition of lipid peroxidation (monitored in terms of malondialdehyde, and both lipoxygenase activity and enzyme-protein levels) might be one of the mechanisms responsible for their antisenescence effects (Borrell et al., 1997). The inhibition of lipid peroxidation by spermine was suggested to be due to direct binding of spermine to microsomal phospholipids (Kitada et al., 1979). Tadolini (1988) proposed a possible molecular mechanism by which spermine may inhibit lipid peroxidation. This is related to the ability of spermine to form a ternary complex with iron, and the phospholipid polar head may change the susceptibility of Fe^{2+} to autoxidation and thus its ability to generate free oxygen radicals (Tadolini, 1988).

Transgenic tomato plants overexpressing yeast SAM decarboxylase under the control of the ethylene- and ripening-inducible E8 promoter have been generated (Mehta et al., 1997). Some transgenic plants showed delayed leaf senescence as well as delayed fruit ripening. The independent transgenic tomato lines are being used as a system to study the interactive roles of polyamines and ethylene in regulating plant senescence (Mehta et al., 1997).

Polyamines and stress

Plant polyamine metabolism is extremely sensitive to adverse environmental conditions (Flores, 1990, 1991; Galston and Kaur-Sawhney, 1990; Rajam, 1997; Bouchereau et al., 1999); thus, polyamines are considered as creditable *stress markers* in plants. Putrescine accumulates in response to several conditions of *abiotic stress*: K^+ and Mg^{2+} deficiency, high NH_4 nutrition, osmotic-, acid-, salt- and water stresses, Cd and SO_2 toxicity, anaerobiosis, low temperature, UV-B and UV-C radiation, ozone fumigation and herbicide application (see Flores 1990, 1991; Galston and Kaur-Sawhney, 1990; Langebarlets et al., 1991;

Kurepa et al., 1988). Some authors have described putrescine accumulation in response to *biotic stress*, e.g., pathogen infection (Rajam, 1997; Torrigiani et al., 1997), but the effects of plant competition on polyamine metabolism have yet received very little attention (e.g., Pedrol et al., 1999).

The accumulation of putrescine is much more pronounced in cereals than in other plants, since putrescine represents an important pool of C and N in cereals that can be used during stress conditions (Slocum and Weinstein, 1990). These authors suggested that putrescine accumulation under stress could be related to a detoxification mechanism preventing NH_4 accumulation.

Plants under stress are characterised by showing high photorespiration rates leading to increased NH_4 levels that are toxic for the plant. The increase of NH_4 would stimulate the synthesis of glutamate, thus supplying ornitine and arginine (via Urea cycle) for the synthesis of putrescine (Slocum and Weinstein, 1990). In spite of this hypothesis, the exact mechanism leading to putrescine accumulation under stress is still under continuos debate.

Under a practical point of view, it should be pointed out that in several cases the damaging effects produced by some types of stresses could be mitigated by exogenous application of polyamines. For example, polyamine treatment is able to reduce the severity of leaf necrosis induced by ozone fumigation (Bors et al., 1989). Treatment with spermidine protects plants against the toxicity of some herbicides (Kurepa et al., 1998).

Pre-treatment of plants with inhibitors of polyamine biosynthesis that prevents the rise in spermidine levels leads to increased sensitivity to UV-C damaging effects (see Tiburcio et al., 2001). In all these cases, it is possible that polyamines may alleviate the effects of *oxidative damage* probably by acting as free radical scavengers, thus avoiding the initiation of cell lysis and necrosis (Prasad and Rengel, 1998).

☆ *Salt stress*

During salt stress, the accumulation of putrescine is a generalised response in order to compensate for the loss of K^+ (Flores, 1990). The sensitivity to salt is related to the inability to maintain the selectivity for K^+ ions. The tolerance of certain species is due to the substitution of Na^+ by K^+ (Cuartero et al., 1992); but other cationic compounds (i.e.

polyamines) can substitute K^+ to maintain cation/anion balance as well as an appropriate pH (Smith, 1985). Thus, free polyamines could act as *compatible solutes* to substitute K^+. In this regard, it has been observed that the accumulation of Na^+ is accompanied by a decrease in both K^+ and free polyamines levels (Santa-Cruz et al., 1997). Furthermore, increased tolerance to water- and salt-stress has been reported in carrot transgenic cultures over-expressing ODC (Minocha and Sun, 1997).

Numerous experiments show that salt sensitivity depends on the *relative concentration of different types of polyamines* (Basu and Gosh, 1991). For example, salt tolerance in rice is associated to a relative increase in the levels of free spermidine and spermine (Krishnamurthy and Bhagwat, 1989; Santa-Cruz et al., 1997). Flores (1990) and Tiburcio et al. (2001) have suggested that only spermidine and spermine, but no putrescine, be implicated in the protective response against stress. On the other hand, salt-sensitive cultivars accumulate putrescine, but no spermidine or spermine, probably due to an inhibition of SAM decarboxylase activity (Erdei et al., 1996). All these observations suggest that stress tolerance is closely related with the antisenescence capacity of polyamines (i.e., delay of the *exhaustion phase* in the stress syndrome? – Larcher, 1995).

Finally, it has to be taken into account that plants may follow different mechanisms in response to environmental stresses. For example, plants may react by preferential accumulation of other osmoprotectants, like proline (Chapter 22). In this regard, unpublished results (Tiburcio et al.) showed that the degree of putrescine accumulation was much higher in monocot than in dicot species. In contrast, the opposite pattern was observed with regard to proline accumulation (Tiburcio et al. Unpublished results). Osmotic stress induces the expression of pyrroline-5-carboxylase synthetase (P5CS), enzyme that is involved in proline biosynthesis (see Chapter 22). An ABA-dependent signalling pathway (Strizhov et al., 1997) regulates the transcription of this gene. Since polyamine metabolism is interconnected with proline biosynthesis, via glutamate, it will be interesting to study in the future whether or not this ABA-dependent signalling pathway also mediates *polyamine/proline interaction*. The combined study of the stress-inducible metabolites polyamines and proline constitutes undoubtedly a useful tool to understand the mechanisms of plant adaptation to the environment.

ANALYZING POLYAMINES IN PLANT TISSUES BY TLC AND HPLC

Separation and quantification of each amine in a given extract requires the use of chromatographic techniques. The polyamines contents of various plant tissues have been analysed using several ion exchange chromatography, thin-layer chromatography (TLC) and high-performance liquid chromatography (HPLC, whose basis are widely described in Chapter 18).

Smith and Davies (1987) offered a descriptive review of methods to separate and quantify polyamines by HPLC. Most common HPLC methods require pre-derivatisation of polyamines. Dansyl derivatised amines –dansylamines- are monitored by fluorescence detectors (e.g., Smith and Davies, 1985; Walter and Geuns, 1987; Corbin et al., 1989; Marcé et al., 1995). Benzoyl derivatives –benzoylamines- are UV monitored (e.g., Flores and Galston, 1982; Slocum et al., 1989). Other methods such as ion-pair reverse phase HPLC or ion-exchange chromatography operate with underivatised polyamines; fluorogenic detection is performed following post-column derivatisation (see Smith and Davies, 1987). HPLC offer the advantage over TLC of a superior resolution and easier quantification of polyamines. Many plant scientists utilise this sensitive method to follow the changes in the levels of polyamines during plant growth and development. Besides the most common polyamines (putrescine, spermidine and spermine) HPLC has been extended to the analysis of a large number of other naturally occurring amines in plant tissues, including polyamine homologues, heterocyclic and aromatic amines (Smith and Davies, 1987).

On the other hand, the progressive optimisation of TLC for polyamine analysis (Seiler, 1971; Smith and Wilshire, 1975; Smith and Best, 1977; Flores and Galston, 1982; Wettlanfer and Weinstein, 1988) have made it a fast and inexpensive tool for many laboratories of Plant Physiology. Contrary to HPLC, TLC allows the concurrent processing of multiple samples, besides the co-processing of internal reference standards on the same layer, simultaneously. Moreover, with the use of HPTLC plates (see above), together with the choice of a good quantification method, we can obtain results that rival HPLC in accuracy and precision. TLC is being used with satisfactory and accurate results in comparative studies, specially following the effects of different stress factors on polyamine levels (e.g., Reggiani et al., 1989; Willadino et al., 1996; Pedrol et al., 2000).

Now we will describe the basic steps for the analysis of polyamines in different plant tissues by means of both HPTLC and HPLC. In the methods here propossed, polyamines are analysed as their fluorescent dansylated derivatives, 'dansylpolyamines'. Fluorescence monitoring of polyamines through their dansyl derivatives is a rapid and highly sensitive method with *limits of detection at the femtomolar level* (Smith and Davies, 1985). Both chromatographic procedures allow high resolutive separations of the polyamines diaminopropane, putrescine, *cadaverine* - another diamine of limited occurrence in higher plants, formed from lysine decarboxylation (see Tiburcio et al., 1990)-, spermidine and spermine. HPLC offers undoubtly easier and more rapid quantifications but, as said before, equipment requirements are highly superior to TLC. *Extraction* and *dansylation* procedures (summarized in Fig. 4), as well as *standard preparation* are common for both chromatographic methods; further analytical steps will be detailed separately.

Extraction of polyamines from plant tissues

Samples of plant material are powdered with liquid nitrogen, using pre-cooled mortars and pestles. Polyamines are extracted by homogenising in 5 % (V/V) cold perchloric acid (PCA) at a ratio of 100 to 300 mg fresh-mass · mL^{-1} PCA in the case of leaf tissue. Homogenates are transferred to cold centrifuge tubes (e.g., Eppendorf tubes) and incubated in an ice bath for 30 min (Tiburcio et al., 1985, Marcé et al., 1995).

The required amount of plant material will depend on the water content of each tissue and species. As examples, *calli* from *in vitro* culture are extracted at 600 mg fresh-mass · mL^{-1} PCA, whereas lyophilised leaf material is extracted at a ratio of 50 mg dry-mass · mL^{-1} PCA. Fresh material must be processed immediately after being harvested or kept in freezer at -80 °C until extraction, regarding polyamines are extremely labile.

After incubation, homogenates are centrifuged at 27,000 g_n for 20 min (Marcé et al., 1995). Slight modifications can be introduced, e.g., the extraction procedure for leaf tissue proposed by Reggiani et al. (1989) includes centrifugation at 12,000 g_n for 15 min at 4 °C.

Both the supernatant (S) and the pellet (P) fractions are collected.

The pellet is resuspended in the original volume with 1 M NaOH. Aliquots (300 μL) each of the pellet suspension and the original supernatant are mixed 1:1 with 12 M HCl into an injectable vial, which is sealed with a flame and heated at 110 °C for 16 to 18 h. The resulting mixtures are filtered through glass wool (to eliminate carbonised material) and dried in vacuum. Then, the dried material is redisolved in 300 μL of 5 % (V/V) PCA (Marcé et al., 1995).

In this way we obtain three fractions: the non-hydrolysed PCA supernatant (S fraction: free polyamines), the hydrolysed PCA supernatant (SH fraction: containing the polyamines liberated from the PCA-soluble conjugates) and the hydrolysed pellet (PH fraction: containing the polyamines liberated from the PCA-insoluble conjugates).

These samples may be kept at 4 °C for several hours, or stored long-term at -20 °C until dansylation.

Preparation of standard solutions

Stock solutions of commercial polyamine standards: diaminopropane, putrescine, cadaverine, spermidine and spermine (in the form of hydrochlorides) are prepared at the concentration of 1 mM in H_2O (or in 0.01 N HCl).

For the HPLC procedure, a stock solution 1 mM of *1,7-diaminoheptane* is also prepared. The unnatural amine 1,7 diaminoheptane is used in HPLC as internal reference compound because it resolves well from derivatives of endogenous amines, elutes near amines of interest, and it is stable under storage conditions (Smith and Davies, 1987).

- **Working standard solutions for HPLC** (0.05 mM) are prepared by 1:20 dilution of the stock [i.e., 50 μL 1 mM stock + 950 μL H_2O (or 0.01 N HCl)].
- **Working standard solutions for TLC** will be the 1 mM stock solutions, undiluted.

Standards are stable for at least 2 months if stored at –20 °C in plastic tubes. Plastic containers (e.g., Eppendorf tubes) must be used for storage because polyamines adsorb to the surface of glass (Smith and Davies, 1987).

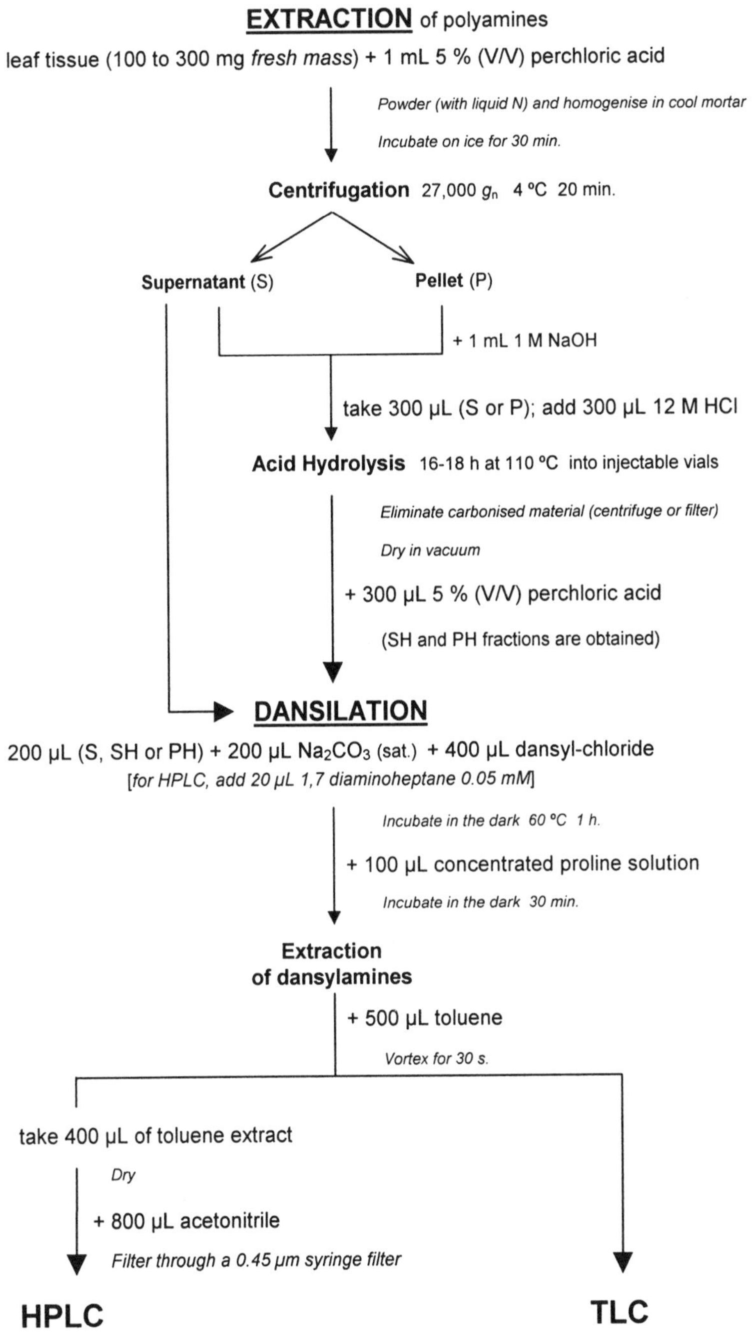

EXTRACTION of polyamines

leaf tissue (100 to 300 mg *fresh mass*) + 1 mL 5 % (V/V) perchloric acid

Powder (with liquid N) and homogenise in cool mortar

Incubate on ice for 30 min.

Centrifugation 27,000 g_n 4 °C 20 min.

Supernatant (S) **Pellet** (P)

+ 1 mL 1 M NaOH

take 300 µL (S or P); add 300 µL 12 M HCl

Acid Hydrolysis 16-18 h at 110 °C into injectable vials

Eliminate carbonised material (centrifuge or filter)

Dry in vacuum

+ 300 µL 5 % (V/V) perchloric acid

(SH and PH fractions are obtained)

DANSILATION

200 µL (S, SH or PH) + 200 µL Na_2CO_3 (sat.) + 400 µL dansyl-chloride

[*for HPLC, add 20 µL 1,7 diaminoheptane 0.05 mM*]

Incubate in the dark 60 °C 1 h.

+ 100 µL concentrated proline solution

Incubate in the dark 30 min.

Extraction of dansylamines

+ 500 µL toluene

Vortex for 30 s.

take 400 µL of toluene extract

Dry

+ 800 µL acetonitrile

Filter through a 0.45 µm syringe filter

HPLC **TLC**

Figure 4. Extraction and dansylation procedures for the analysis of polyamines by TLC and HPLC. According to Marcé et al. (1995).

Derivatisation by dansylation

Dansyl chloride (see Fig. 5) is the most widely used reagent for derivation of polyamines prior to separation by TLC or HPLC (Smith and Davies, 1987).

Dansylation of polyamines is carried out after Seiler (1971), with some modifications (Marcé et al., 1995), as follows: Aliquots (200 μL) of the 3 fractions: S, SH and PH, are transferred to reaction tubes. Then, 400 μL of dansyl chloride (5 mg · mL^{-1} in acetone) are added, besides 200 μL of a saturated Na_2CO_3 solution (in order to produce an alkaline medium for labelling, which is optimal at pH of 9.5 to 10.5; Smith and Davies, 1987).

After vortexing the mixtures for a wile, they are incubated in the dark overnight at room temperature (alternatively for 10 min. at 70 °C). The reaction is stopped by the addition of 100 μL of a proline solution (100 mg · mL^{-1} in H_2O) followed by incubation in the dark for 30 min. This way, the excess of dansyl chloride is removed.

Finally, dansylated polyamines are extracted from the alkaline medium with 500 μL toluene by vortexing intensively for 30 s. In this step, dansylated amino acids remain in the aqueous phase, being discarded.

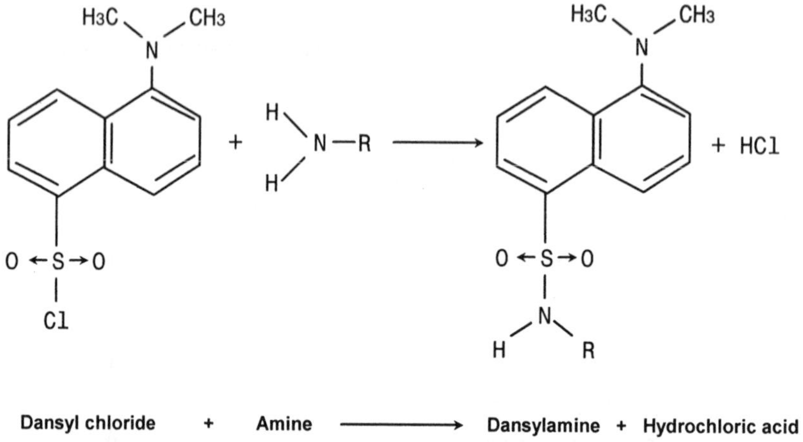

Dansyl chloride + Amine ⟶ Dansylamine + Hydrochloric acid

Figure 5. Dansylation of amines. Dansyl chloride (5-dimethylaminonaphthalene-1-sulfonyl chloride) reacts stoichiometrically with primary and secondary amines; if the substances have more than one amino group, each is derivatised by the reagent (Drawn after Smith and Davies, 1987).

Dansylation of standards is carried out in the same way described for samples, but consider:

- Polyamine **standards for HPLC** are prepared by mixing 40 μL of each 0.05 mM working solutions (diaminopropane, putrescine, cadaverine, 1,7-diaminoheptane, spermidine and spermine), having a final volume of 240 μL. Moreover, 40 μL 1,7-diaminoheptane 0.05 mM are added to each sample before dansylation, as internal reference, thus having equal experimental volumes for samples and standards.
- Polyamine **standards for TLC** are prepared by mixing 20 μL of each 1 mM stock solutions (diaminopropane, putrescine, cadaverine, spermidine and spermine). Final volume of 200 μL (equal to sample aliquots) is achieved by adding 100 μL H_2O (or HCl 0.1 N). Alternatively, different concentrations of standards can be prepared (e.g., 25, 20, 15, 10 μL of each 1mM stock + 75, 100, 125, 150 μL H_2O) in order to obtain a standard curve for each amine (i.e., concentration *vs.* fluorescence).

Dansylation must be performed in the dark since the derivatives are light sensitive, but dim light in the laboratory is tolerable. Solution of dansyl chloride in acetone can be stored for 24 h at 4 °C in the dark. After the dansylation reaction, samples must be kept in the dark or in dark vials.

Along all the described procedures of extraction and dansylation, samples may be kept at 4 °C for several hours after the extraction, after the hydrolysis, after resuspended in PCA, or in toluene.

The TLC procedure

Separation of the dansylamines is performed on a HPTLC plate of silicagel 60 Å (see Fig. 6), using 4:1 (V/V) chloroform/triethylamine as the developing solvent system (Tiburcio et al., 1985). However, a trial-and-error approach would be necessary to adequate the proportions of solvents to local conditions; e.g., 5:1 (V/V) can be used (Slocum et al., 1989).

Other solvent systems can provide accurate separation of dansylpolyamines. For example, Regianni et al. (1989) and Slocum et al. (1989) develop TLC in 5:4 (V/V) cyclohexane/ethyl acetate solvent system. Nontheless, with the latter solvent the diamines putrescine and cadaverine are poorly separated (Slocum et al., 1989); so, its use would be inadequate for cadaverine-rich species (see Tiburcio et al., 1990).

Chromatography is developed under penumbra, into a closed TLC chamber. The chamber containing the solvents mixture (usually 90 to 100 mL) must be prepared some hours before chromatography in order to equilibrate the mixture and provide optimal conditions for migration and separation of dansylamines.

Samples and standards are spotted in the application zone (see Fig. 6) by using a micropipet. Generally 10-40 μL of each toluene extract are applied. Large volumes must be spotted by repeated applications of small aliquots, with solvent evaporation in between increments. After application, samples are dried in a warm air stream. Then, the plate is placed vertically into the chamber, dipping the immersion zone (see Fig. 6) into the developing solvent. The chamber is then closed hermetically.

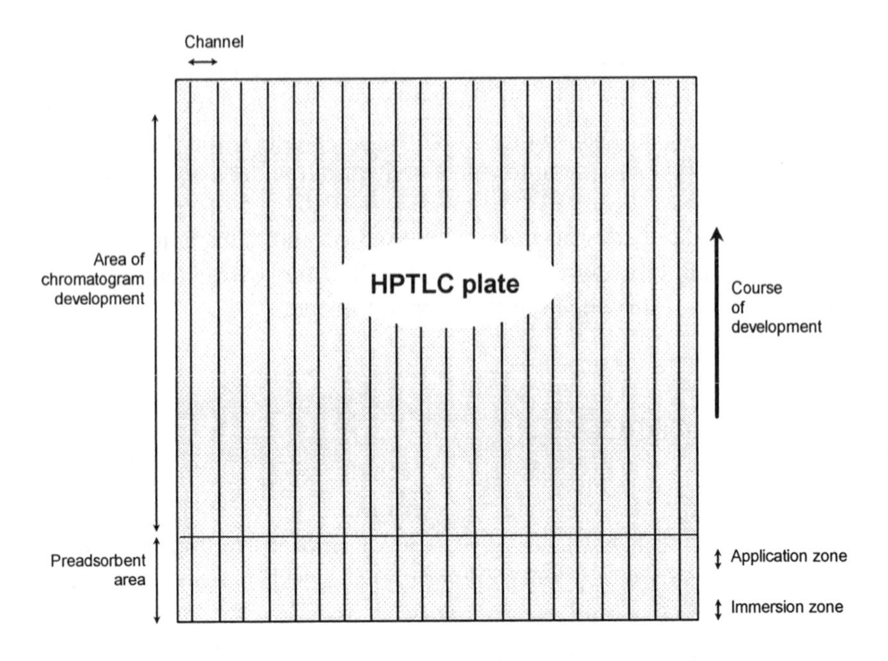

Figure 6. Representation of a HPTLC plate of silicagel with 19 separated channels (to avoid sample mixing), and a preadsorbent or concentration area provided with a thicker layer.

Times of development are different for the different solvent system. Plates are developed for approximately 1 h and 15 min. in 4:1 (V/V) chloroform/triethylamine; and 40 min. in 5:4 (V/V) cyclohexane/ethyl acetate. Time of development can also vary slightly with local conditions too, specially with room temperature.

Once developed, the plate is removed from the chamber and dried quickly with warm air (e.g., dried for 10 min. at 60 °C, Regianni et al., 1989).

Spots corresponding to the different amines are visualised under UV-light, and identified by comparison to standards. When developed in 4:1 (V/V) chloroform/triethylamine, the migration order of the different polyamines is diaminopropane (firstly separated) followed by putrescine, cadaverine, and finally, spermidine and spermine (see Fig. 7A).

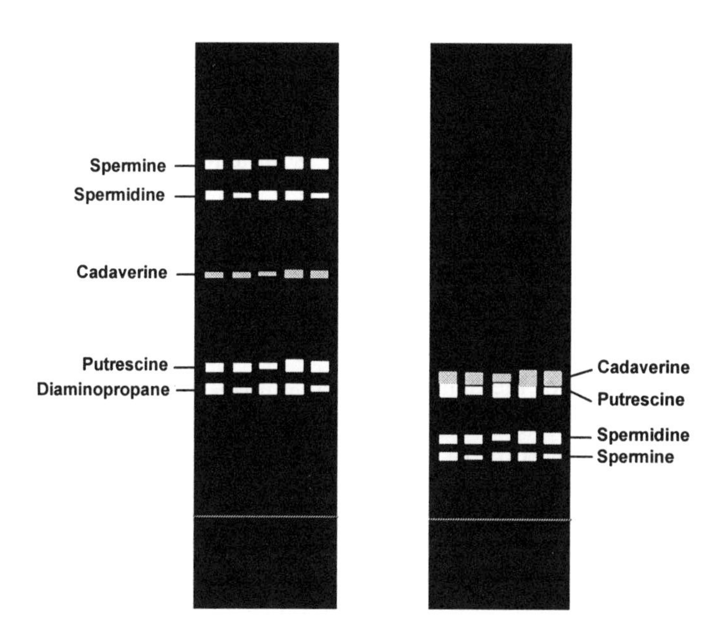

Figure 7. Representation of two chromatograms obtained by HPTLC showing the separated dansylamines, as visualised under UV-light. Dansylamines chromatographed under (A) 4:1 (V/V) chloroform/triethylamine solvent system (1h 15 min.), and (B) 5:4 (V/V) cyclohexane/ethyl acetate solvent system (40 min.), with the diamines putrescine and cadaverine poorly separated. [There have been solely represented the spots corresponding to the target analytes (diaminopropane, putrescine, cadaverine, spermidine and spermine), which are identified by comparison of their R_f values with those of co-processed standards. Other spots that are always visualised (not represented) correspond to products produced under the conditions of dansylation (dansyl methylamine, dansyl dimethylamine, dansyl ammonia and dansyl hydroxide); other dansylated amines should also appear.].

When developed in 5:4 (V/V) cyclohexane/ethyl acetate, the migration order of the different polyamines is spermine (firstly separated) followed by spermidine, and finally, putrescine and cadaverine (if present) poorly separated (see Fig. 7B). Once visualised, the spot bounders are marked with pencil.

Quantification is carried out following the *zone elution* method (see above). The marked silicagel zones corresponding to diaminopropane, putrescine, cadaverine, spermidine and spermine are scraped and transferred carefully to assay tubes. Dansylamines are recovered from the silica gel by eluting in 2 mL ethyl acetate. Analytes and standards are measured with fluorescence spectrophotometer with excitation at 350 nm and emission at 495 nm, using ethylacetate as blank. Sample concentrations of the different amines (expressed as nmol $\cdot$ g^{-1}) are determined by comparison with polyamine standards, taking into account the different dilutions made during the extraction procedure.

As an important final regard, notice that the analysed SH fraction would contain the polyamines liberated from the soluble conjugates but also the free polyamines (E. Bernet, pers. com.); thus, for final quantification S fraction value must be subtracted to the SH fraction value, i.e., [$SH_{real} = SH_{measured} - S$].

The HPLC procedure

After the extraction of the dansylated polyamines, 400 µL of the organic phase (the toluene extract) is removed, dried (e.g., under a stream of N_2) and re-dissolved in 800 µL of acetonitrile (which is compatible with the HPLC column). Finally, the samples are passes through a 0.45 µm pore size syringe filter before inyection in HPLC.

Separation of dansylpolyamines is performed according to the high-performance liquid chromatographic method of Marcé et al. (1995). The chromatographic system used consists of a reversed-phase column and a mobile phase of acetonitrile and water. The separation of 1,3-diaminopropane, putrescine, cadaverine, spermidine and spermine takes only 9 min. This method provides a good resolution between 1,3-diaminopropane and putrescine, which were poorly separated by other previously described rapid HPLC methods (e.g., Walter and Geuns , 1987).

The HPLC system (see Chapter 18) must be equiped with a Brownlee reverse-phase ODS Spheri-5 (C_{18}, 5 μm spherical, 80 Å pore size, 220 × 4.6 mm I.D.), besides a fluorescence detector.

The procedure is carried out as follows: Generally 10-20 μL of each acetonitrile extract are injected into the HPLC system. The sample is subjected to a *gradient elution in acetonitrile and water*, both solvents previously sonicated and filtered through a 0.45 μm filter. The initial conditions are 70 % acetonitrile and 30 % water pumping at a flow-rate of 1.5 mL · min^{-1}. The mixture is pumped for 4 min; then the concentration of acetonitrile is raised to 100 %. This concentration of acetonitrile is kept constant for 4 min., and finally returned to the initial conditions (see Fig. 8).

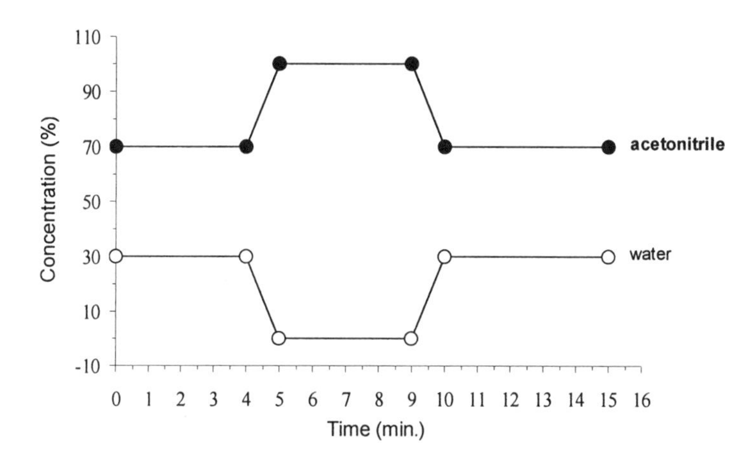

Figure 8. Diagram of the gradient used in the HPLC method: percent concentration acetonitrile and water in the mobile phase *vs.* time in minutes. Drawn after Marcé et al. (1995).

Chromatographic separation is such a way developed in a *reversed-phase* mode (in which the eluent strength decreases as the solvent becomes more polar) thus providing optimal separations of low molecular weight organic compounds (see Harris, 1999, p.733). The proper described conditions were optimised by Marcé et al. (1995) for the elution of spermidine and spermine, which are both strongly retained on the column.

Dansylamines are monitored with a fluorescence detector, with excitation at 252 nm and emission at 500 nm. Under the described conditions, spermidine is the last peak eluted from the column (approximately 8.5 min., see Fig. 9), and the column is then re-equilibrated in the remaining 6.5 min. The *elution order* of the different dansylamines (Fig. 9) is diaminopropane, putrescine, cadaverine, diaminoheptane, spermidine and spermine.

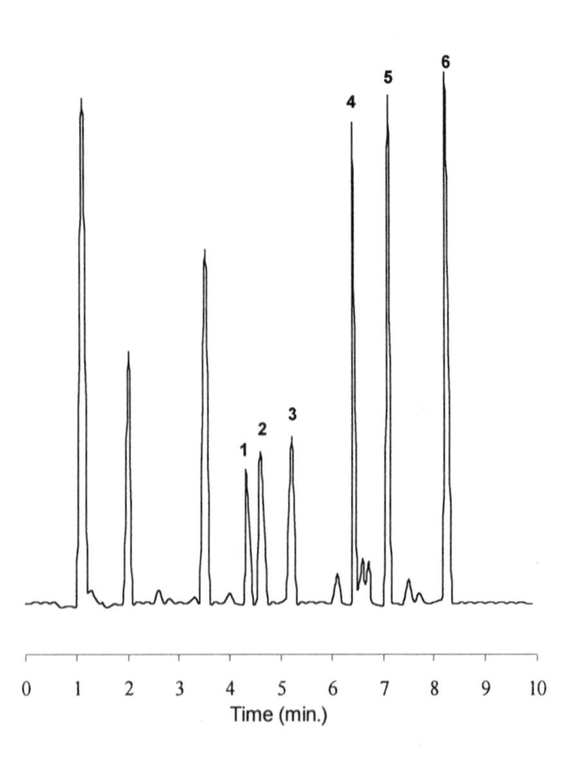

Figure 9. Chromatogram of dansylated standard polyamines. Peaks: (1) diaminopropane, (2) putrescine, (3) cadaverine, (4) diaminoheptane, (5) spermidine, (6) spermine. Drawn after Marcé et al. (1995).

Finally, quantification of the different amines is usually performed by means of a software program coupled to the HPLC system. It consists on integrating the areas of each sample peak, compared to those of polyamine standards of known concentration. Final concentrations (expressed as nmol $\cdot$ g^{-1}) are calculated regarding the different dilutions made during the extraction procedure.

REFERENCES

Altman A., Kaur-Sawhney R., Galston A.-W. Stabilization of oat leaf protoplasts through polyamine-mediated inhibition of senescence. Plant Physiol 1977; 60:570-574

Bagni N. "Transport and Subcellular Compartimentation of Polyamines in Plants." In *Polyamines as Modulators of Plant Development*. International Meetings on Biology. Madrid, España: Fundación Juan March, 1991

Basu R., Gosh B. Polyamines in various rice genotypes with respect to sodium chloride salinity. Physiol Plantarum 1991; 82:575-581

Besford R.-T., Richardson C., Campos J.-L., Tiburcio A.-F. Effect of polyamines on stabilization of molecular complexes in thylakoid membranes of osmotically-stressed oat leaves. Planta 1993; 189:201-206

Borrell A., Carbonell L., Farrás R., Puig-Parellada P., Tiburcio A.-F. Polyamines inhibit lipid peroxidation in senescent oat leaves. Physiol Plantarum 1997; 99:385-390

Bors W., Langebarlets C., Michel C., Sandermann H.Jr. Polyamines as radical scavengers and protectants against ozone damage. Phytochemistry 1989; 25:367-371

Bouchereau A., Aziz A., Larher F., Martin-Tanguy J. Polyamines and environmental challenges: recent development. Plant Sci 1999; 140:103-125

Cimpou C., Hodisan T. Mobile phase optimization in thin-layer chromatography (TLC). Rev Anal Chem 1997; 16:299-321

Corbin J.L., Marsh B.H., Peters G.A. An improved method for analysis of polyamines in plant tissue by precolumn derivatisation with o-phthalaldehyde and separation by High Performance Liquid Chromatography. Plant Physiol 1989; 90:434-439

Cuartero J., Yeo A.R., Flowers T.J. Selection of donors for salt-tolerance in tomato using physiological traits. New Phytol 1992; 121:63-69

Drolet G., Dumbroff E.B., Legge R.L., Thompson J.E. Radical scavenging properties of polyamines. Phytochemistry 1986; 25:367-371

Erdei L., Szegletes Z., Barabás N.K., Pestenácz A. Responses in polyamine titer under osmotic and salt stress in sorghum and maize seedlings. J Plant Physiol 1996; 147:599-603

Evans P.T., Malmberg R.L. Do polyamines have roles in plant development? Annu Rev Plant Physiol Plant Mol Biol 1989; 40:235-269

Federico R., Angelini R. "Polyamine Catabolism." In *Biochemistry and Physiology of Polyamines in Plants*. R.D. Slocum, H.E. Flores, eds. Boca Ratón, Florida: CRC Press, 1991

Flores H.E. "Polyamines and Plant Stress." In *Stress Responses in Plants. Adaptation and Acclimation Mechanisms*. R.G. Alscher, J.R. Cumming, eds. New York, USA: Wiley-Liss Inc., 1990

Flores H.E. "Changes in Polyamine Metabolism in Response to Abiotic Stress." In *Biochemistry and Physiology of Polyamines in Plants*. R.D. Slocum, H.E. Flores, eds. Boca Ratón, Florida: CRC Press, 1991

Flores H.E., Galston A.W. Analysis of polyamines in higher plants by high performance liquid chromatography. Plant Physiol 1982; 69:701-706

Fried B., Sherma J. *Thin-Layer Chromatography.* New York, USA: Marcel Dekker Inc., 1992

Galston A.W. Polyamines as modulators of plant development. Bioscience 1983; 33:382-387

Galston A.W., Flores H.E. "Polyamines and Plant Morphogenesis." In *Biochemistry and Physiology of Polyamines in Plants.* R.D. Slocum, H.E. Flores, eds. Boca Ratón, Florida: CRC Press, 1991

Galston A.W., Kaur-Sawhney R., Altabella T., Tiburcio A.-F. Plant polyamines in reproductive activity and response to abiotic stress. Bot Acta 1997; 110:197-207

Galston A.W., Kaur-Sawhney R.K. Polyamines in plant physiology. Plant Physiol 1990; 94:406-410

Hahn-Deinstrop E. *Applied Thin-Layer Chromatography. Best Practice and Avoidance of Mistakes.* Weinheim, Germany: Wiley-VCH, 2000

Harris D.C. *Quantitative Chemical Analysis.* New York, USA: V Ed.W.H. Freeman and Co., 1999

Katz E. *Quantitative Analysis using Chromatographic Techniques.* E. Katz, ed. New York, USA: John Wiley, 1987

Kitada M., Igarashi K., Hirose S., Kitagawa H. Inhibition by polyamines of lipid peroxide formation in rat liver microsomes. Biochem Bioph Res Co 1979; 87:388-394

Krishnamurthy R., Bhagwat K.A. Polyamines as modulators of salt tolerance in rice cultivars. Plant Physiol 1989; 91:500-504

Kumar A., Altabella T., Taylor M.-A., Tiburcio A.-F. Recent advances in polyamine research. Trends Plant Sci 1997; 2:124-130

Kurepa J., Smalle J., Van Montagu M., Inzé D. Polyamines and paraquat toxicity in *Arabidopsis thaliana.* Plant Cell Physiol 1998; 39:987-992

Kushad M., Dumbroff E. "Metabolic and Physiological Relationships between the Polyamine and Ethylene Biosynthetic Pathways. In *Biochemistry and Physiology of Polyamines in Plants.* R.D. Slocum, H.E. Flores, eds. Boca Ratón, Florida: CRC Press, 1991

Langebarlets C., Kerner K., Leonardi S., Schraudner M., Trost M., Heller W., Sandermann H. Jr. Biochemical plant responses to ozone. I. Differential induction of polyamine and ethylene biosynthesis in tobacco. Plant Physiol 1991; 95:882-889

Larcher W. *Physiological Plant Ecology.* Berlin, Germany: Springer-Verlag, 1995

Malmberg R.L., Watson M.B., Galloway G.L., Yu W. Molecular genetic analysis of plant polyamines. Crit Rev Plant Sci 1998; 17:199-224

Marcé M., Brown D.S., Capell T., Figueras X., Tiburcio A.F. Rapid high-performance liquid chromatographic method for the quantitation of polyamines as their dansyl derivatives: application to plant and animal tissues. J Chromatogr 1995; 666:329-335

Martin-Tanguy J. Conjugated polyamines and reproductive development: biochemical, molecular and physiological approaches. Physiol Plantarum 1997; 100:675-688

Metha R.-A., Handa A., Li N., Mattoo A.-K. Ripening-activated expression of S-adenosylmethionine decarboxylase increases polyamine levels and influences ripening in transgenic tomato fruits (abstract no. 134). Plant Physiol 1997; 114:5-44

Minocha S.C., Sun D.-Y. Stress tolerance in plants through transgenic manipulation of polyamine biosynthesis. Plant Physiol Supp 1997; 114:297

Pedrol N., Ramos P., Reigosa M.J. Ecophysiology of perennial grasses under water deficits and competition. In *Photosynthesis: Mechanisms and Effects*. Proceedings of the XI International Congress on Photosynthesis, Vol. V. G. Garab, ed. Dordrecht, The Netherlands: Kluwer Academic Publishers, 1999

Pedrol N., Ramos P., Reigosa M.J. Phenotypic plasticity and acclimation to water deficits in velvet-grass: a long-term greenhouse experiment. Changes in leaf morphology, photosynthesis and stress-induced metabolites. J Plant Physiol 2000; 157:383-393

Pell E.J., Dann M.S. "Multiple Stress-induced Foliar Senescence and Implications for Whole-Plant Longevity." In *Response of Plants to Multiple Stresses*. H.A. Mooney, W.E. Winner, E.J. Pell, eds. San Diego, California, USA: Academic Press, 1991

Poole C.F., Poole S.K. Instrumental Thin-Layer Chromatography. Anal Chem 1994; 66:27

Prasad M.N.V., Rengel Z. "Plant Acclimation and Adaptation to Natural and Antropogenic Stress." In *Stress of Life: From Molecules to Man*. Annu New York Acad Sci P. Csermely, ed. New York, USA: The New York Academy of Sciences, 1998

Rajam M.V. "Polyamines." In *Plant Ecophysiology* M.N.V. Prasad, ed. New York, USA: John Willey, 1997

Reggiani R., Hochkoeppler A., Bertani A. Polyamines in rize seedlings under oxygen-deficit stress. Plant Physiol 1989; 91:1197-1201

Santa-Cruz A., Acosta M., Pérez-Alfocea F., Bolarin M.C. Changes in free polyamine levels induced by salt stress in leaves of cultivated and wild tomato species. Physiol Plantarum 1997; 101:341-346

Seiler N. Identification and quantitation of amines by thin-layer chromatography. J Chromatogr 1971; 63:97-112

Serafini-Fracassini D. "Cell Cycle-dependent Changes in Plant Polyamine Metabolism." In *Biochemistry and Physiology of Polyamines in Plants*. R.D. Slocum, H.E. Flores, eds. Boca Ratón, Florida: CRC Press, 1991

Sherma J. "Basic Techniques, Materials, and Apparatus." In *Handbook of Thin Layer Chromatography*. J. Sherma, B. Fried, eds. New York, USA: Marcel Dekker Inc., 1990

Slocum R.D., Flores H.E. In *Biochemistry and Physiology of Polyamines in Plants*. R.D. Slocum, H.E. Flores, eds. Boca Ratón, Florida: CRC Press, 1991

Slocum R.D., Flores H.E., Galston A.W., Weinstein L.H. Improved method for HPLC analysis of polyamines, agmatine and aromatic monoamines in plant tissue. Plant Physiol 1989; 89:512-517

Slocum R.D., Weinstein L.H. "Stress-induced Putrescine Accumulation as a Mechanism of Ammonia Detoxification in Cereal Leaves." In *Polyamines and Ethylene: Biochemistry, Physiology, and Interactions*. H.E. Flores, R.N. Arteca, J.C. Shannon, eds. Rockville, Maryland: American Society of Plant Physiologists, 1990

Smith M.A., Davies P.J. "Monitoring Polyamines in Plant Tissues by High Performance Liquid Chromatography." In *High Performance Liquid Chromatography in Plant Sciences*. H.F. Linskens, J.F. Jackson, eds. Berlin, Germany: Springer-Verlag, 1987

Smith T.A. Polyamines. Annu Rev Plant Physiol Plant Mol Biol 1985; 36:117-143

Smith T.A., Best G.R. Polyamines in barley seedlings. Phytochemistry 1977; 16:841-843

Strizhov N., Abrham E., Okresz L., Blickling S., Zilberstein A., Schell J., Koncz C., Szabados L. Differential expression of two P5CS genes controlling proline accumulation during salt stress is regulated by ABA1, ABAII and AXR2 in *Arabidopsis*. Plant J 1997; 12:557-569

Tadolini B. Polyamine inhibition of lipoperoxidation. The influence of polyamines on iron oxidation in the presence of compounds mimicking phospholipid polar heads. Biochem J 1988; 249:33-36

Tiburcio A.F., Altabella T., Masgrau M. "Polyamines." In: *Plant Hormone Research*. K. Palme, J. Schell, eds. Springer-Verlag, 2001. In press.

Tiburcio A.F., Altabella T., Borrell A., Masgrau C. Polyamine metabolism and its regulation. Physiol Plantarum 1997; 100:664-674

Tiburcio A.-F., Besford R.-T., Capell T., Borrell A., Testillano P.-S., Risueño M.-C. Mechanisms of polyamine action during senescence responses induced by osmotic stress. J Exp Bot 1994; 45:1789-1800

Tiburcio A.F., Campos J.L., Figueras X., Besford R.T. Recent advances in the understanding of polyamine functions during plant development. Plant Growth Regul 1993; 12:331-340

Tiburcio A.F., Kaur-Sawhney R., Galston A.W. "Polyamine Metabolism." In *The Biochemistry of Plants, Intermediary Nitrogen Fixation*. B.J. Miflin, P.J. Lea, eds. New York, USA: Academic Press, 1990

Tiburcio A.F., Kaur-Sawhney R., Ingersoll R.B., Galston A.W. Correlation between polyamines and pyrrolidine alkaloids in developing tobacco callus. Plant Physiol 1985; 78:323-326

Torrigiani P., Rabiti A.L., Bortolotti C., Betti L., Marani F., Canova A., Bagni N. Polyamine synthesis and accumulation in the hypersensitive responses to TMV in *Nicotiana tabacum*. New Phytol 1997; 135:467-473

Walden R., Cordeiro A., Tiburcio A.F. Polyamines: small molecules triggering pathways in plant growth and development. Plant Physiol 1997; 113:1009-1013

Walter H.J.P., Geuns J.M.C. High speed HPLC analysis of polyamines in plant tissues. Plant Physiol 1987; 83:232-234

Wettlaufer S., Weinstein L.H. Quantitation of polyamines using thin-layer chromatography and image analysis. J Chromatogr 1988; 441:361-366

Willadino L., Camara T., Boget N., Claparols I., Santos M., Torné J.M. Polyamine and free amino acid variations in NaCl-treated embryogenic maize callus from sensitive and resistant cultivars. J Plant Physiol 1996; 149:179-185

Interesting web sites: http:/cobra.mdc.net/cyberguides/lcgc/cat11419.html (equipment and supplies for TLC and HPLC); http://www.chemistry.adelaide.edu.au/external/Soc-Rel/Content/tlc.htm (TLC procedures); http://analytical.chemweb.com (Analytical chemistry)

CHAPTER 22

FREE PROLINE QUANTIFICATION

Pilar Ramos Tamayo and Nuria Pedrol Bonjoch
Dpto Bioloxía Vexetal e Ciencia do Solo. Universidade de Vigo. Spain

INTRODUCTION

Free proline accumulation is one of the most frequently reported metabolic modifications induced by different stresses in plants. In spite of this, the precise role of proline in stress physiology, as well as the metabolic adjustments associated with its biosynthesis, still remain matters of controversy. Still today, there is no definitive evidence for the adaptive value of proline itself under adverse conditions (Gibon et al., 2000). Nonetheless, many researchers continue being attracted to the study of free proline regarding its significance in the integrated metabolic response to stress in plants.

A colorimetric method for measuring free proline

The most often used analytical method for proline quantification is the rapid colorimetric procedure described by **Bates et al. (1973)**. These authors described the technique as being adequate for its use in studies regarding water stress. Nonetheless, application to any type of stress has been widely demonstrated, including the less studied biotic stresses such as allelopathy (Sánchez-Moreiras, 1996) and resource competition (Pedrol et al., 1999).

M.J. Reigosa Roger, Handbook of Plant Ecophysiology Techniques, 365–382.

Chemical basis

The method proposed by Bates et al. (1973) is based in the reaction that takes place between *ninhydrin* and amino acids. Ninhydrin is a powerful oxidant that produces the oxidative deamination of the α-amino group, releasing ammonium, CO_2, the corresponding aldehide and ninhydrin in reduced form. Released ammonium reacts with an additional mol of ninhydrin and with reduced ninhydrin, thus producing a *coloured complex*. For the amino acids, this purple complex has an adsorption maximum at 570 nm, being such absorption nearly a lineal function of the quantity of amino groups present in the solution. Therefore, this reaction is a very convenient probe for quantitative colorimetric amino acids determination.

Considering that proline is strictly an *imino acid* (partially substituted amino group), the reaction with ninhydrin will be different to the rest of proteic amino acids, forming a coloured complex whose absorption maximum is near to 440 nm (Maler and Cordes, 1971).

Accuracy

Although several amino acids can interfere with proline quantification, it is noteworthy to state that the concentration of other free amino acids normally present in stressed plants are very low if compared to proline, so adsorption interference is considered as unimportant.

The *detection range* of the method is considered between 0.1 and 36 µmol (*free proline*) per g of fresh weight (plant tissue) (Bates et al. 1973). Following the base of this colorimetric reaction several authors have done *different modifications* in function of the type of stress, used material (species and tissue), possible interference of several reactives, or considering the need for pre-treatments when using *in vitro* callus cultures (method of Troll and Lindsley, 1955; modified by Magné and Larher, 1992; Lutts et al., 1996; Trotel-Aziz et al., 2000).

Before explaining the detailed procedure for free proline quantification in plant tissues, it is better to remember some ecophysiological questions regarding this amino acid in plants. They will be undoubtedly very useful to interpret our own results.

THE ECOPHYSIOLOGICAL SIGNIFICANCE OF FREE PROLINE

Plants and microorganisms accumulate frequently some solutes as a response to water and salt stress. When water potential in the soil decreases, the plant increases the metabolic response known as '**osmotic adjustment**', accumulating *cytoplasm solutes* that produce a decrease in inner water potential, thus allowing the plant to absorb water from the soil. These stress metabolites known as *compatible osmolytes*, are neutral organic compounds non toxic to the cell, that do not inhibit enzymatic activity even at high concentrations (Aspinall and Paleg, 1981).

These organic compatible solutes are also named '*osmoprotectants*' (Yancey, 1994, and see Chapter 20), because they have functions in minimising the impact of abiotic stresses on plants. These solutes are typically hydrophilic, which suggests they could replace water at the surface of proteins, protein complexes, or membranes, thus acting as non-enzymatic low-molecular-weight chaperones (Hasegawa et al., 2000; and see Chapter 20).

Compatible solutes include *diverse organic compounds* such as free amino acids (proline, alanine, β-alanine, taurine), quaternary ammonium compounds (prolinebetaine, glycinebetaine, β-alaninebetaine, glycerophosphorylcholine), tertiary sulphonium compounds (β-dimethylsulphoniopropionate) and carbohydrates (trehalose, sorbitol, mannitol, glycerol, pinitol). Glycinebetaine seems to be the best osmoprotectant, with prolinebetaine and hydroxyprolinebetaine presenting better capacity than proline. The acquisition of enzymatic activity to catalyse the conversion from proline to betaine derivatives seems to be an adaptive strategy that would allow some species to convert proline *pool* in better osmoprotectants (Hanson et al., 1994). Synthesis and accumulation of organic osmolytes are widespread in plants, but the *distribution* of specific compatible solutes varies among plant species. The amino acid proline is accumulated by a taxonomically diverse set of plants, whereas accumulation of the quaternary ammonium compound β-alanine betaine appears to be confined to representatives of a few genera in the Plumbaginaceae (Bray et al., 2000).

With respect to their *intracellular localisation*, free proline can be mainly found in cytoplasm (including also some organelles) while it is practically absent in vacuoles (Moftah and Michel, 1987). High concentrations of glycinebetaine have been found in chloroplasts isolated from several species of plants submitted to saline stress (McNeil et al., 1999).

Why measuring free proline?

Free proline accumulation in the leaves, shoots and roots in Angiosperms has been considered as one of the most widespread **stress induced response**. So, its measurement would be an excellent stress detector.

Free proline contents increases have been found under water and osmotic stress (Singh et al., 1972; Bhaskaran et al., 1985) and also when plants were subjected to salt, cold or freezing stress (Chu et al., 1974, 1976; Pérez-Alfocea et al., 1994). Proline can also accumulate in some non-stressed plant structures that dehydrate, like pollen and seeds (Chiang and Dandekar, 1995; Hua et al., 1997). This fact avails its role in the tolerance to water stress acting as osmoprotectant.

The accumulation of free proline will depend on the *type and intensity of stress*. Ten to 100-fold increments in free proline concentrations in leaves submitted to water deficits have been described, attaining up to 1 % of leaf fresh weight in some species (Aspinall and Paleg, 1981). Those enlargements are even more marked that those produced by salt stress (e.g., Chiang and Dardekar, 1995). Free proline concentration increase can be dramatic under severe stress when the initial proline content is low; however, it is possible that such an increase is not evident under mild stress (e.g., Pedrol et al. 2000). Those changes are only possible when the tissue carbohydrates concentrations are adequate and they can be reverted if optimal conditions are restored (Hsiao, 1973).

Many evidences relate proline metabolism with ***stress tolerance***, and many papers have studied proline accumulation as a parameter whose genetical transmission has been used in selection for tolerance to wilting or salinity in a great number of plant species (Dörffing et al., 1997; Lutts et al., 1999; Gibon et al., 2000).

Biosynthesis and metabolism of proline

Proline biosynthesis lies in *a central crossroads between carbon and nitrogen assimilation* metabolic pathways. Any change in proline concentration (be it stress related or not) will be accompanied with changes in the general nitrogen metabolism (Hanson and Hitz, 1982; Pearson and Stewart, 1987; Erskine et al., 1996), affecting also other molecules metabolism, like *proteins* (see Chapters 18 and 20) and *polyamines* (see Chapter 19). In higher plants there are several pathways leading to proline biosynthesis and they are interconnected (Samaras et al, 1995; see Fig. 1):

- Proline is synthesised in the cytoplasm mainly ***from glutamate***. Glutamate is produced by the action of glutamate synthetase, or by the transformation of the α-ketoglutarate produced in the tricarboxylic acids cycle (TCA) via glutamate dehydrogenase. First, the enzyme P5CS (Δ^1-pyrroline-5-carboxylate synthetase) catalyses in two steps the transformation to glutamic-γ-semialdehyde, which spontaneously produces Δ^1-pyrroline-5-carboxylate; in the second step, P5CR (Δ^1-pyrroline-5-carboxylate reductase) catalyses final conversion to proline (see Fig. 1).
- Other synthesis pathways come ***from ornithine*** (Fig. 1): both forms of OAT enzyme (ornithine aminotransferase) catalyse the reactions that are previous to the Δ^1-pyrroline-5-carboxylate or Δ^1-pyrroline-2-carboxylate formation, which will derive to proline by the action of P5CR (Δ^1-pyrroline-5-carboxylate reductase) and P2CR (Δ^1-pyrroline-2-carboxylate reductase) enzymes, respectively. It is noteworthy that ornithine also proceeds from glutamate, and it is also a common precursor of polyamine synthesis; in this way, derivation of these amino acids towards one or other synthesis route will be partially determinant of *the relative levels proline and polyamines* in plant tissues (Slocum and Weinstein, 1990; Lutts et al, 1999).

Proline *catabolism* is less known at least its regulation processes. The first step consists in the proline oxidation to Δ^1-pyrroline-5-carboxylate, reaction that is catalysed by proline-dehydrogenase in inner mitocondrial membrane. In plants this enzyme is linked to the electron transport chain, thus meaning that proline degradation is coupled to ATP synthesis (Elthon and Stewart, 1982). Beginning with Δ^1-pyrroline-5-carboxylate, other enzyme-mediated reactions finally lead to the glutamate formation (see Samaras et al., 1995). Most part of proline that accumulated during water stress is rapidly metabolised (oxidised till glutamate) as soon as stress has ceased; only a relatively small quantity of proline derives to protein synthesis (Stewart, 1981).

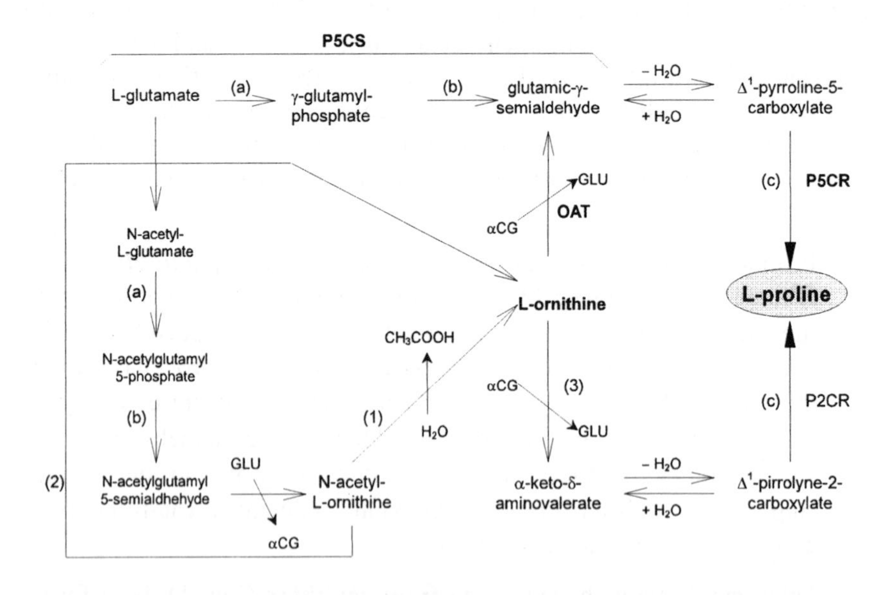

Figure 1. Alternative pathways of synthesis of proline in higher plants. P5CS, Δ^1-pyrroline-5-carboxylate synthetase; P5CR, Δ^1-pyrroline-5-carboxylate reductase; P2CR, Δ^1-pyrroline-2-carboxylate reductase; αCG, α-ketoglutarate; GLU, L-glutamate; (1) N^2-acetyl-L-ornithine-amidohydrolase; (2) N^2-acetyl-L-ornithine:L-glutamate N-acetyltransferase; OAT, ornithine-δ-aminotransferase; (3) ornithine-α-aminotransferase (a) +ATP→ADP; (b) +NADPH+H^+→$NADP^+$+Pi; (c) +NADPH+H^+→$NADP^+$. After Samaras et al. (1995).

Proline accumulation: stress symptom or adaptative response?

Some authors have considered proline accumulation in tissues as a *pathological consequence of the stress* better than an adaptive response (Bhaskaran et al., 1985; Ibarra-Caballero et al., 1988; Pérez-Alfocea et al., 1994). When a plant has a low tolerance or does not possess acclimation mechanisms, a long submission to severe stress conditions will finally lead to senescence. Initially there will be a shortening in protein synthesis and an increase in protein degradation, thus increasing the free amino acids cytoplasm concentrations (Wen et al., 1996) including proline. Nevertheless, although the latter hypothesis seems reasonable, other authors have described important simultaneous increases in free proline and soluble proteins during more or less long water deficits (Pearson and Stewart, 1990; Ashraf and Yasmin, 1995; Pedrol et al., 2000), without finding any relationship between free proline accumulation and protein metabolism damage.

In some occasions, under the point of view of cultivated plants breeding, and especially in the selection drought and salt tolerant cereals, free proline accumulation has been considered as *a symptom of sensibility* to water stress, without considering this has an adaptive value (Dutta Gupta et al, 1995; Lutts et al, 1996, 1999). This is because selection of tolerant genotypes is done relatively to the maximum yield under stress conditions, whilst proline accumulation is frequently accompanied by a growth decrease (photosynthetic assimilates are mainly used for osmoprotection and storage -Bellinger et al., 1991-). But, under our *ecophysiological point of view*, we must take into account that, in wild plants, a reduction of transpirating biomass (i.e., leaves) or a lower invest to 'expensive' organs (e.g., fruits) can be rather an adaptative strategy, far from a stress symptom.

Hanson et al. (1979) proposed to select barley cultivars using the lowest proline concentration as a tolerance index, stating that those tolerant genotypes are the ones that present a positive correlation between proline accumulation and crop yield, while in resistant genotypes that correlation would be negative. But we cannot forget that plants can accumulate preferentially other compatible solutes (K^+, putrescine, poliols, etc.) and this also can decrease *correlation between resistance/tolerance and proline*. In the middle of controversy, the truth is that there is not a clear consensus about the interpretation of proline accumulation as a possible mechanism of *resistance* or *tolerance*. Some authors conclude that high proline accumulation in strong stress is associated to water deficit tolerant genotypes (Rhodes and Hanson, 1993), whilst other authors find a positive correlation of free proline concentration and resistance indexes (Singh et al., 1972).

Notwithstanding, proline accumulation has been observed that is variable between cultivars independently of their salt tolerance, and also in the different development phases (Lutts et al., 1995, in *Oryza sativa*); so, many researchers continue considering adequate the use of proline super-production and the genetic manipulation of proline synthesis as a mean to confer water stress tolerance (Van Rensburg et al., 1993; Zhang et al., 1995).

Besides the previous considerations, several physiological causes of proline accumulation have been considered related to ABA mediated regulation, thus availing a clear ***adaptive significance*** as response to several kinds of stress. Let us see some of them:

Possible causes of proline accumulation during stress

Induction of *de novo* synthesis

Heyser et al. (1989) and Chiang and Dardekar (1995) proved using marked molecules the incorporation of glutamate, ornithine, and arginine to the *de novo* proline synthesis in plants subjected to water stress, in a process dependent of ABA concentration. The gene codifying P5CR (Figure 1) can be super-expressed as a response to salinity (Delauney and Verma, 1990; Verbruggen et al., 1993) and drought (Hare and Cress, 1996); P5CS super-expression under water deficit has also been described (Yoshiba et al., 1995). OAT activity can increase three-fold due to water (Sundaresan and Sudhakaran, 1995) and salinity can induce the expression of its codifying gene (Lutts et al., 1999).

Induction of transport mechanisms

Rentsch et al. (1996) found that nitrogen distribution in plants during stress was dependent of proline *specific permeases*: water stress induces the genetic expression of *proline carriers*, while in the same conditions other amino acid transporters synthesis are inhibited; besides this, salinity and water deficit regulate genetic expression of permeases differently. Recently, as previously cited authors had supposed, it has been proved that proline accumulation in tissues under low water potentials can be explained by the *phloem transport from other parts of the plant*, and not necessarily by *de novo* synthesis (Verslues and Sharp, 1999).

Oxidation inhibition

Decreases up to 50% in proline dehydrogenase activity have been observed in plants subjected to water deficit, as well in isolated mitochondria (Rapayani and Stewart, 1991) as in whole plants (Sundaresan and Sudhakaran, 1995). Water stress inhibits genetic expression of that enzyme (Kiyouse et al., 1996; Taji et al., 1999), thus suggesting that proline mitochondrial degradation is inhibited by the same conditions that induce its cytoplasm synthesis.

The ecophysiological roles of proline

As previously stated, the role of osmoprotectant against low water potential has been frequently assigned to proline. Nevertheless, proline contribution to osmoregulation is small in most cultivated species in stress conditions. Several *non-osmotic* important roles have been assigned recently to proline water stress accumulation; in some cases *local proline concentration* is hypothetically more important than absolute quantity (Hasegawa et al., 2000).

As stated before, there are now two different trends in the interpretation of proline in stress conditions:

- Although as matter of controversy (see above), proline accumulation can be considered as a pathological consequence of stress, that is, a ***damage symptom*** better than a resistance indicator (Bhaskaran et al., 1985; Moftah and Michel, 1987; Ibarra-Caballero et al., 1988; García et al., 1997; Lutts et al., 1999).
- Proline can also be considered as an ***adaptive response to stress***, with recent special emphasis in proline adaptative roles without needing massive proline accumulation (Taylor, 1996; Hasegawa et al., 2000). Several general regulation functions have been cited:

   1. **Osmoregulation**, as explained before.
   2. **Metabolism protection against stress conditions**. In this sense, proline is implicated in the protection of cytosolic enzymes as nitrate-reductase and rubisco (Bandurska, 1993; Solomon et al., 1994, respectively), stabilisation of tertiary protein structure against the inactivation produced by heat (Solomon et al., 1994), or protection of intra-cellular structures, stabilising membranes by interaction with phospholipids, thus affecting membrane integrity and chloroplast ultrastructure (Van Rensburg et al., 1993).
   3. **Cytosol pH regulation**

      It has been suggested that glucides and photosynthetic energy are both implied in stress proline accumulation. Although light is not used to produce proline synthesis it stimulates this synthesis. Sugars would serve as a source of carbonated structures, and light energy would give the NAD(P)H and ATP needed to synthesise proline from glutamate. Proline synthesis depends on the sugar catabolism and respiration proved by the fact that inhibitors of Krebs cycle (providing α-cetoglutaric acid) or of glucolisis, as well as anoxia, avoid proline accumulation. Under water stress respiratory and photorespiratory organic acids production is stimulated, thus acidificating cytosol. pH regulation related to proline synthesis has been suggested, due to the fact that in that process 3 molecules of $NAD(P)^+$ per synthesised molecule of proline are produced, thus allowing a shift in cytosolic pH (Venekamp, 1989).

4. **Carbon and Nitrogen storage**. Proline could serve as a storage compound that is easily mobilised once growth conditions are restored (Jäger and Meyer, 1977). Besides, after water deficit ceases, degradation of accumulated free proline produces energy (ATP) usable for the recuperation of cell functions.
5. **Antioxidant function.** Proline intervenes in the prevention of oxygen radical production or it can act as scavenger of reactive oxygen species (ROS) and other free radicals (Smirnoff and Cumbes, 1989; Alia et al., 1995).
6. Finally, the role of proline in the plant development is still unknown, although there are some proofs of its participation in several **morphogenetic phenomena** (Nanjo et al., 1998).

A PROTOCOL FOR PROLINE QUANTIFICATION IN PLANT TISSUES (Bates et al. 1973)

Required reagents:

Sulphosalicylic acid 3 % (w/v)

Phosphoric acid 6 M

Glacial acetic acid

Acid-ninhydrin, prepared by warming 1.25 g ninhydrin in 30 mL glacial acetic acid, and adding 20 mL phosphoric acid 6 M (Take care: Acid ninhydrin will keep stable only for 24 hours, at 4°C).

Purified proline

Toluene

Preparation of the standard curve

Sample absorbances can be converted to proline concentration by comparison to a proline standard curve obtained using known proline concentrations. Every time proline is calculated, a standard curve must be prepared.

A trial-and-error assay will be often necessary to *select our adequate standard curve points*. Several proline solutions must be prepared at different concentrations; they must cover the expected range of unknown sample concentrations. Different solutions are prepared by adding distilled water to an 'stock' solution of commercial purified proline.

All tubes in standard curve will follow the same procedure used with the samples, according to ninhydrin reaction (see Step 2).

Steps of the method

Here we describe the method for free proline quantification in leaf tissues, which includes four key steps (see Figure 2):

1. Homogenisation and filtration

250 to 500 mg of fresh leaf tissue (obtained from the youngest fully expanded leaves) are powdered in a cold mortar with liquid nitrogen until pulverisation and homogenised with 5 mL of sulphosalicylic acid 3 % (to get protein precipitation). Then, the homogenate is filtered through a n· 2 filter paper (e.g., Whatman® n· 2).

The required amount of fresh sample will depend on the tissue's water content. A trial-and-error approach will be often necessary (see Chapter 21 for further explanation).

2. Reaction with ninhydrin

2 ml of the filtrated sample are mixed with 2 ml of glacial acetic acid and 2 ml of acid-ninhydrin (previously prepared) in test tubes. After agitation, the sample is incubated at 100 °C during 1 hour, and this produces the coloured complex formation. After one hour, reaction is stopped by putting the tubes in an ice bath.

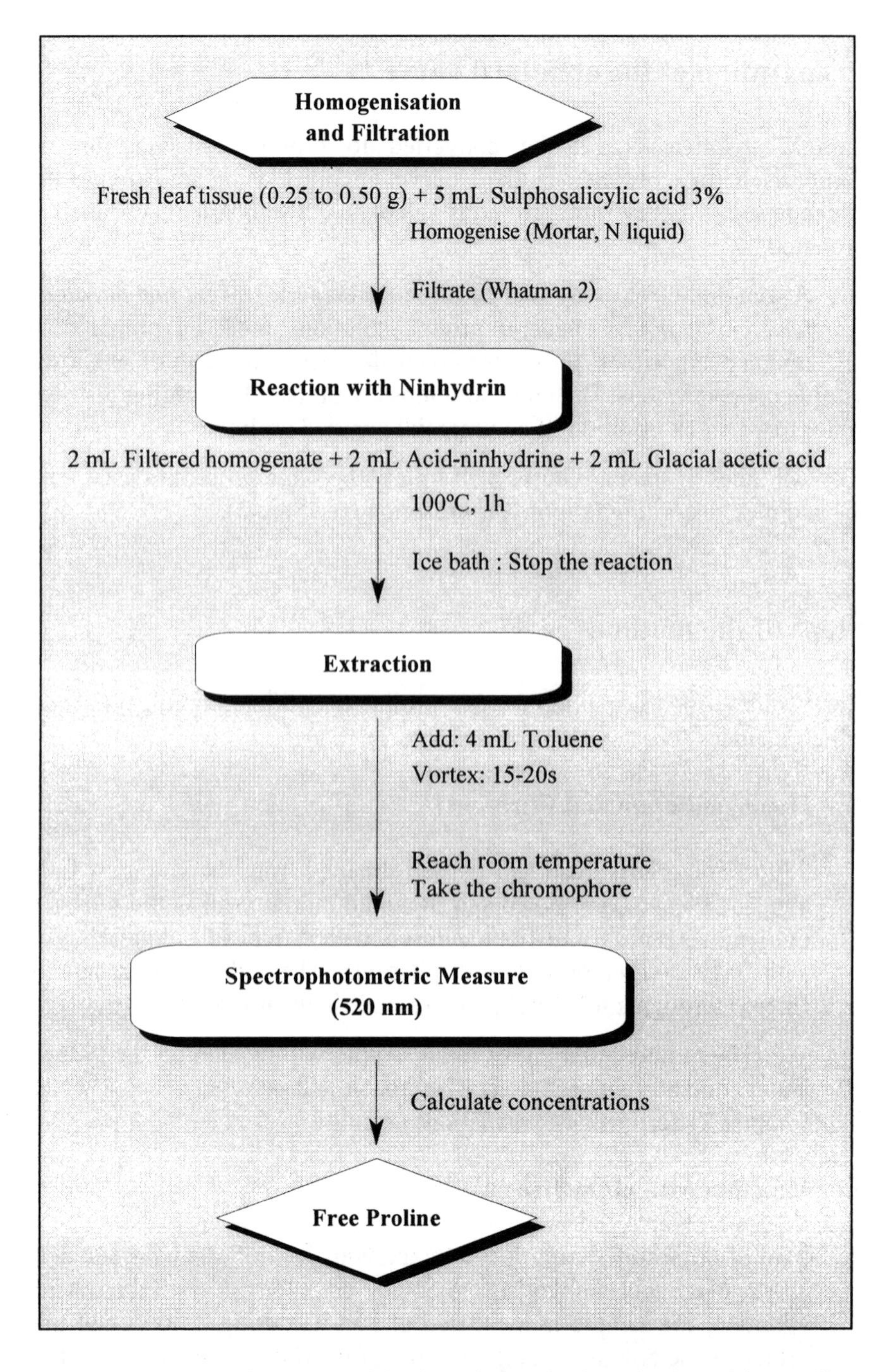

Figure 2. Main steps of the protocol in the colorimetric determination of free proline (Bates et al., 1973).

3. Extraction

4 mL toluene are added to each tube, and vortexed for 15-20 s. Then, organic and inorganic phases are separated, obtaining the cromophore dissolved in toluene.

4. Spectrophotometric measurement

The maximum absorbance of the chromophore when dissolved in toluene is attained at a wavelength of 520 nm. Thus, the organic phase of each sample and standard containing the chromophore is collected, and absorbance at 520 nm is measured spectrophotometrically.

Calculations and data analysis

Using the data obtained from the standard curve prepared at known concentrations, a lineal regression is done (comparing absorbance *vs.* proline concentration). Thus, we obtain the equation that will allow sample concentration calculations.

Finally, it is advisable to express free proline concentration relative to dry weight (μmol/g of dry weight of plant material).

Further considerations

As previously affirmed, abiotic stress factors as salinity or drought can induce the expression or super-expression of genes that codify proline synthetic enzymes, as OAT (ornithine aminotransferase), P5CR (Δ^1-pyrroline-5-carboxylate reductase), or P5CS (Δ^1-pyrroline-5-carboxylate synthetase) (Verbruggen et al., 1993; Hare and Cress, 1996; Sundaresan & Sudhakaran, 1995). So, another kind of studies that could provide valuable information on the physiological significance of proline accumulation is investigating about the effects of stress on enzyme activities involved in proline metabolism (Yoshiba et al., 1995; Zhang et al., 1995).

In vitro tissue culture constitutes a useful tool to study the cellular mechanisms of stress resistance since it allows a control of the homogeneity of the stress and its application at the cellular level, independently of regulatory mechanisms occurring at the whole plant level (Lutts et al., 1996). Cell level studies using callus cultures have provided knowledge about synthesis, regulation, accumulation and possible functions of proline.

However, Plant Ecophysiology is based in integrated studies in whole plant, considering the different plant parts, relationships between different metabolites, and between different parameters affecting plant physiology.

REFERENCES

Alia Prasad K.V.S.K., Pardha Saradhi P. Effect of zinc on free radicals and proline in *Brassica* and *Cajanus*. Phytochemistry 1995; 39:45-47

Ashraf M., Yasmin N. Responses of 4 arid zone grass species from varying habitats to drought stress. Biol Plantarum 1995; 37:567-575

Aspinall D., Paleg L.G. "Proline Accumulation: Physiological Aspects." In *The Physiology and Biochemistry of Drought Resistance in Plants*. L.G. Paleg, D. Aspinal, eds. New York, USA: Academic Press, 1981

Bandurska H. *In vivo* and *in vitro* effect of proline on nitrate reductase activity under osmotic stress in barley. Acta Physiol Plantarum 1993; 15:83-88

Bates L., Waldren R.P., Teare I.D. Rapid determination of free proline for water stress studies. Plant Soil 1973; 39:205-207

Bellinger Y., Bensaoud A., Larher F. "Physiological Significance of Proline Accumulation, a Trait of Use to breeding for Stress Tolerance." In *Physiology-Breeding of Winter Cereals for Stressed Mediterranean Environments*. Paris: INRA Les Colloques 55, 1991

Bhaskaran S., Smith R.H., Newton R.J. Physiological changes in cultured sorghum cells in response to induced water stress. I. Free proline. Plant Physiol 1985; 79:266-269

Bray E.A., Bailey-Serres J., Weretilnyk E. "Responses to Abiotic Stresses." In *Biochemistry and Molecular Biology of Plants*. B.B. Buchanan, W. Gruissem, R.L. Jones, eds. Maryland: American Society of Plant Physiologists, 2000

Chiang H.H., Dandekar A.M. Regulation of proline accumulation in *Arabidopsis thaliana* (L.) Heynh during development and in response to dessication. Plant Cell Environ 1995; 18:1280-1290

Chu T.M., Aspinall D., Paleg F.J. Stress metabolism. VI. Temperature stress and the accumulation of proline in barley and radish. Aust J Plant Physiol 1974; 1:87-97

Chu T.M., Aspinall D., Paleg F.J. Stress metabolism. VII. Salinity and proline accumulation in barley. Aust J Plant Physiol 1976; 3:219-228

Delauney A.J., Verma D.P.S. A soybean gene encoding Δ^1-pyrroline-5-carboxylate reductase was isolated by functional complementation in *Escherichia coli* and is found to be osmoregulated. Mol Gen Genet 1990; 221:299-305

Dörffling K., Dörffling H., Lesselich G., Luck E., Zimmermman C., Melz G. Heritable improvement of frost tolerance in winter wheat by *in vitro*-selection of hydroxyproline resistant proline overproducing mutants. Euphytica 1997; 93:1-10

Dutta Gupta S., Auge R.M., Denchev P.D., Conger B.V. Growth, proline accumulation and water relations of NaCl-selected and non-selected callus lines of *Dactylis glomerata* L. Environ Exp Bot 1995; 35:83-92

Elthon T.E., Stewart C.R. Proline oxidation in corn mitochondria. Plant Physiol 1982; 70:567-572

Erskine P.D., Stewart G.R., Schmidt S., Turnbull M.H., Unkovich M., Pate J.S. Water availability - a physiological constraint on nitrate utilization in plants of Australian semi-arid mulga woodlands. Plant Cell Environ 1996; 19:1149-1159

García A.B., de Almeida E.J., Iyer S., Gerats T., Van Montagu M., Caplan A.B. Effects of osmoprotectants upon NaCl stress in rice. Plant Physiol 1997; 115:159-169

Gibon Y., Ronan S., Larher F. Proline accumulation in canola leaf discs subjected to osmotic stress is related to the loss of chlorophylls and to the decrease of mitochondrial activity. Physiol Plantarum 2000; 110:469-476

Hanson A.D., Hitz W.D. Metabolic responses of mesophytes to plant water deficits. Annu Rev Plant Physiol Plant Mol Biol 1982; 33:163-203

Hanson A.D., Nelsen C.E., Pedersen A.R., Everson E.H. Capacity for proline accumulation during water stress in barley and its implications for breeding for drought tolerance. Crop Sci 1979; 19:489-493

Hanson A.D., Rathinasabapathi B., Rivoal J., Burnet M., Dillon M.O., Gage D.A. Osmoprotective compounds from the Plumbaginaceae: a natural experiment in metabolic engineering of stress tolerance. Proc Natl Acad Sci USA 1994; 91:306-310

Hare P.D., Cress W.A. Tissue-specific accumulation of transcript encoding Delta(1)-pyrroline-5-carboxylate reductase in *Arabidopsis thaliana*. Plant Growth Regul 1996; 19:249-256

Hasegawa P.M., Bressan R.A., Zhu J-K., Bohnert H.J. Plant cellular and molecular responses to high salinity. Annu Rev Plant Physiol Plant Mol Biol 2000; 51:463-499

Heyser J.W., Chacon M.J., Warren R.C. Characterization of L-[5-^{13}C]-proline biosynthesis in halophytic and nonhalophytic suspension cultures by ^{13}C NMR. J Plant Physiol 1989; 135:459-466

Hsiao T. C. Plant responses to water stress. Annu Rev Plant Phys 1973; 24:519-570

Hua X.J., Van de Cotte B., Vann Montagu M., Verbruggen N. Developmental regulation of pyrroline-5-carboxylate reductase gene expression in *Arabidopsis*. Plant Physiol 1997; 114:1215-1224

Ibarra-Caballero J., Villanueva-Verduzco C., Molina-Galán J., Sánchez-De Giménez E. Proline accumulation as a sympton of drought stress in maize: a tissue differentiation requirement. J Exp Bot 1988; 39:889-897

Jäger H., Meyer H.R. Effect of water stress on growth and proline metabolism of *Phaseolus vulgaris* L. Oecologia 1977; 30:83-86

Kiyouse T., Yoshiba Y., Yamaguchi-Shinozaki K., Shinozaki K. A nuclear gene encoding mitochondrial proline dehydrogenase, an enzyme involved in proline metabolism, is upregulated by proline but downregulated by dehydration in *Arabidopsis*. Plant Cell 1996; 8:1323-1335

Lutts S., Kinet J.M., Bouharmont J. Changes in plant responses to NaCl during development of rice (*Oryza sativa* L.) varieties differing in salinity resistance. J Exp Bot 1995; 46:1843-1852

Lutts S., Kinet J.M., Bouharmont J. Effects of various salts and of mannitol on ion and proline accumulation in relation to osmotic adjustment in rice (*Oryza sativa* L.) callus cultures. J Plant Physiol 1996; 149:186-195

Lutts S., Majerus V., Kinet J.-M. NaCl effects on proline metabolism in rice (*Oryza sativa*) seedlings. Physiol Plantarum 1999; 105:450-458

Magné C., Larher F. High sugar content interferes with colorimetric determination of amino acids and free proline. Anal Biochem 1992; 200:115-118

Maler H.R., Cordes E.H. *Química Biológica*. Barcelona: Omega, 1971

McNeil S.D., Nuccio M.L., Hanson A.D. Betaines and related osmoprotectants. Targets for metabolic engineering of stress resistance. Plant Physiol 1999; 120:945-949

Moftah A.E., Michel B.E. The effect of sodium chloride on solute potential and proline accumulation in soybean leaves. Plant Physiol 1987; 83:238-240

Nanjo T., Yoshiba Y., Sanada Y., Wada K., Tsukaya H., Kakubari Y., Yamaguchi-Shinozaki K., Shinozaki K. Roles of proline in osmotic stress tolerance and morphogenesis of *Arabidopsis thaliana.* Plant Cell Physiol Supp 1998; 39:104

Pearson J., Stewart G.R. "Aspects of Nitrogen Metabolism in Barley in Relation to Drought Stress." In *Drought Resistance in Plants: Physiological and Genetic Aspects*. Report EUR 10700, L. Monti, E. Porceddu, eds. Brussels: Commission of the European Communities, 1987

Pearson J., Stewart G.R. Free proline and prolamin protein in the grain of three barley varieties to a gradient of water supply. J Exp Bot 1990; 226:515-519

Pedrol N., Ramos P., Reigosa M.J. "Ecophysiology of perennial grasses under water deficits and competition." In *Photosynthesis: Mechanisms and Effects*. Proceedings of the XIth International Congress on Photosynthesis Vol. V, G. Garab, ed. Dordrecht, The Netherlands, Kluwer Academic Publishers, 1999

Pedrol N., Ramos P., Reigosa M.J. Phenotypic plasticity and acclimation to water deficits in velvet-grass: a long-term greenhouse experiment. Changes in leaf morphology, photosynthesis and stress-induced metabolites. J Plant Physiol 2000; 157:383-393

Pérez-Alfocea F., Santa-Cruz A., Guerrier G., Bolarin M.C. NaCl stress-induced organic solute changes on leaves and calli of *Lycopersicon esculentum* L. pennelli and their interspecific hybrid. J Plant Physiol 1994; 143:106-111

Rapayani P.J., Stewart C.R. Solubilization of a proline dehydrogenase from maize (*Zea mays* L.) mitochondria. Plant Physiol 1991; 95:787-791

Rentsch D., Hirner B., Schmelzer E., Frommer W.B. Salt stress-induced proline transporters and salt stress-repressed broad-specificity amino-acid permeases identified by suppresion of a yeast amino-acid permease-targeting mutant. Plant Cell 1996; 8:1437-1446

Rhodes D., Hanson A.D. Quaternary ammonium and tertiary sulphonium compounds in higher plants. Annu Rev Plant Physiol Plant Mol Biol 1993; 44:357-384

Samaras Y., Bressan R.A., Csonka L.N., García-Ríos M.G., Paino D'urzo M., Rhodes D. "Proline Accumulation during Drought and Salinity." In *Environment and Plant Metabolism: Flexibility and Acclimation.* N. Smirnof, ed. Oxford: Bios Scientific, 1995

Sánchez-Moreiras A.M. Efecto de compuestos fenólicos en *Lactuca sativa* L. Tesis de Licenciatura. Vigo, Spain: University of Vigo, 1996

Singh T.N., Aspinall D., Paleg L.G. Proline accumulation and varietal adaptability to drought in barley: a potential metabolic measure of drought resistance. Nat New Biol 1972; 236:188-190

Slocum R.D., Weinstein L.H. "Stress-induced Putrescine Accumulation as a Mechanism of Ammonia Detoxification in Cereal Leaves." In *Polyamines and Ethylene: Biochemistry, Physiology, and Interactions.* H.E. Flores, R.N. Arteca, J.C. Shannon, eds. Maryland: American Society of Plant Physiologists, 1990

Smirnoff N., Cumbes Q.J. Hydroxyl radical scavenging activity of compatible solutes. Phytochemistry 1989; 28:1057-1060

Solomon A., Beer S., Waisel Y., Jones G.P., Paleg L.G. Effects of NaCl on the carboxylating activity of Rubisco from *Tamaxis jordanis* in the presence and absence of proline-related compatible solutes. Physiol Plantarum 1994; 90:198-204

Stewart G.R. "Proline Accumulation: Biochemical Aspects." In *The Physiology and Biochemistry of Drought Resistance in Plants.* L.G. Paleg, D. Aspinal, eds. Sydney: Academic Press, 1981

Sundaresan S., Sudhakaran P.R. Water stress induced alterations in the proline metabolism of drought-susceptible and drought-tolerant cassava (*Manihot esculenta*) cultivars. Physiol Plantarum 1995; 94:635-642

Taji T., Seki M., Yamaguchi-Shinozaki K., Kamada H., Giraudat J., Shinozaki K. Mapping of 25 drought-inducible genes, RD and ERD, in *Arabidopsis thaliana.* Plant Cell Physiol 1999; 40:119-123

Taylor C.B. Proline and water deficit. Ups, downs, ins and outs. Plant Cell 1996; 8:1221-1224

Troll W., Lindlsey J. A photometric method for determination of proline. J Biol Chem 1955; 215:655-660

Trotel-Aziz P., Niogret M.-F., Larher F. Proline level is partly under the control of abscisic acid in canola leaf discs during recovery from hyper-osmotic stress. Physiol Plantarum 2000; 110:376-383

Van Rensburg L., Krüger G.H., Krüger. H. Proline accumulation as drought-tolerance selection criterion: its relationship to membrane integrity and chloroplast ultrastructure in *Nicotiana tabacum* L. J Plant Physiol 1993; 141:188-194

Venekamp J.H. Regulation of cytosol acidity in plants under conditions of drought. Physiol Plantarum 1989; 76:112-117

Verbruggen N., Villarrole R., Van Montagu M. Osmoregulation of a pyrroline-5-carboxylate reductase gene in *Arabidopsis thaliana.* Plant Physiol 1993; 103:771-781

Verslues P.E., Sharp R.E. Proline accumulation in maize (*Zea mays* L.) primary roots at low water potentials. II. Metabolic source of increased proline deposition in the elongation zone. Plant Physiol 1999; 119:1349-1360

Wen J.Q., Tan B.C., Liang H.G. Changes in protein and amino acid levels during growth and senescence of *Nicotiana rustica* callus. J Plant Physiol 1996; 148:707-710

Yancey P.H. "Compatible and counteracting solutes". In *Cellular and molecular physiology of cell volume regulation.* K Strange, ed. Boca Raton, Florida, USA, CRC Press, 1994

Yoshiba Y., Kiyouse T., Katagiri T., Udea H., Mizoguchi T., Yamaguchi-Shinozaki K., Wada K., Harada Y., Shinozaki K. Correlation between the induction of a gene for Δ^1-pyrroline-5-carboxylase synthetase and the accumulation of proline in *Arabidopsis thaliana* under osmotic stress. Plant J 1995; 7:751-760

Zhang C.S., Lu Q., Verma D.P.S. Removal of feedback inhibition of Delta(1)-pyrroline-5-carboxylate synthetase, a bifunctional enzyme catalyzing the first two steps of proline biosynthesis in plants. J Biol Chem 1995; 270:20491-20496

General References

Aspinall D., Paleg L.G. "Proline Accumulation: Physiological Aspects." In *The Physiology and Biochemistry of Drought Resistance in Plants.* L.G. Paleg, D. Aspinal, eds. New York, USA: Academic Press, 1981

Bray E.A., Bailey-Serres J., Weretilnyk E. "Responses to Abiotic Stresses." In *Biochemistry and Molecular Biology of Plants.* B.B. Buchanan, W. Gruissem, R.L. Jones, eds. Maryland: American Society of Plant Physiologists, 2000

Samaras Y., Bressan R.A., Csonka L.N., García-Ríos M.G., Paino D'urzo M., Rhodes D. "Proline Accumulation during Drought and Salinity." In *Environment and Plant Metabolism: Flexibility and Acclimation.* N. Smirnof, ed. Oxford: Bios Scientific, 1995

Stewart G.R. "Proline Accumulation: Biochemical Aspects." In *The Physiology and Biochemistry of Drought Resistance in Plants.* L.G. Paleg, D. Aspinal, eds. Sydney: Academic Press, 1981

Taylor C.B. Proline and water deficit. Ups, downs, ins and outs. Plant Cell 1996; 8:1221-1224

CHAPTER 23

ASSESSMENT OF D-RIBULOSE-1,5-BISPHOSPHATE CARBOXYLASE/OXYGENASE (RUBISCO) ENZYMATIC ACTIVITY

Teodoro Coba de la Peña[1], Adela M. Sánchez-Moreiras[2], Xoan Xosé Santos Costa[2], and Ana Martínez Otero[2]
[1]*Dpto Fisiología y Bioquímica Vegetal. Centro de Ciencias Medioambientales. Consejo Superior de Investigaciones Científicas. Madrid. Spain*
[2]*Dpto Bioloxía Vexetal e Ciencia do Solo. Universidade de Vigo. Spain*

INTRODUCTION

D-ribulose-1,5-bisphosphate carboxylase/oxygenase (Rubisco, EC 4.1.1.39) catalyses atmospheric CO_2 uptake in plant leaves. This reaction allows a following net synthesis of sugars through the Calvin cycle. Thereby, this carboxylase activity of Rubisco has a key importance in photosynthesis (Portis, 1992).

This enzyme is located on the stromal surface of the thylacoid membrane, and it represents more than 16 % of total proteins of the chloroplast. That means that it is the most abundant enzyme in plants (Jensen, 1977).

In higher plants and some other organisms, Rubisco is an enzymatic complex composed by 8 small (S) chains or subunits (codified by nuclear DNA) with a molecular weight of 13 kDa each, and 8 large (L) subunits (codified by chloroplastic DNA) with a molecular weight of 55 kDa each. Assembly of all these subunits occurs in the chloroplast stroma, building the whole holoenzyme (L_8S_8), also called 'Form I'.

M.J. Reigosa Roger, Handbook of Plant Ecophysiology Techniques, 383–397.

Each L chain has both a catalytic site and a regulatory site, and it is in the L chains were principal enzymatic activity is located. S chain stimulates the catalytic activity of L chain (Hartman and Harpel, 1994; Gutteridge and Gatenby, 1995).

The simplest Rubisco found in some photosynthetic bacteria is a dimer of L-subunits and it is called ‘Form II’.

Rubisco has two substrates: plant ribulose-1,5-bisphosphate (which binds first to the enzyme; Lorimer et al., 1976) and atmospheric carbon dioxide. Reaction catalysed by Rubisco is:

$$\text{Ribulose 1,5-bisphosphate (RuBP)} + CO_2 \rightarrow 2\ \text{(3-phosphoglycerate)}$$

The whole reaction that induces the synthesis of one sugar (hexose) molecule involves 6 rounds of the Calvin cycle, and it takes place during the biochemical phase of photosynthesis:

$$6\ \text{RuBP} + 6\ CO_2 + 18\ \text{ATP} + 12\ \text{NADPH} + 12\ H^+ + 12\ H_2O \rightarrow$$
$$6\ \text{RuBP} + \text{Hexose} + 18\ \text{ADP} + 18\ \text{Pi} + 12\ \text{NADP}^+$$

Rubisco is a bifunctional enzyme: it has also an oxygenase activity, catalysed by the same active site. This catalytic activity is responsible for photorespiration, which decreases the yield of photosynthesis about 30 to 50 %. Under normal atmospheric conditions and 25 °C, 3 of every 4 molecules of RuBP are carboxylated, and 1 of every 4 is oxygenated by Rubisco (Portis, 1992; Hartman and Harpel, 1994).

Up to now it is not clear why Rubisco owns this oxygenase activity. It has been tried to obtain an enzyme devoid of this activity by enzymatic engineering (site-directed mutagenesis). This achievement would involve a very important increase in photosynthesis efficiency. But regulation of Rubisco gene expression, translation and enzyme activity have revealed to be very complex, rendering very difficult to obtain the desired modified enzyme (Hartman and Harpel, 1994).

Recently, Sivakumar et al. (1998) have observed an *in vitro* suppression of purified Rubisco activity induced by proline accumulation (due to salt stress) in some higher plants, like *Brassica juncea, Sesbania sesban* and *Oryza sativa.* This was a surprising result, given that a protective role of proline accumulation on enzyme and protein functions and integrity during stress was supposed. In the same way, these results suggest that salt or water stress, and maybe allelochemicals that induce different types of stress, can alter Rubisco activity *in vivo* and thereby photosynthetic activity in higher plants. If so, Rubisco could be used as a stress indicator.

In our lab, we have observed proline accumulation following stress or allelochemical treatment (Sánchez Moreiras, 1996; Reigosa et al., 1999; González-Vilar et al., unpublished). It will be very interesting to verify whether *in vivo* Rubisco activity is altered or suppressed under these conditions.

Rubisco has several particular characteristics to be taken into account in the enzymatic assay:

The enzyme exists in several forms (Portis, 1992). The three main forms are:

- Inactive form of Rubisco (E), devoid of catalytic activity.
- Inactive and carbamylated form (EC), obtained after incorporation of CO_2 to a non-charged ε-amino of a lysine residue in the E form of the enzyme. This reaction induces the formation of a carbamate residue, negatively charged. The CO_2 involved in this carbamylation is called 'CO_2 activator', which is different from the CO_2 substrate of the carboxylation enzymatic reaction. Carbamylated Rubisco is still inactive.
- Active form of Rubisco, with catalytic potential (ECM), is characterised by the presence of a metal chelate resulting from the union of the EC form to a divalent metal ion (Mg^{2+} or Mn^{2+}). It is probable that this metal ion bound to the enzyme acts as an 'electron sink' during catalysis. Oxygenase reaction also requires the ECM form of Rubisco.

Carbamylation and chelate formation depend upon stromatic pH and CO_2 and Mg^{2+} concentrations.

Determination of Rubisco activity involves the assessment of two types of enzymatic activities:

1. **'Initial Activity'** is the enzymatic activity of Rubisco just as it is initially present in the sample. This sample can be either previously

purified Rubisco or a leaf extract, which contain a mixture of active and inactive forms of the enzyme.

2. **'Final Activity'**, obtained by adding CO_2 and Mg^{2+} to the leaf extract (or previously purified Rubisco) in which initial activity has been previously determined. This addition induces the activation of inactive forms of the enzyme that are present in the sample. So, final activity is a total or maximum Rubisco enzymatic activity of the sample (Portis, 1992).

Activation state of Rubisco (usually expressed in %) is the ratio of initial activity to final activity, which allows an estimation of Rubisco carbamylation.

REGULATION OF RUBISCO ACTIVITY

It has been shown that Rubisco is submitted to a complex regulation (Portis, 1992; Hartman and Harpel, 1994). Some important regulator factors are:

- RuBP acts simultaneously as substrate and as inhibitor of Rubisco activity, because RuBP binds with stronger affinity to the E form than to the ECM form of Rubisco, inhibiting by this way carbamylation and activation of the E form when CO_2 activator and Mg^{2+} are supplied. Binding of RuBP to Rubisco is always affected by pH and presence of bicarbonate HCO_3^-.

- Other sugar-phosphates bind to Rubisco in a biphasic way. Many of these phosphorylated compounds can bind to all the three forms of the enzyme, but the effects of any particular sugar-phosphate are not clear.

- There are some evidences of light regulation (Hartman and Harpel, 1994; Parry et al., 1997). Higher amounts of activated Rubisco have been observed in leaves exposed to the light. A pH and Mg^{2+} - dependent carbamylation, induced by light, has been observed. In fact, it was found that the 'initial activity' of the enzyme varies upon photosynthetic activity (light phase) and with the proton flux density (PFD). It involves that, in comparative studies of Rubisco activity between control and treated plants, the same conditions of illumination must be maintained.

- A soluble chloroplastic protein, activator of Rubisco, has been identified. It has been called Rubisco-activase. It seems that Rubisco-activase acts preferably on the RuBP - E form of Rubisco. Regulation mechanisms and factors that influence the action of Rubisco-activase are, at present, one important subject of research (Lan and Mott, 1991; Eckardt and Portis, 1997; Spreitzer, 1999).

- One interesting case of sugar phosphate is 2-carboxyarabinitol 1-phosphate (CA1P), which reduces Rubisco activity *in vivo*. Contrary to RuBP, CA1P binds with stronger affinity to the carbamylated form of the enzyme, just in the active site, diminishing directly by this way the catalytic activity. In natural conditions, CA1P degrades gradually in the measure that PFD increases, and it disappears in less than 10 minutes if leaves are illuminated with saturating PFD. So, CA1P is involved in Rubisco regulation by light. In some species, CA1P concentration increases during night and inhibits by this way Rubisco activity during night. CA1P and another derivative, carboxyarabinitol-1,5 biphosphate (CABP), are two inhibitors synthesised and used in some labs in studies of inhibition kinetics. They are not commercially available (Portis, 1992; Hartman and Harpel, 1994).

Methods for measuring Rubisco enzymatic activity

The two principal methods are:

1. *Radioisotopic method*: $^{14}CO_2$ incorporation into steady acid compounds is quantified. This methodology is described in Vu et al. (1984) and Sage et al. (1993).
2. *Spectrophotometric method*: NADH oxidation to NAD^+ in a coupled enzyme system is quantified. Absorbance variation at 340 nm is recorded in a spectrophotometer (Lilley and Walker, 1974). This method has the advantage of being simple and easily available, it does not involve the use of radioactive material, and it allows a continuous recording of Rubisco activity over time (however, radioactive method does not).

Comparisons of both methods are showed in Ward and Keys (1989) and Reid et al. (1997).

Owing to these advantages, our lab decided to adapt the spectrophotometric method for measuring Rubisco activity in *Lactuca sativa* L. and other plant species that are interesting in Allelopathy. This approach will allow a study of comparative activity in control and allelochemicals-treated plants.

The first references of this method are dated from the sixties and seventies (Lilley and Walker, 1974, and references therein), but it was not broadly used owing to the involvement of a long lag time (from one to more minutes) from Rubisco carboxylation to NADH oxidation in the coupled enzyme system. This event reduced considerably the resolution of this method.

Lilley and Walker (1974) performed a modification of this method, obtaining a considerably shorter lag time by adding an ATP regeneration system in the reaction mixture. This modification has allowed a much broader use of this method. At present, all protocols using the spectrophotometric method are based on slight modifications from Lilley and Walker, reducing even more the lag time.

The coupled enzyme system is:

1) The substrate RuBP is added in the reaction mixture containing the leaf extract (or purified Rubisco), and Rubisco catalyses this reaction:

$$RuBP + CO_2 \rightarrow 2\ (3\text{-phosphoglycerate})$$

2) Another added enzyme is **3-phosphoglyceric phosphokinase (PGK)**, and catalyses the next reaction in the coupled system:

$$2\ (3\text{-phosphoglycerate}) + 2\ ATP \rightarrow 2\ (1,3\text{-phosphoglycerate}) + 2\ ADP$$

3) **Glyceraldehyde-3-phosphate-dehydrogenase (G3P-DH)** catalyses the next reaction:

$$2\ (1,3\text{-phosphoglycerate}) + 2\ \underline{NADH} \rightarrow 2\ G3P + 2\ \underline{NAD^+} + 2\ Pi$$

4) ADP is a very sensitive inhibitor of PGK, and its accumulation is one of the key responsible factors for the lag time. Lilley and Walker (1974) diminished the lag time and got the efficiency of the method to be much higher by adding an ATP regeneration system, the enzyme **creatine phosphokinase (PCK)**, which catalyses the reaction:

$$Phosphocreatine + ADP \rightarrow creatine + ATP$$

5) Rubisco enzymatic activity is measured by recording NADH oxidation at 340 nm in a spectrophotometer. So, the kinetics of NADH absorbance decrease along time is measured.

Some authors have modified the concentrations of some of the three coupled enzymes, principally increasing the concentrations of PGK and G3P-DH. Sometimes PCK concentration is also increased to reduce ADP concentration.

By another way, the substrate RuBP can isomerise spontaneously to xylulose-1,5-bisphosphate, which is an inhibitor of Rubisco. In fact, commercially available RuBP is relatively expensive, it is partially isomerized to xylulose-1,5-bisphosphate and it seems to support a relatively low carboxylase activity.

Owing to this, some authors make their own RuBP from ribose-5-phosphate (methodology described in Sharkey et al., 1991, and in Du et al., 1996). RuBP obtained by this way results much cheaper, stable and supports higher Rubisco carboxylase activity, than the commercial one. Lilley and Walker had suggested the use directly of ribose-5-phosphate as Rubisco substrate, because it is more stable and induces higher Rubisco activities, but it usually involves longer lag times.

The Japanese team of Du et al. (1996), working on sugar cane, got to diminish, and even to cancel, the lag time, simply by increasing PCK concentration. They used both RuBP purchased from Sigma and fresh RuBP synthesised in their lab from ribose-5-phosphate. They observed that fresh RuBP allows to double the recorded enzymatic activity compared with commercial RuBP, but a lag time is still present. In contrast, commercial RuBP involves lower enzymatic activities, but the lag time is cancelled. We have chosen this last solution.

SPECTROPHOTOMETRIC ENZYMATIC ASSAY. TECHNICAL INSTRUMENTATION AND PERFORMANCE

We have used the spectrophotometric method of Lilley and Walker (1974) for measuring Rubisco activity in *Lactuca sativa* L. (v. Great Lakes; California) leaves, with some modifications of those exposed in Reid et al. (1997), who measured Rubisco activity in leaf extracts from *Glycine max*, and they do comparative stress studies. Details of this method are:

Plant material

Seeds of *Lactuca sativa* L. (v. Great Lakes, California) germinate in hydroponic conditions, and after few days, young plants are transferred to vermiculite, where they grow from 90 days to 2 months before being used for Rubisco assay. Both germination and growth occur in these environmental conditions: 80 % of moisture, 100 $\mu mol \cdot m^{-2} \cdot s^{-1}$ and a light/dark cycle of 18 h light (25 °C) and 6 h dark (18 °C).

Sixteen leaf dishes (corresponding to 40.7 cm^2) are extracted randomly with a punch (18 mm diameter) from leaves of *Lactuca sativa* L. (v. Great Lakes, California), beginning with the youngest one. These dishes are rapidly (approximately 10 seconds) enveloped in kitchen foil, and frozen in liquid nitrogen. Samples are stored at –80 °C, for up to ten days or longer, until the enzymatic assay (Reid et al., 1997).

Rubisco extraction

The extraction procedure follows Sage et al. (1993) and Reid et al. (1997). Frozen leaf dishes are ground to a fine powder in a mortar pre-cooled with liquid nitrogen. Then, 4 ml of extraction buffer (see composition in Appendix I) are added over the obtained powder, and the obtained solution is homogenised in a glass Heidolph type RZR1 glass homogenizer, for 2-3 min, setting speed position 3. All steps of this procedure must be performed at 0-4 °C.

The obtained homogenate is centrifuged for 1 min. at 12000 rpm in a cooled Hettich EBA 12R microcentrifuge. The obtained supernatant will be used for the Rubisco activity assay. The extraction step must be performed rapidly (5 min, approximately).

Rubisco activity assay

The next step is to assess the 'initial activity' of the sample, and it must be done immediately after obtaining the supernatant. For the spectrophotometric measure, 100 µL of supernatant are added to 870 µL of assay buffer (Appendix II), into a spectrophotometric cuvette.

Finally, 0.5 mM (final concentration) substrate RuBP (Sigma R0878) is added (30 µL) to the mix, and NADH oxidation is followed by measuring absorbance decrease for 5 min at 340 nm and 24.5 ± 0.5 °C, in a HP spectrophotometer.

In our case, lag time is below 30 seconds, and a decrease in absorbance of about 35 % over 5 min is observed. Later on, only the lineal section of the obtained graph must be used for activity estimations.

Another aliquot from the sample is incubated, in parallel, 15 min at 25 °C in the assay buffer supplemented with $NaHCO_3$ up to a final concentration of 10 mM, for assessment of the 'final activity' of Rubisco in the sample.

An average calibration curve of equivalence between absorbance decrease and NADH concentration changes must be performed, simply by recording absorbance, at 340 nm, of the assay buffer with different initial concentrations of NADH, and supplemented with either the extract and water (instead of the substrate RuBP), or with substrate and water (instead of the extract). This procedure allows verifying if there is an unspecific oxidation of NADH

The subsequent use of purified Rubisco from spinach (Sigma product R8000, with 1 M Tris pH 8.0 as solvent), instead of the leaf extract, will allow obtaining a calibration graph for conversion of NADH concentration changes in Rubisco enzymatic activity units.

The amount of mg of total proteins that are present in the leaf extract are quantified using the methodology described in Bradford (1976), allowing the assessment of Rubisco specific activity per mg of total protein. It is also recommended to record the fresh and dry weight of the leaf sample, in addition to leaf surface. Rubisco activity is usually expressed in μmol CO_2/m^2 s.

One example of stress studies is that of Reid et al. (1997), who have compared Rubisco activity values in leaf extracts of control and ozone-treated plants of field-grown *Glycine max*. Control plants have initial activities of 42.6 ± 2.99 μmol CO_2/m^2 s. Ozone-treated plans have a Rubisco initial activity of 31.3 ± 1.41 μmol CO_2/m^2 s. Respective final activity values are very close from those of initial activities.

One way to estimate Rubisco content in a leaf extract is performing, in the conditions mentioned above, an inhibition kinetics of Rubisco in the presence of increasing concentrations of the inhibitor CABP (Reid et al., 1997). This will allow the assessment of specific enzymatic activity per mg of Rubisco present in the leaf extract.

Furthermore, radioactive and anion exchange chromatography methods to accurately measure the content of Rubisco in plant tissues have been described (Ferreira et al., 2000, and references therein).

A schematic protocol of this methodology is shown in Figure 1.

Modifications of the described methodology

Several labs use some modifications of the conditions described above, depending on the experimental system and plant species under study:

Du et al. (1996) use high concentrations of PCK (20 U/ml), comparing to the other two used enzymes of the coupled system (6 U/ml), with the goal of diminishing or cancelling the lag time.

Kane et al. (1998) usually add, to the three indicated enzymes, an additional system with glycerol-phosphate-dehydrogenase and triose-phosphate-isomerase (both of them from rabbit muscle), which allows to eliminate remaining phosphoglycerate from the reactions referred above. Phosphoglycerate, and the whole enzyme coupled system, are very affected by Pi produced during the reaction. By this way, phosphoglycerate is converted to glycerol phosphate.

Lan et al. (1991) add phosphoenolpyruvate and pyruvate kinase to the three enzymes used by Lilley and Walker. This is an additional system to regenerate ATP, with the advantage of inducing a more irreversible reaction than does PCK. But this system can involve a problem: it is known that PCK does not influence Rubisco activation by itself, but the possibility that pyruvate kinase can activate Rubisco, and thereby altering activity values, is not discarded.

Finally, Eckardt and Portis (1997) use G3P-DH and PGK, but they use pyruvate kinase, glycerol-phosphate-dehydrogenase and triose-phosphate-isomerase instead of PCK.

Some **constraints** must been taken into account (using suitable controls) for the correct assessment of Rubisco activity in leaf extracts:

- These extracts can induce an unspecific residual oxidation of NADH, in absence of the substrate RuBP.
- A high initial absorbance can be recorded, due to the presence of chlorophyll and other compounds in the leaf extracts.

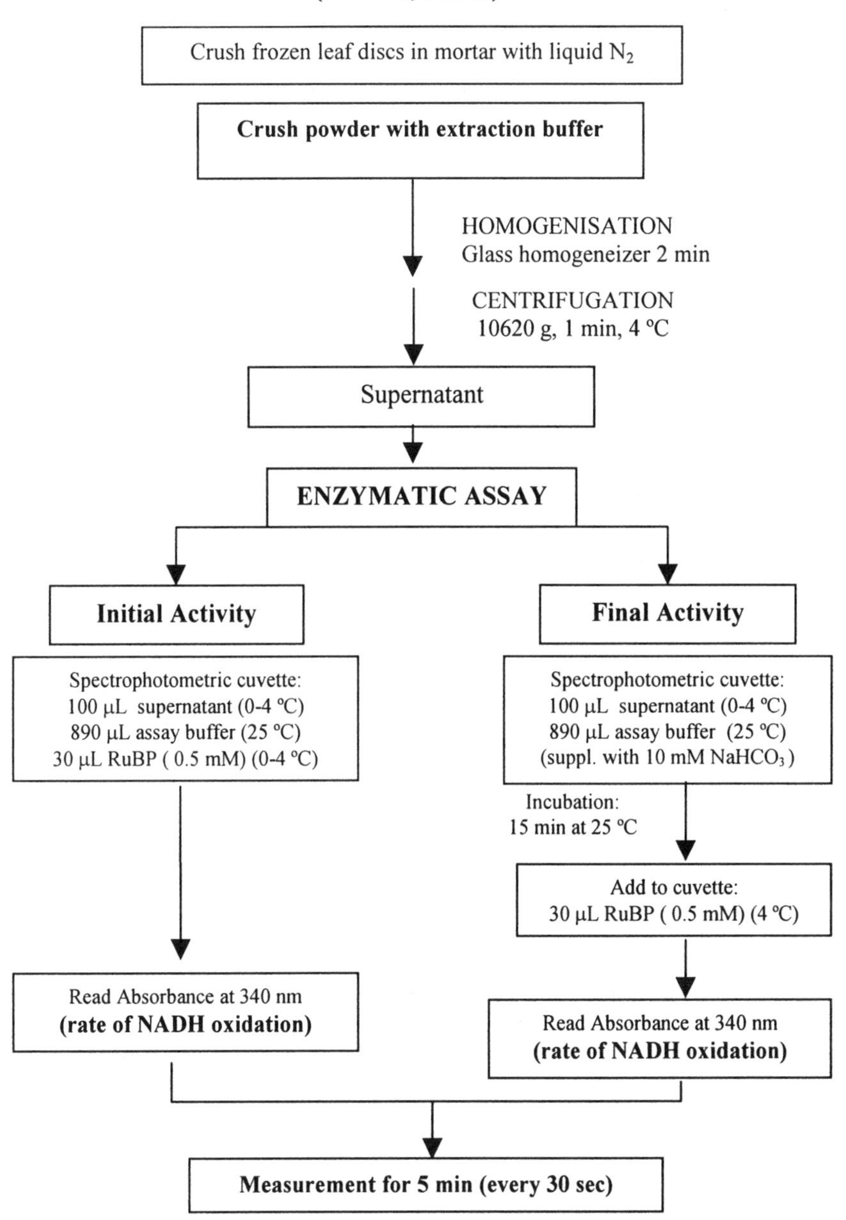

Figure 1. Scheme of the procedure for measurement of Rubisco activity

- The presence of phenolic compounds, and other substances, in the extract can interfere with the coupled enzyme system and/or Rubisco activity.
- Strong decreases in Rubisco initial activity can be done, owing to the presence of proteases and slow binding inhibitors.

APPENDIX I

The composition of the **extraction buffer** is:

100 mM Bicine, pH 8.0 [Sigma product B3876]

1 mM DTT[a] [antioxidant agent; Sigma product D0632]

0.1 % BSA (Sigma A7906)

1.5 % (w/v) polyvinylpilorridone[b] (Sigma PVP-40)

3.3 mM amino-n-caproic acid [plasmin inhibitor; Sigma A2504]

0.7 mM Benzamide [inhibitor of an NO-activated enzyme; Sigma B2009]

20 mM $MgCl_2$ (Sigma M8266)

1 mM EDTA (Panreac, 131669)

150 mM $NaHCO_3$[c] (Sigma S8875)

a) DTT must be added to the buffer just before use, owing to its rapid oxidation.

b) PVP-40 is insoluble and of high molecular weight. It is an enzyme stabiliser by eliminating phenolic impurities.

c) Bubbling gaseous N_2 into the solution for 20 minutes, before adding $NaHCO_3$, BSA and PVP-40, is recommended. By this way, solubilization of gaseous CO_2 into the buffer (which could partially activate the enzyme) is avoided.

APPENDIX II

The composition of the **assay buffer** is:

100 mM Bicine pH 8.0

25 mM $KHCO_3$ (Sigma P9144)

20 mM $MgCl_2$

3.5 mM ATP[a], disodium salt grade I, minimum 99 % (Sigma A2383)

5 mM Phosphocreatine (Fluka 27920)

0.25 mM NADH[b] (Roche 107 727)

80 nkat[c] glyceraldehyde-3-phosphate-dehydrogenase[d,e] (EC 1.2.1.12, from rabbit muscle, lyophilized powder, Sigma G2267)

80 nkat 3-phosphoglyceric phosphokinase[d,e] (EC 2.7.2.3, from Baker's yeast, sulphate free powder, Sigma P1136)

80 nkat creatine phosphokinase[d,f], type I (EC 2.7.3.2, from rabbit muscle, salt free powder, Sigma C3755)

a) ATP powder is added directly to the assay buffer just before use, because of its instability and rapid degradation in solution.

b) NADH stock solution is obtained by dissolving in sodium bicarbonate pH 9.0, at a final concentration of 10 mg/ml. This solution is stable for one month at 4 °C.

c) One katal is the amount of enzyme that induces the formation of 1 mol of product per second. One unit of enzymatic activity is equal to 16.67 nkat.

d) We use sulphate-free powder forms of the enzymes. Other commercial and available forms are precipitated or dissolved in ammonium sulphate. Sulphate is a Rubisco inhibitor, because it competes with RuBP for the active site of the enzyme. In this case, sulphate must be eliminated of the enzyme product. Ammonium sulphate is removed by centrifugation and elimination of the supernatant, and the pellet is resuspended and submitted to dialysis against 50 volumes of water. DTT is added for avoiding enzyme denaturation during dialysis. The product is stored at 0-4 °C.

e) Glyceraldehyde-3-phosphate-dehydrogenase and 3-phosphoglyceric phosphokinase are dissolved in distilled water, at a preferable protein concentration of 5 mg/ml and frozen in aliquots at –20 °C. The enzymes should be stable greater than 6 months.

f) Creatine phosphokinase is dissolved in 10 mM Glycine (Sigma G7403) pH 8.0 at a protein concentration or greater than 1 mg/ml (preferable 5 mg/ml). The enzyme should be stable more than six months frozen in aliquots at –20 °C.

REFERENCES

Bradford M.M. A rapid and sensitive method for quantitation of microgram quantities of protein utilizing the principle of protein-dye binding. Anal Biochem 1976; 72:248-254

Du Y.C., Nose A., Kawamitsu Y., Murayama S., Wasano K., Uchida Y. An improved spectrophotometric determination of the activity of ribulose 1,5-bisphosphate carboxylase. Jpn J Crop Sci 1976; 65:714-721

Eckardt N.A., Portis A.R. Jr. Heat denaturation profiles of ribulose-1,5-bisphosphate carboxylase/oxygenase (Rubisco) and Rubisco activase and the inability of Rubisco activase to restore activity of heat-denatured Rubisco. Plant Physiol 1997; 113:243-248

Ferreira R.B., Esquivel M.G., Teixeira A.R. An accurate method to quantify ribulose bisphosphate carboxylase content in plant tissue. Plant Cell Environ 2000; 23:1329-1340

González-Vilar M., González L., Reigosa M.J. Ecophysiological responses to light intensity and water stress in two citotypes of cocksfoot (*Dactylis glomerata* L.) in Galicia, NW of Spain Submitted

Gutteridge S., Gatenby A.A. Rubisco synthesis, assembly, mechanism, and regulation. Plant Cell 1995; 7:809-819

Hall N.P., Tolbert N.E. A rapid procedure for the isolation of ribulose bisphosphate carboxylase/oxygenase from spinach leaves. FEBS Lett 1978; 96:167-169

Hartman F.C., Harpel M.R. Structure, function, regulation, and assembly of D-ribulose-1,5-bisphosphate carboxylase/oxygenase. Annu Rev Biochem 1994; 63:197-234

Jensen R.G. Ribulose 1,5-bisphosphate carboxylase-oxygenase. Annu Rev Plant Physiol 1977; 28:379-400

Kane H.J., Wilkin J.M., Portis A.R.Jr., Andrews T.J. Potent inhibition of ribulose-bisphosphate carboxylase by an oxydized impurity in Ribulose-1,5-bisphosphate. Plant Physiol 1998; 117:1059-1069

Lan Y., Mott K.A. Determination of apparent Km values for ribulose 1,5-bisphosphate carboxylase/oxygenase (Rubisco) activase using the spectrophotometric assay of Rubisco activity. Plant Physiol 1991; 95:604-609

Lilley R.McC., Walker D.A. An improved spectrophotometric assay for ribulose bisphosphate carboxylase. Biochim Biophys Acta 1974; 358:226-229

Lorimer G.H., Badger M.R., Andrews T.J. The activation of ribulose-1,5-bisphosphate carboxylase by carbon dioxide and magnesium ions. Equilibria, kinetics, a suggested mechanism, and physiological implications. Biochemistry 1976; 15:529-536

Parry M.A.J., Andralojc P.J., Parmar S., Keys A.J., Habash D., Paul M.J., Alred R., Quick W.P., Servaites J.C. Regulation of Rubisco by inhibitors in the light. Plant Cell Environ 1997; 20:528-534

Portis A.R. Jr. Regulation of ribulose 1,5-bisphosphate carboxylase/oxygenase activity. Annu Rev Plant Physiol Plant Mol Biol 1992; 43:415-437

Reid C.D., Tissue D.T., Fiscus E.L., Strain B.R. Comparison of spectrophotometric and radioisotopic methods for the assay of Rubisco in ozone-treated plants. Physiol Plantarum 1997; 101:398-404

Reigosa M.J., Sánchez-Moreiras A.M., González L. Ecophysiological approach in Allelopathy. Crit Rev Plant Sci 1999; 18:577-608

Sage R.F., Reid C.D., Moore B., Seemann J.R. Long-term kinetics of the light dependent regulation of ribulose-1,5-bisphosphate carboxylase/oxygenase activity in plants with and without 2-carboxyarabinitol 1-phosphate. Planta 1993; 191:222-230

Sánchez-Moreiras A.M. Efecto de compuestos fenólicos en *Lactuca sativa* L. Minor Thesis. Vigo, Spain: University of Vigo, 1996

Seftor R.E., Bahr J.T., Jensen R.G. Measurement of the enzyme-CO_2-Mg^{2+} form of spinach ribulose 1,5-bisphosphate carboxylase/oxygenase. Plant Physiol 1986; 80:599-600

Sharkey T.D., Savitch L.V., Butz N.D. Photometric method for routine determination of K_{cat} and carbamylation of Rubisco. Photosynth Res 1991; 28:41-48

Sivakumar P., Sharmila P., Saradhi P.P. Proline suppresses rubisco activity in higher plants. Biochem Bioph Res Co 1998; 252:428-432

Spreitzer R.J. Questions about the complexity of chloroplast ribulose-1,5-bisphosphate carboxylase/oxygenase. Photosynth Res 1999; 60:29-42

Vu C.V., Allen L.H.Jr., Garrard L.A. Effects of enhanced UV-B radiation (280-320 nm) on ribulose-1,5-bisphosphate carboxylase in pea and soybean. Environ Exp Bot 1984; 24:131-134

Ward D.A., Keys A.J. A comparison between the coupled spectrophotometric and uncoupled radiometric assays for RuBP carboxylase. Photosynth Res 1989; 22:167-171

CHAPTER 24

ATP PHOSPHOHYDROLASE ACTIVITY

Adela M. Sánchez-Moreiras
Depto Bioloxía Vexetal e Ciencia do Solo. Universidade de Vigo. Spain

ATPase. WHAT IS THIS?

It is well known that proton pumps play an important role in plant physiology at cellular and organ level, developing different functions at different plant stages, but all of them with vital importance for the plant.

Cells use energy to absorb charged solutes which can not be transported by a passive absorption and which need an active transport using the energy stored in the ATP molecules. This energy is available for the cell when a molecule of ATP is hydrolysed in its terminal phosphate, releasing an ADP molecule and an inorganic phosphate (P_i).

$$ATP (Mg) + H_2O \leftrightarrow ADP(Mg) + P_i$$

This strongly exergonic reaction (7.6 kcal per mole of ATP) allows protons to go out of the cytosol thus creating an electrochemical gradient. In the same way, proton pumps help to ion transport from one side of the membrane to the other side into the cell.

The enzyme ATP phosphohydrolase (ATPase) - EC 3.6.1.35 - catalyses this reaction. It is an integral protein large enough to cross one or more membranes. This physical characteristic is essential in its function, which, in a reaction mediated by Mg^{2+} in combination with the ATP molecule, is able to change its shape and thus allowing proton transport (see Fig. 1, 2).

M.J. Reigosa Roger, Handbook of Plant Ecophysiology Techniques, 399–412.

Later, with the help of carrier proteins and using the energy offered by the ATP, these protons will be used in the cross inward of anions by cotransport (symport) and in the cross outward of cations by counter transport (antiport).

We can found ATPases seemingly in every membrane of every cell and from every plant alive. Therefore, there are specific ATPases in the plasma membrane (H^+ and Ca^{2+} pumps), mitochondria, chloroplast, or in the vacuole, but perhaps the most studied and better known enzyme is the plasma membrane H^+-ATPase.

All these ATPase types differ markedly in their protein chemistry, their reaction mechanism and their evolutionary origins (Buchanan et al., 2000). Each enzyme catalyses its reaction in a different way. For example, the plasma membrane H^+-ATPase is a single polypeptide of about 100 kDa that allows the crossing of a single proton out of the cell for each hydrolysed ATP molecule (see Fig. 2), and it is K^+-dependent. By contrary, the ATPases found in tonoplast are not K^+-dependent and they allow the cross into the vacuole of two protons per each ATP hydrolysed. The chloroplastic and mitochondrial ATPases present even more differences.

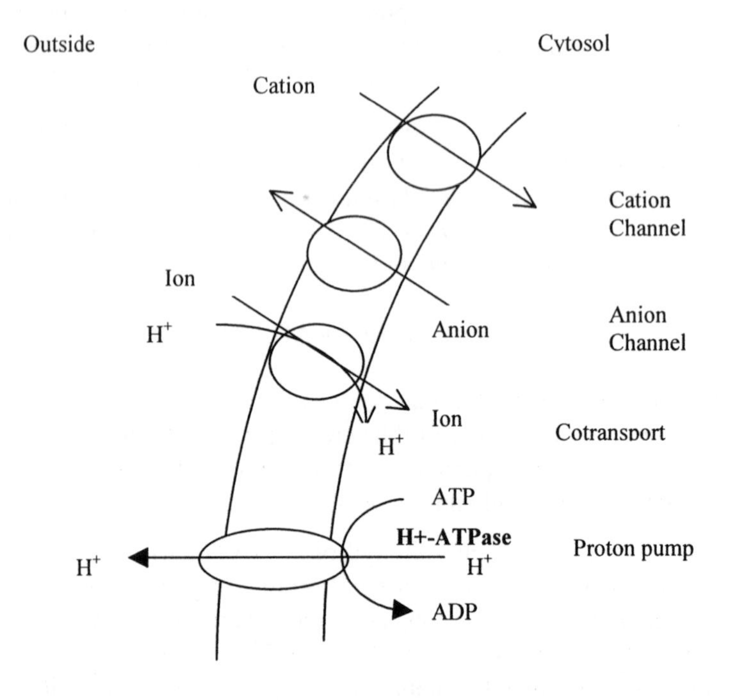

Figure 1. Ion active transport across the plasma membrane

In any case, proton pumping ATPase has a major role in many functions in the cell "controlling directly intracellular and extracellular pH, constituting what has been called the biophysical pH-state" (Serrano, 1989). Their presence is essential in the generation of an electrochemical proton gradient, needed for basic plant functions as nutrient uptake or turgor (Poole, 1978; Serrano, 1985; 1989). This gradient drives the transport of many ions and molecules through the plant cell membranes, see Fig. 1, (Marré and Ballarin-Denti, 1985; Serrano, 1988; 1989; Palmgren, 1998).

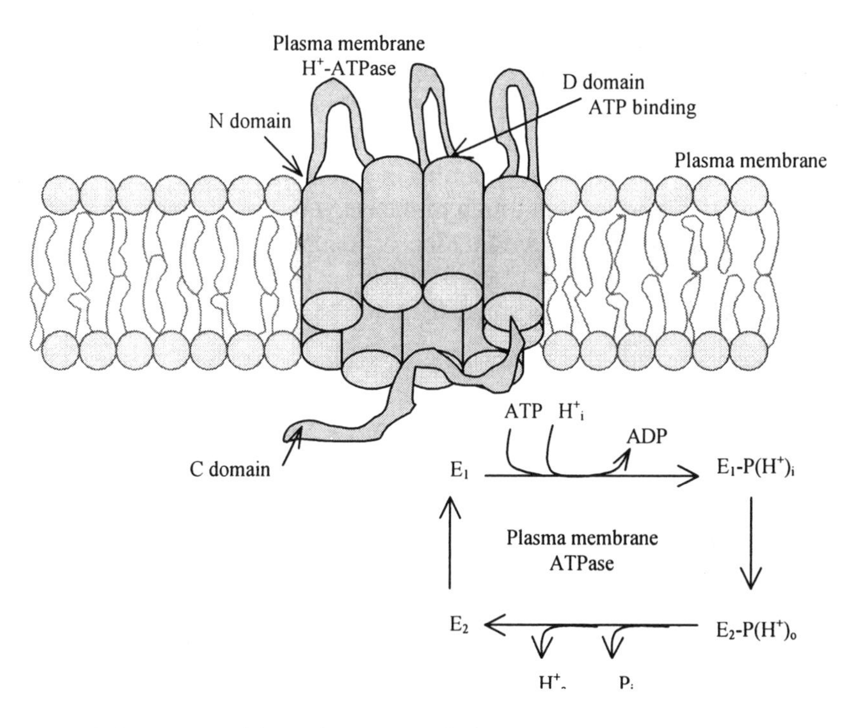

Figure 2. Structure, position and reaction cycle for a plasma membrane ATPase, a P-type H^+-ATPase. All plasma membrane ATPases share a common domain structure (N, D, C domains). After proton binding to the enzyme in the E_1 conformation, the ATP hydrolysis (that bind to ATPase in the D domain), and the enzymatic phosphorylation cause a change to the E_2 conformation. Once that the enzyme acquires this conformation, the lower affinity of it for the H^+ allows the release of the protons in the other side of the membrane. Inorganic phosphate is also released and the enzyme returns to the E_1 conformation. Redrawn from Buchanan et al. (2000).

Of course, the nutrient transport is a basic requirement of growth. But ATPase enzyme was found also to be related with cell division, cell differentiation, polar growth in some parts of the plant, processes of dormancy during germination or cell elongation according with the "acid growth theory" (Rayle and Cleland, 1977; Marrè, 1979; Marrè et al. 1988; Altabella et al. 1990). Changes on intracellular pH are directly related with a lot of processes occurring in the cell and therefore they are directly related to growth, development and vigour of the organism.

It was also shown that stimulation of proton pumps, presents in the guard cells, is implicated in the stomatal movements, when ATPase activity is suppressed (Zhao et al., 2000).

The most important factors in plant physiology participate in the regulation of ATPase activity, because of all vital functions in the cell, which were described above. Stimulation of this enzyme can be induced by the light receptors (stomata opening), by a decrease in the cell turgor (osmoregulation), by almost all plant hormones that are interacting with it (auxin in the cell elongation, cytokinins in cotyledons, etc.), or also by the action of protein kinases. Inhibition of the enzyme was shown in presence of abscisic acid (inhibiting germination or inducing stomata closing) and after injury in the tissues for example after a exposition to strong thermal differences (Marré, 1979; Hanson and Trewavas, 1982; Serrano, 1989).

Considering all these reasons, it seems necessary to use this parameter when we want to see if some of these plant processes are affected.

MEASUREMENT SIGNIFICANCE IN ALLELOPATHY

When secondary compounds are released into the environment, they are able to affect plants or other organisms in many vital processes. So, in some circumstances the development of the organisms is subjected to the presence of these secondary metabolites in the medium. Allelochemicals can affect many processes as radicle growth, respiration, photosynthesis, growth, elongation, etc (Einhellig, 1986), but when an altered development is found, more extensive biochemical studies are needed in order to see the real site and mode of action of this or these compound/s.

When an alteration of growth is detected and we can see an effect on the integrity of plasma membrane, it is necessary to know which is the reason for these first observed effects. Perhaps occurs an increase of lipid peroxidation? Are the carrier proteins affected? Maybe the proteins related with oxygen metabolism are disturbed? But it is also possible to have an altered activity of enzymes responsible of proton pumps in the cell? We can find one or all of these effects as the mode/s of action of a single allelochemical compound within the cell. So, to be sure that the exact site/s of action is/are identified we must test all of these possible effects.

ATPase requires a determined lipid environment to act correctly (Serrano, 1988). An increase in lipid peroxidation in the plasma membrane (induced perhaps by the accumulation of superoxide, hydrogen and hydroxyl radicals) in presence of an allelopathic compound can come accompanied by a subsequent damage in the membrane inducing finally the loss on selective membrane permeability (Baziramakenga et al. 1995; Politycka, 1996; Zhang and Kirkham, 1996). And also the properties of membrane enzymes such as plasma membrane ATPase can be changed. The knowledge of its enzymatic activity can explain something more about the state of this chain of effects.

Sensitivity of ATPase enzyme to abiotic (Ahn et al., 2000) and biotic stress in presence of hydroxamic acids (Queirolo et al., 1983; Friebe et al., 1997); resin glycoside mixture (Calera et al., 1995); monoterpene compounds (Cruz Ortega et al., 1990) or flavonoid compounds (Balke, 1985) was found in the last years. In the most part of these studies a first germination and/or growth alteration was observed, and ATPase activity quantification comes to confirm one of the modes of action of these mentioned compounds.

In 1983 Queirolo et al. described a fast and reversible inhibition of the ATPase from chloroplasts from Gramineae in presence of the cyclic hydroxamic acid DIMBOA (2,4-dihydroxy-7-methoxy-1,4-benzoxazin-3-one), a methoxy derivative of DIBOA (2,4-dihydroxy-1,4-benzoxazin-3-one). They suggested that this reactivity could be caused by the reaction of DIMBOA with the sulfhydryl group on the enzyme.

In the same way, Friebe et al. (1997) studied the effects on *Avena sativa* of some hydroxamic acids, and observed a significant decrease in the activity of plasma membrane H^+-ATPase in presence of DIMBOA. They could demonstrate in *A. sativa* the correlation between a decrease on plasma membrane H^+-ATPase activity in presence of DIMBOA and the inhibition of radicle growth. So, the development of the plant could be determined by the appearance or not of this compound.

Friebe et al. in 1997 suggested that sulfhydryl group of the enzyme could play an important role in this inhibition, as was previously suggested by Queirolo et al. (1983).

Many works with new allelopathic compounds, testing their effect on plant ATPases, were published in the last years. In 1990 Cruz Ortega et al.

suggested that the action of diacetyl-piquerol *in vivo* could be related with the effect on the plasma membrane H^+-ATPase activity. And Hager et al. (1991) has shown that an inhibition in the elongation of maize coleoptiles was correlated with a decrease in the amount of plasma membrane H^+-ATPase.

By other hand, the phytotoxicity of allelochemicals, like phenolic compounds has been demonstrated along the years. This kind of phytotoxicity was primarily shown in their influence over the cellular membranes. In presence of phenols the H^+-gradient through the inner mitochondrial membrane was suppressed and the synthesis of ATP was inhibited (Moreland and Novitzky, 1988). The default of this ATP in the cell can suppress also ATPase activity.

The possibility to relate ATPase activity with germination, development and vigour of plants makes that this parameter is being used more and more in the field of allelopathy.

ATPase DETERMINATION

At the beginning, primary pumps were principally identified with "in vivo" experiments seeing the alteration on the functions attributed to ATPase proton pumps activity. Along the years new techniques allowed the localisation, isolation of ATPase enzyme and the direct measurement of its activity (Hodges et al. 1972; Hodges, 1976; Calera et al., 1995; Friebe et al. 1997; Jahn et al., 1998). Now, after several physiological works we can study the molecular and genetic aspects of this enzyme and also the complicated and not very much known regulatory mechanisms of this master enzyme within the cell.

At this time we have several techniques to purify ATPase, to study the cell and subcellular localisation of this enzyme, and to know directly its activity. We must select the appropriate technique depending on our objective.

The **immunohistochemical methods**, utilising for example a plasma membrane-H^+-ATPase specific monoclonal antibody to localise the enzyme, will be an useful technique if we want to see an effect in the shape, amount or distribution of the enzyme in the cell (Baur et al., 1996; Jahn et al. 1998). Immunostaining for immunofluorescence microscopy and electrophoretic separation of bands, which are related to the presence of a type of ATPase, are two techniques used in this field. The monoclonal antibody binds to the multiple H^+-ATPase isoforms and then, after a fractionation, the enriched extraction is subjected to electrophoretic separation. The bands that appear are analysed. This mixture is fixed and observed by immunofluorescence microscopy to know the different aspects of the enzyme.

Perhaps the most employed methods to study the activity of the enzyme present in the sample, and not the shape or the amount of the enzyme, are the **colorimetric methods**. The most frequent procedures are based in the detection of inorganic phosphate released in the enzymatic reaction (Cruz Ortega et al., 1990; Calera et al., 1995; Friebe et al. 1997).

Approaches of colorimetric methods were made to find the most sensitive and colour and time stable procedure. Extraction and purification of ATPase enzyme is one of the most important parts in the technique for ensuring a good spectrophotometric measurement. Alternatives for a major purification of the enzymes are given in the literature (Hager et al., 1991; Friebe et al. 1997).

The use of different specific inhibitors for plasma membrane, vacuolar, mitochondrial, or chloroplast ATPases in the enzymatic incubation allows the measurement of a specific ATPase activity in one only sample.

An assay for the detection of inorganic phosphate of high sensitivity and simplicity without losses by long and complicated handling is needed in this technique, too. The selected method must also be able to detect amounts of less than 1 μg/ml of inorganic phosphate in the sample (Penney, 1976).

The colorimetric method described below (see Fig. 3) was selected for the measurement of the effects of allelopathic compounds on ATPase activity because it was considered to be a single method that gives a real measurement of this enzymatic activity.

PROCEDURE

ATPases extraction from a plant sample

In the extraction process we must be able to get the most purified microsomal fraction (ATPase is an integral protein of membranes). To obtain the microsomal fraction, 3-4 g plant leaves are homogenised in an extraction buffer (see below Preparation of buffers: *Extraction Buffer*) in a ratio of 1 g fresh weight plant : 1 ml buffer. The suspension is filtered and centrifuged at 10000 g_n for 20 min at 4 °C. After centrifugation, the pellet is discarded and the supernatant (crude extract) is centrifuged again at 80000 g_n for 30 min at 4 °C. Then, the supernatant (soluble fraction) is discarded and the resulting pellet (microsomal fraction) is resuspended in a new buffer (see below Preparation of buffers: *Resuspension Buffer*).

Total proteins are assayed with Bradford's method (1976), using bovine seroalbumine (BSA) as standard (see below Preparation of buffers: *Bradford Reagent*).

Enzymatic assay for ATPase activity

Once that microsomal fraction is purified we have to test the ATPase enzymatic activity in the sample. Therefore, we must provide the enzyme with the all-necessary cofactors for the reaction. So, we have to add to the enzymatic buffer: ATP (the substrate of this hydrolytic reaction to obtain ADP and P_i), magnesium combined with ATP (ATPase present in higher plants is magnesium-dependent), and potassium in form of KCl (plasma membrane ATPase is potassium-stimulated) (Friebe et al. 1997).

But, we will have all ATPase types living in the cell, in this microsomal fraction. So, when we obtain the enzymatic activity, this activity is a mixture of mitochondrial ATPase activity, vacuolar ATPase activity, plasma membrane ATPase activity, etc.

So, if we want to see only a specific ATPase type, the following inhibitors will be added to the enzymatic buffer when it is required (see below Preparation of buffers: *Enzymatic Assay Buffer*): NaN_3 (mitochondrial ATPase inhibitor); Na_2VO_4 (plasma membrane ATPase inhibitor); $NaNO_3$ (vacuolar ATPase inhibitor) KNO_3 (tonoplast ATPase inhibitor), CCCP (for uncoupling the oxidative phosphorylation in mitochondrial system) and ammonium molybdate that inhibit acid phosphate (Calera et al. 1995; Friebe et al., 1997).

Enzymatic buffer is mixed in this ratio: 500 μL buffer with 10-15 μg protein and incubate 20 min at 37 °C. After this time the enzymatic reaction must be stopped by the addition of SDS at 6 %.

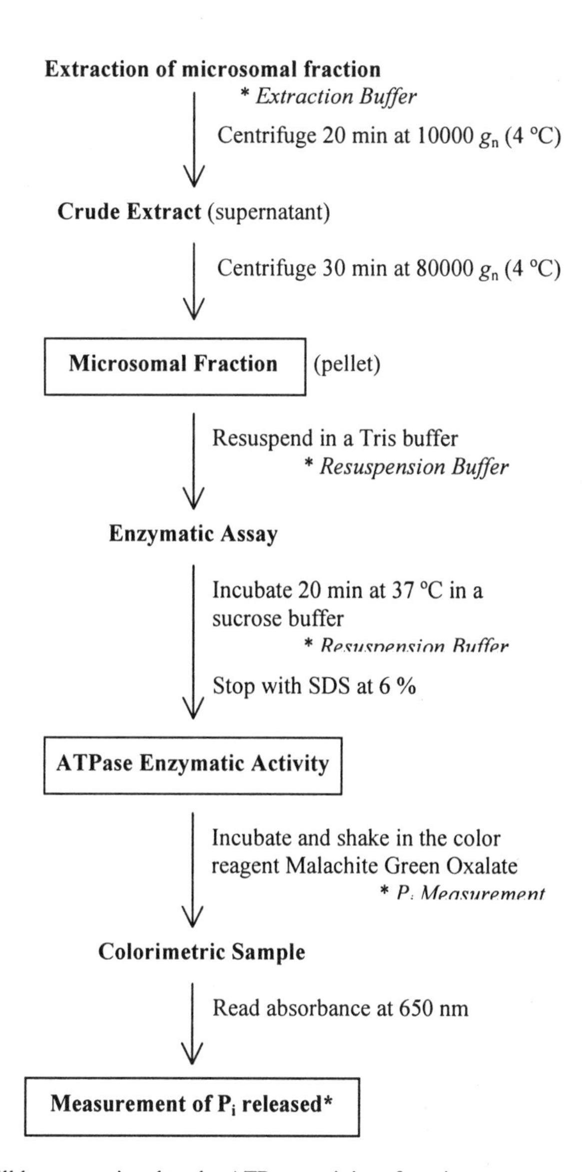

Extraction of microsomal fraction
* Extraction Buffer
Centrifuge 20 min at 10000 g_n (4 °C)
Crude Extract (supernatant)
Centrifuge 30 min at 80000 g_n (4 °C)
Microsomal Fraction
(pellet)
Resuspend in a Tris buffer
* Resuspension Buffer
Enzymatic Assay
Incubate 20 min at 37 °C in a sucrose buffer
Stop with SDS at 6 %
ATPase Enzymatic Activity
Incubate and shake in the color reagent Malachite Green Oxalate
Colorimetric Sample
Read absorbance at 650 nm
Measurement of P_i released*

* This value will be proportional to the ATPase activity of our interest

Figure 3. Scheme of the procedure for measurement of ATPase activity

Measurement of P_i release

Inorganic phosphate released in the ATPase reaction can be measured by different procedures (Penney, 1976; Lanzetta et al., 1979; Chifflet et al., 1988) as spectrophotometric analyses, gas chromatographic techniques, or fluorimetric methods. For this procedure a sensitive, fast, single and reproducible colorimetric method (Penney, 1976) was selected.

Dried KH_2PO_4 was used as a standard. Malachite green oxalate was the selected dye for this method. The concentration used of this dye (126 mg / 100 ml) was tested to be very sensitive (Penney, 1976). Just before the absorbance reading, all components required (see below Preparation of buffers: *P_i Measurement*) are mixed in the spectrophotometer cuvette and shaken.

The absorbance was read 2 min after the dye addition, at 650 nm. The value of P_i appeared in the sample and released in the reaction catalysed by the ATPase enzyme will be directly proportional to the enzymatic activity.

PREPARATION OF BUFFERS

Extraction buffer (Friebe et al., 1997):

Compound	*Concentration*	*Amount (to 100 ml buffer)*
Sucrose	20 % (p/v)	20 g
Tris-HCl (pH 7.5)	250 mM	3.94 g
EDTA	25 mM	0.931 g
DTT	5 mM	77.12 mg
Pefabloc	0.05 mM	1.2 mg
PVPP insoluble	2 % (p/v)	2 g
Ascorbic acid	10 mM	0.176 g
Distilled water		to 100 ml

Resuspension buffer (Friebe et al., 1997):

Compound	*Concentration*	*Amount (to 100 ml buffer)*
PMSF	1 mM	17.42 mg
Sucrose	250 mM	8.56 g
Tris-HCl (pH 7.0)	5 mM	78.8 mg
BSA	0.2 % (p/v)	0.2 g
DTT	1 mM	15.42 mg
Glycerol	3.7 % (v/v)	3.7 ml
Distilled water		to 100 ml

Bradford reagent (Bradford 1976):

Compound	*Concentration*	*Amount (to 1L)*
Coomasie Blue	0.01 %	100 mg
Ethanol	4.7 %	50 ml
Phosphoric acid 85 %	8.5 %	100 ml

Dilute the dye in ethanol, add the phosphoric acid and distilled water to get 1 L reagent.

Enzymatic assay buffer (Calera et al., 1995; Friebe et al., 1997):

First, prepare a basic solution with:

Compound	*Concentration*	*Amount (to 50 ml buffer)*
Sucrose	125 mM	2.15 g
KCl	50 mM	0.186 g
Triton X-100 (pH 6.5)	0.015 %	7.5 μL
Distilled water		to 50 ml

Now, add the other compounds (to 500 μL final enzymatic buffer):

Compound	*Concentration*
Basic solution	150 μL
Mg ATP	5 mM
NaN_3	1 mM
$(NH_4)_6Mo_7O_{24}$	0.2 mM
$NaNO_3$	50 mM
CCCP	7 μM

P_i **Measurement** (Penney, 1976):

(1) KH_2PO_4 (heat 1 h at 100 °C).

100 mg/L phosphate stock solution → 0.143 g/L KH_2PO_4. Store at room T

(2) 2.5 N HCl

(3) Na_2MoO_4 $2H_2O$ → 26 g/L sodium molybdate stock solution. Store at 4 °C, 2-3 weeks.

(4) Malachite Green Oxalate M·290 → 126 mg/100 ml. Store at room T

Mix 1 vol. (2) and 1 vol. (3) to obtain molybdate acid (just before to the analyses).

Colour reagent:

Mix all these components in the spectrophotometer cuvette:

- 1.8 ml molybdate acid
- 0.5 ml distilled water
- 200 μL sample
- 50 μL Malachite Green Oxalate

REFERENCES

Ahn S.-J., Im Y.-J., Chung G.-C., Seong K.-Y. Sensitivity of plasma membrane H^+-ATPase of cucumber root system in response to low root temperature. Plant Cell Rep 2000; 19:831-835

Altabella T., Palazón J., Ibarz E., Pinol T., Serrano R. Effect of auxin concentration and growth phase on the plasma membrane H^+-ATPase of tobacco calli. Plant Sci 1990; 70:209-214

Balke N. "Effects of Allelochemicals on Mineral Uptake and Associated Physiological Processes." In *The Chemistry of Allelopathy. Biochemical Interactions among Plants,* A.C. Thompson, ed. Washington, D.C., 1985

Baur M., Meyer A.J., Heumann H.-G., Lützelschwab M., Michalke W. Distribution of plasma membrane H^+-ATPase and polar current patterns in leaves and stems of *Elodea canadensis*. Bot Acta 1996; 109:382-387

Baziramakenga R., Leroux G.D., Simard R.R. Effects of benzoic acids on membrane permeability of soybean roots. J Chem Ecol 1995; 21:1271-1285

Bradford M.M. A rapid and sensitive method for the quantification of microgram quantities of protein utilizing the principle of protein-dye binding. Anal Biochem 1976; 72:248-254

Buchanan B.B., Gruissem W., Jones R.L. *Biochemistry and Molecular Biology of Plants*. Rockville, Maryland: American Society of Plant Physiologists, 2000.

Calera M.R., Anaya A.L., Gavilanes-Ruiz M. Effect of phytotoxic resin glycoside on activity of H^+-ATPase from plasma membrane. J Chem Ecol 1995; 21:289-297

Chifflet S., Torriglia A., Chiesa R., Tolosa S. A method for the determination of inorganic phosphate in the presence of labile organic phosphate and high concentration of protein: Application to lens ATPases. Anal Biochem 1988; 168:1-4

Cruz Ortega R., Anaya A.L., Gavilanes-Ruiz M., Sánchez Nieto S., Jiménez Estrada M. Effect of diacetyl piquerol on H^+-ATPase activity of microsomes from *Ipomoea purpurea*. J Chem Ecol 1990; 16:2253-2261

Einhellig F.A. "Mechanism and Modes of Action of Allelochemicals." In *The Science of Allelopathy*. R.A. Putnam, Ch-Sh Tang, eds. New York, John Wiley and Sons, 1986

Friebe A., Roth U., Kück P., Schnabl H., Schulz M. Effects of 2,4-dihydroxy-1,4-benzoxazin-3-ones on the activity of plasma membrane H^+-ATPase. Phytochemistry 1997; 44:979-983

Hager A., Debus G., Edel H.G., Stransky H., Serrano R. Auxin conduces exocytosis and the rapid synthesis of a high turnover pool of plasma membrane H^+-ATPase. Planta 1991; 185:527-537

Hanson J.B., Trewavas A.J. Regulation of plant cell growth: the changing perspective. New Phytol 1982; 90:1-18

Hodges T.K. "ATPases Associated with Membranes of Plant Cells." In *Encyclopedia of Plant Physiology*. U. Luttge, M.G. Pitnam, eds. Berlin: Springer-Verlag, 1976

Hodges T.K., Leonard R.T., Bracker C.E., Keenan T.W. Purification of an ion-stimulated ATPase from plant roots: association with plasma membranes. Proc Natl Acad Sci USA 1972; 69:3307-3311

Jahn T., Baluska F., Michalke W., Harper J.F., Volkmann D. Plasma membrane H^+-ATPase in the root apex: Evidence for strong expression in xylem parenchyma and asymmetric localization within cortical and epidermal cells. Physiol Plantarum 1998; 104:311-316

Lanzetta P.A., Alvarez L.J., Reinach P.S., Candia O.A. An improved assay for nanomole amounts of inorganic phosphate. Anal Biochem 1979; 100:95-97

Marré E. "Integration of Solute Transport in Cereals." In *Recent Advances in the Biochemistry of Cereals*. D.L. Laidman, R.G. Wyn Jones, eds. New York: Academic Press, 1979

Marré E., Ballarin-Denti A. The proton pumps of the plasmalemma and the tonoplast of higher plants. J Bioenerg Biomembr 1985; 17:1-21

Marré M.T., Moroni A., Albergoni F., Marré E. Plasmalemma redox activity and H^+ extrusion. I. Activation of the H^+-pump by ferricyanide-induced potential and cytoplasmic acidification. Plant Physiol 1988; 87:25-29

Palmgren M.G. Proton gradients and plant growth: Role of the plasma membrane H^+-ATPase. Adv Bot Res 1998; 28:1-70

Penney C.L. A simple micro-assay for inorganic phosphate. Anal Biochem 1976; 75:201-210

Politycka B. Peroxidase activity and lipid peroxidation in roots of cucumber seedlings influenced by derivatives of cinnamic and benzoic acids. Acta Physiol Plant 1996; 18:365-370

Poole R.J. Energy coupling for membrane transport. Annu Rev Plant Physiol 1978; 29:437-460

Queirolo C.B., Andreo C.S., Niemeyer H.M., Corcuera L.J. Inhibition of ATPase from chloroplasts by a hydroxamic acid from the Gramineae. Phytochemistry 1983; 22:2455-2458

Rayle D.L., Cleland R. Control of plant cell enlargement by hydrogen ion. Curr Top Dev Biol 1977; 11:187-214

Serrano R. *Plasma Membrane ATPase of Plants and Fungi*. Boca Raton, Florida: CRC Press, 1985

Serrano R. Structure and function of proton translocating ATPase in plasma membranes of plants and fungi. Biochem Biophys Acta 1988; 947:1-28

Serrano R. Structure and function of plasma membrane ATPase. Annu Rev Plant Physiol Plant Mol Biol 1989; 40:61-94

Zhang J., Kirkham M.B. Lipid peroxidation in sorghum and sunflower seedlings as affected by ascorbic acid, benzoic acid, and propyl gallate. Plant Physiol 1996; 149:489-493

Zhao R., Dielen V., Kinet J.-M., Boutry M. Cosupression of a plasma membrane H^+-ATPase isoform impairs sucrose translocation, stomatal opening, plant growth, and male fertility. Plant Cell 2000; 12:535-546

CHAPTER 25

ROOT UPTAKE AND RELEASE OF IONS

Adela M. Sánchez-Moreiras[1], and Oliver Weiss[2]

[1]*Depto Bioloxía Vexetal e Ciencia do Solo. Universidade de Vigo. Spain.*
[2]*Institut für Pflanzenbau. Rheinische Friedrich-Wilhelms-Universität. Germany*

INTRODUCTION

Ion membrane transport

The root is the first plant organ in contact with the soil, and so, it is usually also the first organ in contact with nutrients present in the soil. This makes that roots have the vital function of absorbing and starting the transport of these solutes through the plant.

The mineral salts amount absorbed by the root surface will be depending on several factors as the ion concentration, the soil pH, the ion availability in the volume soil and the ion requirements for the plant (Lambers et al. 1998). Once that these solutes are within the plant, they are transported to the leaves.

The cell is a dynamic system with a dynamic equilibrium of entries and exits, and with several thermodynamic reactions. Biomembranes are limiting the cells, and generally they are 7.5 to more than 10 nm thick. They are constituted by two layers of different nature, a lipidic bilayer (giving stability to the membrane) and an inner fatty acids layer. Proteins across this structure are also present in the membrane. Since lipid bilayers are essentially impermeable to most metabolically important substances, all transport properties of cell membranes may be attributed to specific

M.J. Reigosa Roger, Handbook of Plant Ecophysiology Techniques, 413–427.

transport proteins inserted in the bilayer (Poole, 1988). A selective anion and cation exchange will take place through these biomembranes.

Ions are in continuous flux, as well in the environment as within the plant, and in migration through the soil-plant-environment.

Because of the chemical and physical cell characteristics, the entry in the cell can be difficult. But there are special structures with the function of ion transport (Larcher, 1994). How do they work?

Ion transport can be classified in either passive or active transport as follows. If there is free energy, or an electrochemical-potential gradient, ion absorption by passive transport is allowed. But if the cell needs to make enter an ion against any gradient, then will be necessary to spend some energy. This is called active transport.

Energetic point-of-view

Active transport	Passive transport
• Proton pump	• Uniport
• Ca^{2+} Efflux pump	• Co-transport (antiport, symport)

Structural point-of-view

• Ion channels	• Carriers

Sattelmacher et al. (1998) reviewed the apoplastic properties and processes in relation to plant mineral nutrition.

Because of irregular diffusion of cations and anions and because the activity of H^+-pumps, electrical potentials and pH gradients are build between the membrane outer and inner side. Therefore, substances with an energy charge could enter into the cell following an electrochemical gradient. Ion transport could be passive until the electrochemical equilibrium will be reached (Larcher, 1994).

Passive ion transport is well known and classified as uniport if only one substance is translocated. The transport of two ions is called co-transport. When these two ions are transported in the same direction it is called symport and when the transport occurs in the opposite directions it is called antiport.

To get an idea about the transport rates of the different kind of translocators see table 1. The pumps are used for active ion transport, the carriers and ion channels are used for passive membrane translocation and

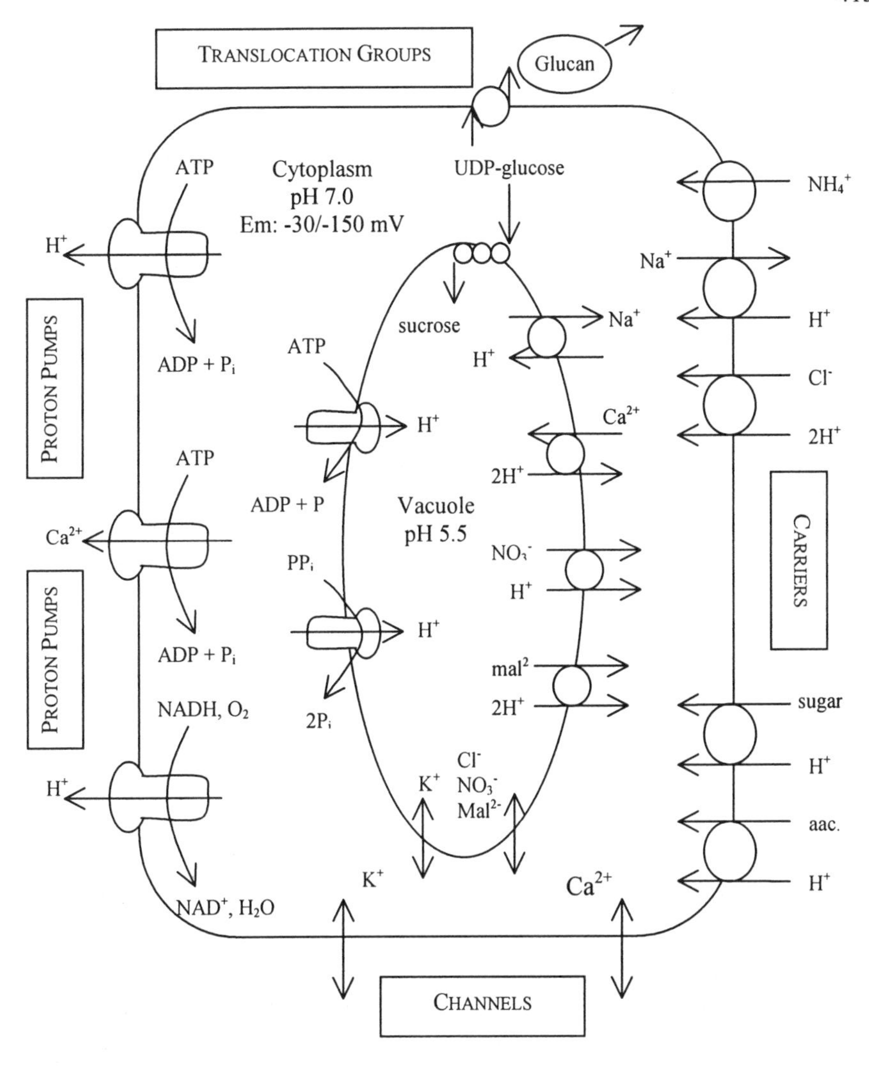

Figure 1. Transport systems in the plasma membrane and tonoplast. Cell functions as nutrition, osmoregulation, etc,, require a coordinated activity of these two biomembranes (redrawn from Poole, 1988).

are highly substrate specific (see Fig. 1). While passive transport follows to electrochemical gradients, active transport can work also against an electrochemical gradient (uphill transport). The transmembrane complex ATPases are called "key enzyme of the active ion uptake" (see chapter X). They also could change cytoplasm pH.

Table 1. Transport rates (Satter and Moran, 1988; Hedrich and Schroeder, 1989)

Transport form	ions s^{-1}
Pump	100 – 400
Carrier	1.000 – 10.000
Channel	10^{6-8}

The proton gradient is a basic requirement for the ions and other particles transport processes through the plasma and vacuole membrane. In the ion uptake and leakage study it is necessary to take in account three compartments interrelated in the cell: the apparent free space (AFS), the protoplasm, and the vacuole. Although, in the transport dynamics these three compartments seem consecutively united, each one has ion uptake and release times really different, from seconds in the AFS to days in the vacuole (Buchanan et al., 2000).

Ion selective transport

As previously said, in these spaces the transport is mediated by passive or active mechanisms, depending on the transported ionic species (see Fig. 1).

The plasmalemma permeability is typically high for K^+ while for Mg^{2+}, and especially for Ca^{2+}, this permeability is lower. Na^+ permeability varies between different plant species (Mengel, 1991). So, the potassium and sodium import is usually in a passive way (although cells are able to make it actively), while sodium efflux generally is actively pumped through the two limiting membranes (Marre, 1979).

Anion uptake generally occurs against electrochemical gradient (uphill transport). The transport of Cl^-, NO_3^-, $H_2PO_4^-$, and SO_4^{2-} is by active accumulation in the cell. The uptake of these anions is generally related to a secondary co-transport with H^+. This co-transport can be simultaneously present with a passive diffusion of these anions through the cell (Poole, 1988).

POSSIBLE SIGNIFICANCES FOR ION MEASUREMENT

The modification of membrane properties is a known response of plants to some biotic or abiotic stress. Several studies have been performed to elucidate the changes in the properties of the membranes and other physiological processes occurred in a plant exposed to some kind of stress (Balke, 1985; Cakmak and Horst, 1991; Pandey, 1994; Yu and Matsui, 1997). The number of cellular phenomena directly related with the membrane activity makes that an alteration at this level can disrupt so important physiological processes as cell enlargement, seed germination or stomata opening (see Fig. 2).

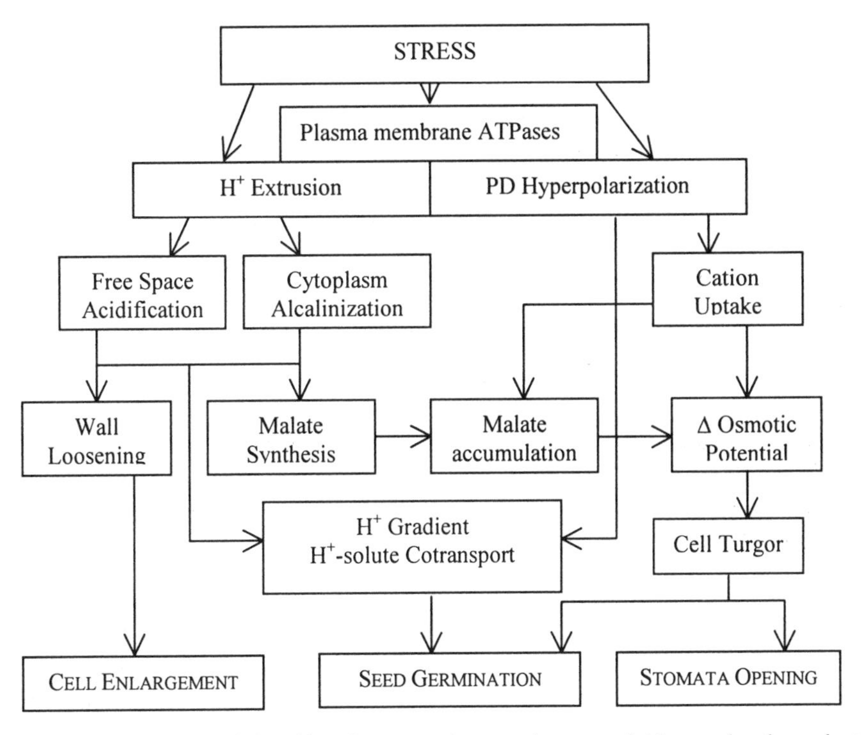

Figure 2. Complex relationships between the membrane activities and other plant physiological processes. Redrawn from Marre, 1979.

Several published works tried to explain the role that allelopathic plant exudates, phenolic acids or their derivatives, or other allelochemical compounds play in the membrane permeability disruption in higher plants (Glass and Dunlop, 1974; Balke, 1985; Radecke and Glenn, 1988; Mitiku, 1991; Baziramakenga et al., 1994; Pandey, 1994; Polyticka, 1996).

The inhibitory effects of vanillic, p-hydroxycinnamic and ferulic acids on the $H_2PO_4^-$ uptake by cucumber seedlings were examined by Lyu et al. in 1990.

In this way, Yu and Matsui (1997) examined the effect of root exudates of cucumber plants on the uptake of different ions (NO_3^-, $H_2PO_4^-$, SO_4^{2-}, K^+, Ca^{2+}, Mg^{2+} and Fe^{2+}). In their experiment they could observe an uptake inhibition by these root exudates for all studied ions except for $H_2PO_4^-$ uptake, as well as a stimulated leakage for specific ions (as K^+). The effect of cinnamic acid (a main constitute of cucumber exudates) and other aromatic acids on ion uptake and leakage was also reported in this study. Similar effects to those reported with a direct addition of cucumber exudates and a pH-dependent phytotoxicity for the allelochemical activity were shown. Yu and Matsui concluded that the acids within the membrane could induce a change in the membrane permeability resulting in an increase of ion leakage and a decrease of ion uptake.

Balke, in 1985, studied how some phenolic acids are able to increase membrane permeability, and so, how they are able to induce the selective efflux or to inhibit the selective absorption of certain anions and cations in plant roots. As Balke proposed in its work, there are different hypothesis for this alteration, all possible in plant roots and perhaps all occurring at the same time, because the mode of action of an allelopathic compound can be diverse in the place and in the time.

The possible solubilisation of these compounds into the cellular membrane with the loss of the membrane structure and the consequent mineral leakage across it, the membrane injury by an increased lipid peroxidation, or the alteration of proton active transport by an allelopathic effect at protein level, are some of the proposed hypothesis to explain the observed disruption in the membrane activity.

Allelochemicals can affect the lipid composition of the plant membranes generating strong membrane disorganisation (see Fig. 3), and inducing finally the loss on selective membrane permeability (Baziramakenga et al. 1995; Politycka, 1996; Zhang and Kirkham, 1996). This increase can be induced by the accumulation of superoxide, hydrogen and hydroxyl radicals as a result of the enzymatic activity alteration.

In this way, Baziramakenga et al., in 1994, found that benzoic and cinnamic acids tempted an alteration on the nutrient uptake by roots of *Glycine max* L. cv. Maple Bell. The absorption of P, K, Mg and Mn was specially depressed by cinnamic acid. After these observations, Baziramakenga et al. in 1995 continued the study and observed an alteration of peroxidase, catalase and superoxidodismutase activities (implied in the oxygen metabolism at cellular level) in presence of allelochemicals (see Fig. 3).

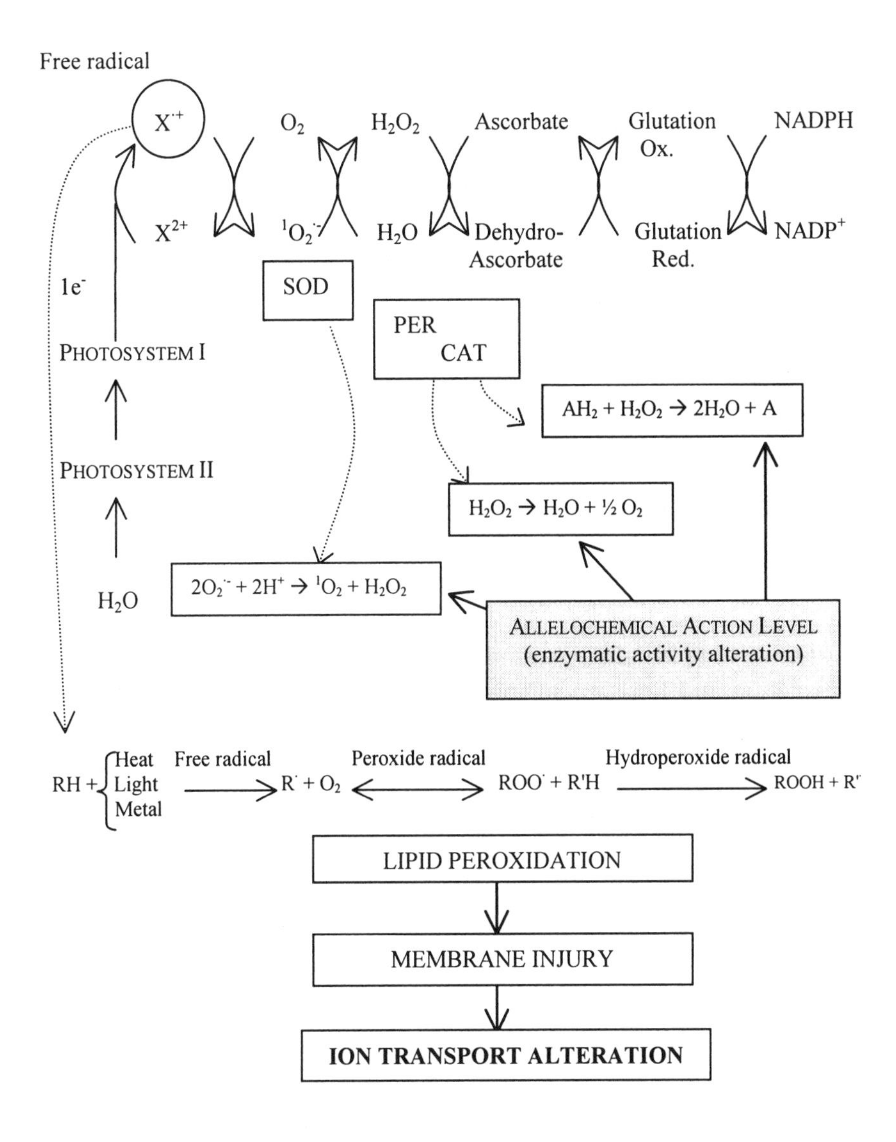

Figure 3. Possible allelochemical mode of action at enzymatic level with the subsequent ion uptake alteration. If the allelopathic compound causes an alteration in the activity of the enzymes implied in the oxygen metabolism, an accumulation of hydroperoxide radicals can occur. These radicals can be toxic to the membrane by a direct damage or by the formation of free radicals that induce an increase in lipid peroxidation.

This alteration stimulated the accumulation of H_2O_2 in the membrane. The accumulation of free radicals can increase also the lipid peroxidation in the plasma membrane with its subsequent damage, inhibiting the correct nutrient uptake by these roots.

Two years later, Politycka (1996) observed also an increase in lipid peroxidation of cucumber roots submerged in benzoic and cinnamic acid solutions. The presence of these allelochemical compounds caused an alteration at membrane level increasing the ion permeability.

Allelochemicals of different nature, as the hydroxamic acids, were also studied for alteration of ion uptake in plants. So, in 1997 Pethö et al. performed an experiment to know the role of cyclic hydroxamic acids in the iron uptake of maize.

Investigations carried out in our laboratory (Sánchez-Moreiras et al., 2001; Reigosa et al., 2001) to determine the effect of 2-benzoxazolinone (BOA, an hydroxamic acid) on membrane permeability of lettuce roots shown an increase in the release of the most part of the studied ions (see Fig. 4). This disruption was time dependent for anions and for cations and was reflected like a continuous release along the time and not like a strong and punctual effect. Significant increases in the NO_3^-, PO_4^{3-}, SO_4^{2-}, Ca^{2+} or NH_4^+ release suggested an alteration on membrane permeability of lettuce root cells.

The membrane disorganisation caused by an increase in the lipid peroxidation and the subsequent damage in the membrane permeability and nutrient uptake can be an important mode of action of the plant secondary metabolites. Zhang and Kirkham (1996) also found effects on lipid peroxidation when sorghum and sunflower seedlings were exposed to ascorbic acid.

Other phenomena can be also important factors in an incorrect membrane activity. So, a decrease in the medium pH (Harper and Balke, 1981; Baziramakenga et al., 1995) combined with the presence of an allelopathic compound could cause a strong cellular disorganisation inhibiting some ion uptake.

Likewise, an alteration in the plasma membrane ATPase activity could be the reason for an abnormal nutrient absorption in plants exposed to allelopathic compounds, because of the role that this enzyme play in the ion gradient and so also in the ion transport through the membrane (Friebe et al., 1997). The effects on this master enzyme are broadly exposed in this handbook in the Chapter 24.

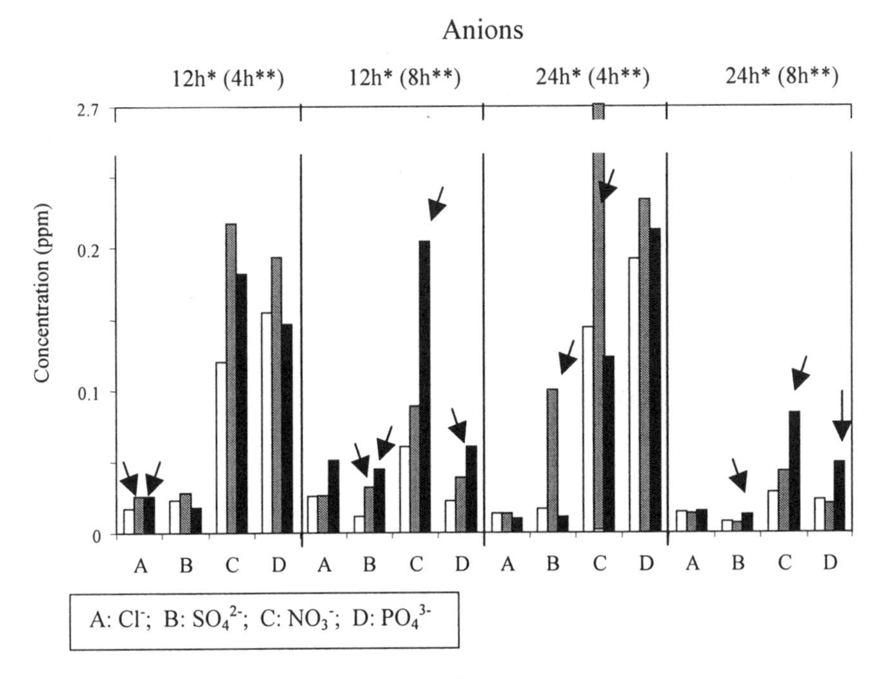

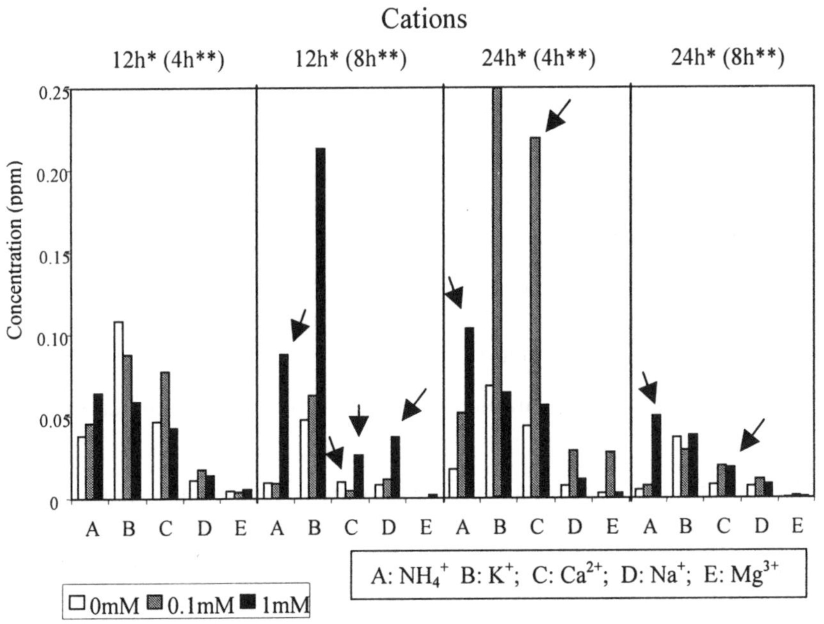

Figure 4. Effect of BOA application on root release of anions and cations from lettuce plants (Reigosa et al., 2001).
* Root exposure time (h) ** Root washing time (h) ↓ Arrow shows significant inhibition

By other hand a disruption in the intrinsic membrane protein activities can be occurring in allelochemical-treated-plants (Vaughan and Ord, 1991; Booker et al., 1992).

An effect on the sulfhydryl groups of carrier proteins was reported to be also responsible of altered ion uptake and membrane permeability damage in plants that were in contact with phenolic compounds (Baziramakenga et al., 1995).

All of these possible allelochemical modes of action in the cell require a detailed and more complex study about the membrane activities and physiological related processes when an alteration on membrane permeability is observed.

ION MEASUREMENTS

Different ways for measurement ion content in a sample were used along the years in the literature (Baziramakenga et al., 1995, Yu and Matsui 1997), but perhaps the capillary electrophoresis is one of the most used techniques to separate for measuring the anions and cations present in an intact and alive plant sample.

The principle of separation and identification in the ion chromatography is based in the capture by the column of the injected ion mixture and the release by the moving phase of the ions separated in the time. So, the mineralised and the organic ions are dosed and separated according to their motility.

Once that the sample is injected in the head of the capillary, a difference of potential is applied and the ions migrate to the opposite polarity. Its velocity is related with its motility.

The injection of the sample can be performed in three different ways:

- Hydrodynamic injection based in the dynamic entry of the sample in the capillary. The capillary and the electrode are submerged in the sample flask and a gas pressure is applied for a determined time to facilitate the entry of some nL in the capillary head. This method allows the entry of important volumes in the system, but is subjected to possible losses.
- Hydrostatic injection based in the entry of the sample in the capillary without using external pressures. The capillary is placed

in the sample flask and both are elevated (in Capillary Ionic Analysis: CIA, h is 10 cm) for a determined time (5-60 sec). The gravity allows the pass of the sample from one extreme to the other. This is a single and reproducible method, which avoid any loss.

- Electrocinetic injection based in the migration in the capillary head of certain ions by applying a difference of potential in the sample. This is an accurate but also complex system.

Once that the sample is injected in the capillary, it is necessary to detect the ions in movement. So, the capillary has a triple function: injection, separation, and detection.

The separation by Capillary Ionic Analysis (CIA) has a high resolution, and the ion migration time and the ion migration order are foregone. In this method the electrolyte is selected according to the ion mixture subjected to separation. It is important to take into account the absorbance, the pH, and the motility in the electrolyte selection. The form and the size of the obtained peaks depend on the ion motility in the electrolyte (see Fig. 6).

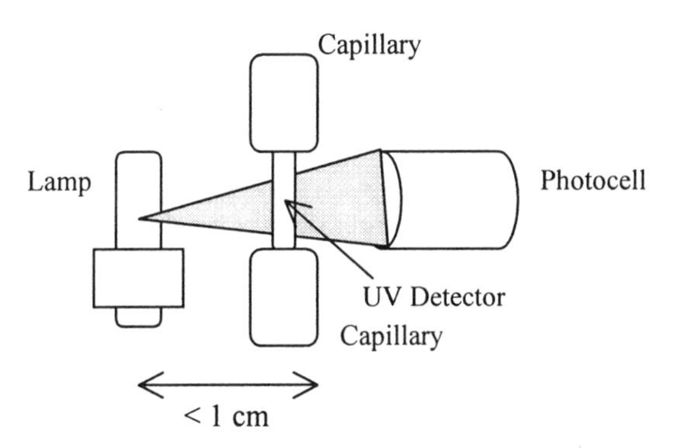

Figure 5. CIA equipment with a high performance UV detector

The detection must be single and routine, a must be able to detect a very little amount of sample. Therefore, a high performance ultraviolet detector will be a good election in the Capillary Ionic Analysis, because it is a single detector with a high sensibility.

After each analysis a pattern curve is made for every analysed ion at different concentrations. This allows the quantification of the exact amount of the ions present in the sample. Because we know the exit time and the exit order for each ion it is possible to obtain a detailed composition of our sample.

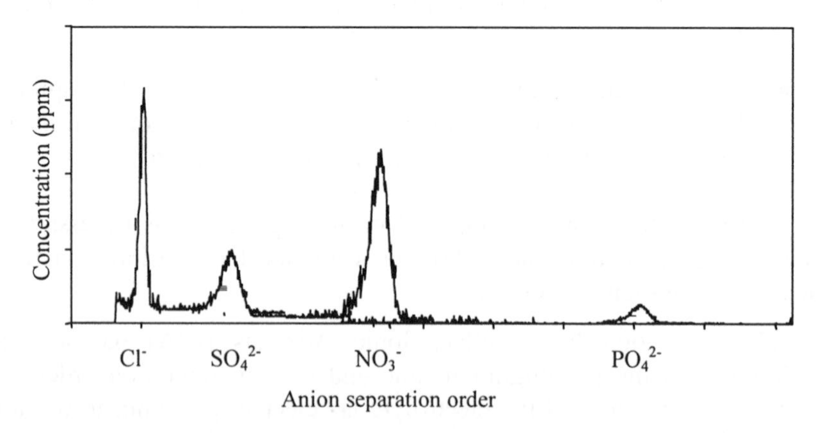

Figure 6. Anion appearance in a sample after a capillary ionic analysis

So, the analytical parameters in the Capillary Ionic Analysis can be resumed in:

- Electrolyte nature, concentration and pH
- Detection wavelength
- UV detection method (direct or inverse detection)
- Injection method
- Capillary size and inner diameter
- Analysis temperature and potential

With this analysis we will obtain a high resolution, a fast analysis, new criteria for ion separation (motility, pK, etc.), single and abundant parameter, an easy management, and finally, a cheap technique and easy maintenance.

PROCEDURE

Root membrane permeability can be studied with Capillary Ionic Analysis by analysing the ion amounts released by the plant in a determined condition.

So, in our methodology, mature plants are transferred to glass individual tubes with the treatment solution to expose roots to the allelochemical. Therefore, the root will be in total contact with the treatment solution. The tubes are placed in an environment chamber with the controlled light and temperature regimes.

After exposure time (some hours) roots are quickly washed to eliminate possible solution rests. Then, roots are placed in new tubes with water for long washing. This washing is performed to allow ionic release through roots. The resultant water is stored at –20°C for later analysis.

Anions and cations are quantified by Capillary Ionic Analysis with a 100x100 capillary and filters of 254 nm wavelength for anions and 185 nm for cations.

Take in account that for relating these results with a possible allelopathic effect on the root membrane permeability, it would necessary to measure pH differences in the samples along the experiment.

LITERATURE

Balke N. "Effects of Allelochemicals on Mineral Uptake and Associated Physiological Processes." In The Chemistry of Allelopathy. Biochemical Interactions among Plants, A.C. Thompson, ed. Washington, D.C., 1985.

Baziramakenga R., Leroux G.D., Simard R.R. Effects of benzoic and cinnamic acids on membrane permeability of soybean roots. J Chem Ecol 1995; 21:1271-1285.

Baziramakenga R., Simard R.R., Leroux G.D. Effects of benzoic and cinnamic acids on growth chlorophyll and mineral contents of soybean. J Chem Ecol 1994; 20:2821-2833.

Booker F.L., Blum U., Fiscus E.L. Short-term effects of ferulic acid on ion uptake and water relations in cucumber seedlings. J Exp Bot 1992; 43:649-655.

Buchanan B.B., Gruissem W., Jones R.L. Biochemistry and Molecular Biology of Plants. Rockville, Maryland: American Society of Plant Physiologists, 2000.

Cacmak I., Horst W.J. Effect of aluminium on lipid peroxidation, superoxide dismutase, catalase, and peroxidase activities in root tips of soybean (Glycine max). Physiol Plant 1991; 83:463-468.

Einhellig F.A. "Mechanism and Modes of Action of Allelochemicals." In The Science of Allelopathy. R.A. Putnam, Ch-Sh Tang, eds. John Wiley and Sons, New York, 1986.

Friebe A., Roth U., Kück P., Schnabl H., Schulz M. Effects of 2,4-dihydroxy-1,4-benzoxazin-3-ones on the activity of plasma membrane H+-ATPase. Phytochem 1997; 44:979-983.

Glass A.D.M., Dunlop J. Influence of phenolic acids on ion uptake: IV. Depolarization of membrane potentials. Plant Physiol 1974; 54:855-858.

Harper J.R., Balke N.E. Characterization of the inhibition of K+ absorption in oat roots by salicylic acid. Plant Physiol 1981; 68:1349-1353.

Hedrich R., Schroeder J.I. The physiology of ion channels and electrogenic pumps in higher plants. Ann Rev Plant Physiol Plant Mol Biol 1989; 40:539-569.

Lambers H., Chapin F.S., Pons T.L. "Mineral Nutrition." In Plant Physiological Ecology. Springer Verlag, New York, 1998.

Larcher W. Ökophysiologie der Pflanzen. 5. Auflage. Verlag Eugen Ulmer, Stuttgart, Germany, 1994.

Marré E. "Integration of Solute Transport in Cereals." In Recent Advances in the Biochemistry of Cereals. D.L. Laidman, R.G. Wyn Jones, eds. Academic, New York, 1979.

Mengel K. Ernährung und Stoffwechsel der Pflanze. Gustav Fischer Verlag Jena, Germany, 1991.

Mitiku G. Action of allelochemicals ferulic and p-coumaric acids on membrane potential, IAA-oxidase/peroxidase and growth of soybean seedlings (Glycine max L.). Diss Abstracts Int B Sci Eng 1991; 51:3215.

Pandey D.K. Inhibition of salvinia (Salvinia molesta Mitchell) by parthenium (Parthenium hysterophorus L.). I. effect of leaf residue and allelochemicals. J Chem Ecol 1994; 20:3111-3131.

Pethó M., Lévai L., Römheld V. The possible role of cyclic hydroxamic acids in the iron uptake of maize. Növénytermelés 1997; 46:139-144.

Politycka B. Peroxidase activity and lipid peroxidation in roots of cucumber seedlings influenced by derivatives of cinnamic and benzoic acids. Acta Physiol Plantarum 1996; 18:365-370.

Poole R.J. "Plasma Membrane and Tonoplast." In Solute Transport in Cells and Tissues. D.A. Baker, J.L. Hall, eds. Monographes and Surreys in the bioscience Langman Scientific and Technical, 1988.

Radecke M.E., Glenn S. Alteration of membrane permeability by allelochemicals in suspended cell cultures. Proceed Ann Meet Northeastern Weed Sci Soc 61, 1988.

Reigosa M.J., González L., Sánchez-Moreiras A.M., Durán B., Puime O., Fernández A., Bolaño C. Comparison of physiological effects of allelochemicals and commercial herbicides. Allelopath J 2001; in press.

Sánchez-Moreiras A.M., Weiss O., Reigosa M.J., Pellissier F. Membrane permeability of lettuce can be affected by some allelochemicals. Submitted 2001.

Sattelmacher B., Mühling K.H., Pennewiss K. The apoplast – its significance for the nutrition of higher plants. Z Pflanz Bodenkunde 1998; 161:485-498.

Satter R.L., Moran N. Ionic channels in plant cell membranes. Physiol Plant 1988; 72:816-820.

Vaughan D., Ord B.G. "Extraction of Potential Allelochemicals and their Effects on Root Morphology and Nutrient Content. In *Plant Root Growth: An Ecological Perspective*. D. Atkinson, ed. Blackwell Scientific Publishers, London, 1991.

Yu J.Q., Matsui Y. Effects of root exudates of cucumber (*Cucumis sativus*) and allelochemicals on ion uptake by cucumber seedlings. J Chem Ecol 1997; 23:817-827.

Zhang J., *Kirkham* M.B. Lipid peroxidation in sorghum and sunflower seedlings as affected by ascorbic acid, benzoic acid, and propyl gallate. Plant Physiol 1996; 149:489-493.

CHAPTER 26

RADIOIMMUNOASSAY OF ABSCISIC ACID

Mark A. Bacon
Biological Sciences Department, Institute of Environmental and Natural Sciences, Lancaster University. United Kingdom.

INTRODUCTION

Radioimmunoassay of abscisic acid (ABA) is a relatively cheap and fast way in which to determine the approximate concentration of this plant growth regulator in a wide range of plant samples. The simplicity of the assay makes it possible to analyse a large number of samples and this remains one of the main reasons why the radioimmunoassay system is a popular choice over other methods such as GC-MS or ELISA.

The use of this assay has advanced our understanding of the role of abscisic acid in regulating plant growth and gas exchange. So much so, that the initial perception that an accumulation of the hormone was needed in order to demonstrate a physiological role, particularly during drought, has been called into question (Wilkinson and Davies; 1997; Bacon et al., 1998). The flux and delivery of ABA via the xylem and its distribution between different compartments within the different tissues of the plant, now seem as, if not more important than the net synthesis and accumulation of this hormone.

This chapter demonstrates the ease with which the radioimmunoassay can be undertaken and considers what plant tissues may be sampled, and how, in order to gain useful information on the physiological significance of the concentrations that are determined. The protocol presented is based on the experiences from our own laboratory in running the assay developed by Quarrie et al., (1988) using the monoclonal antibody AFRC MAC 262 specific for (+)- *cis-trans* abscisic acid.

M.J. Reigosa Roger, Handbook of Plant Ecophysiology Techniques, 429–442.

WHY MEASURE ABA?

Abscisic acid (Fig. 1) was first implicated in abscission and seed dormancy over 30 years ago (Addicot, 1983). Its role in maintaining seed and bud dormancy is now known to be only one of a whole range of physiological phenomenon implicated to involve ABA (see Jones and Davies, 1991). In recent years, the role of ABA as a signalling molecule has received much attention. It is now clear that ABA, acting as a xylem borne chemical messenger, plays a key role in communicating soil water status, perceived by the roots, to all other parts of the plant.

An unequivocal role for ABA in mediating stomatal closure in response to soil drying has been demonstrated in a wide range of different species. Zhang & Davies (1991) demonstrated that removal of ABA from the xylem sap of maize plants using an immunoaffinity column, removed anti-transpirant activity.

However, Munns & King (1988) were unable to show any reductions in transpiration when ABA was fed to plants in distilled water, at a concentration comparable to that found in the xylem sap of droughted wheat plants. The authors concluded that 100 times more ABA than that apparently present in the sap of droughted plants was required to induce similar reductions in transpiration. Some have therefore questioned a physiological role for ABA in controlling leaf expansion during drought (e.g. Munns & Cramer, 1996).

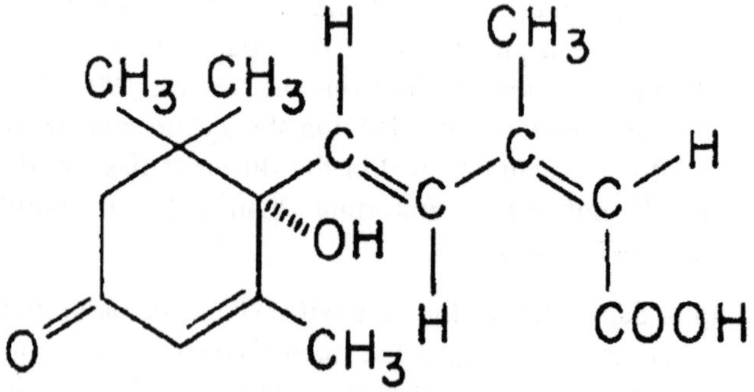

Figure 1. Molecular structure of abscisic acid (ABA).

More recent work has finally been able to dispel much of this scepticism. In a series of recent papers, stomatal closure and a restriction of leaf expansion, in several species, has been shown to be inextricably linked to the presence of ABA at physiologically-relevant concentrations (Wilkinson & Davies, 1997; Wilkinson et al., 1998; Bacon et al., 1998).

A significant body of work (see Davies and Jones, 1991) now demonstrates that ABA has a central role in mediating a plant's response to variation in soil water status. If we are to exploit plants this information for commercial benefit and use it to understand the responses of natural vegetation to their environment, particularly in the context of climate change, the study and measurement of ABA and the physiological responses it governs, remains compelling.

WHERE SHOULD ABA BE MEASURED?

The question posed may appear straightforward. Indeed ABA can be determined in any plant material or other substance, such as soil, without to much difficulty. The main problem is encountered when trying to determine the physiological relevance of the concentration determined. Bulk tissue concentrations of ABA from leaves and roots are easy to determine providing tissue can be ground to a state that allows extraction using deionised distilled water. But what does a determination of ABA concentration in a particular tissue tell us about its physiological role? In roots, an extensive body of evidence now suggests that an accumulation of ABA in growing cells can explain the change in root to shoot ratio observed when plants encounter a whole range of stresses, most notably drought (see Spollen et al., 1993).

The accumulation of solutes, activity of putative cell wall loosening and stiffening enzymes and the accumulation of other plant growth regulators, have all been inextricably linked to an accumulation of ABA (Sharp et al., 1989; Saab et al., 1992; Spollen et al., 2000; Voetberg et al., 1991; Wu et al., 1994; 1996). Determination of bulk levels of ABA in leaves however, has proved less successful in explaining the reduced growth rates observed when plants grow in drying soil, even though ABA is implicated as a primary mediator in this response. Nowhere is this more evident than in a recent set of our own recent experiments which demonstrate that a plant devoid of ABA, as a result of genetic lesion, has a significantly reduced ability to respond to soil drying (Wiklinson et al., 1998; Bacon et al., 1998). Soil drying results in significant increases in the xylem sap pH of several species (see Wilkinson, 1999). Bacon et al.,

(1998) was able to demonstrate that this increase in pH could significantly inhibit leaf expansion.

Measurement of bulk ABA concentration is the growing region of the leaf, failed to explain the response. However, the response to increasing pH could not be found in plants deficient in ABA due to genetic lesion. However, adding low concentrations of ABA (those typically found in well-watered plants) back to the plant, restored the response. The work concluded that ABA must show a greater distribution into an active compartment, as the xylem sap pH increased. As a consequence of ABA being a weakly dissociating acid, as the pH of the xylem sap and ultimately the apoplast increased, ABA would reside in this relatively more alkaline compartment (and the compartment presumed to contain the active site for ABA) before being sequestered into the symplast. While bulk ABA concentrations in the leaf elongation zone proved fruitless in explaining the response (as no differences were seen) measurement within this compartment of the leaf would provide additional evidence to support these conclusions.

It is currently impossible to analyse ABA concentrations within the apoplast, with any accuracy or lack of artifactual risk. Several other pieces of work have also demonstrated the necessity to consider the active compartment of ABA within the plant. Trejo et al., (1994) was able to illustrate an apparent difference in the sensitivity of stomata to ABA was a function of the mesophyll, by determining the ABA concentration of the different compartments of the leaf (vascular bundles, mesophyll and epidermis) rather than the bulk leaf concentration.

In many cases however, most notably within the xylem sap, is there still a valid reason to determine changes in the concentration of ABA in response to environmental perturbation. Collection of xylem sap, which is representative of that found in an intact transpiring plant, still remains a controversial area. The essence of the controversy lies in the fact that most conventional methods of sap collection may result in a concentration or dilution of ABA as a result of collecting sap at rates not comparable to the transpirational fluxes measured within intact transpiring plants. An excellent review of this controversy and opportunities for overcoming the criticisms associated with determining ABA concentrations in collected sap is given by Schurr (1998).

COLLECTION OF SAMPLES FOR ABA ANALYSIS

Although extraction procedures may vary, depending on the specific tissue type, the one general principle is to ensure that the samples are frozen within liquid nitrogen as soon as they are collected. Significant increases in the concentration of ABA may occur within minutes of excising tissue from plants for ABA analysis. Tissue should be quickly removed, wrapped in aluminium foil and immersed in liquid nitrogen. Samples can then be stored indefinitely in liquid nitrogen or transferred to a –80°C freezer. Samples of xylem, cell or phloem sap can be collected into small microfuge tubes, immersed into liquid nitrogen and stored in the same way as other tissue samples.

There is increasing interest in measuring ABA concentrations in soil (Hartung et al., 1996). Soil samples should be collected in the same way as other tissue and frozen in liquid nitrogen to halt microbial activity and potential changes in ABA concentration.

SAMPLE PREPARATION

Prior to extraction in deionised distilled water, tissue samples should be freeze-dried in order to allow extraction of ABA in deionised distilled water. Samples should be dried to constant weight (which typically takes 24-48 hrs for the majority of samples), and stored in the dark in a dessicator until analysis takes place.

ABA can be extracted crudely from plant tissues such as leaves and roots by shaking ground, freeze-dried and weighed samples in deionised distilled water at 4°C overnight. The ratio of water to sample will differ depending on the tissue, species and level of stress the plant has received. Typical extraction ratios vary between 1:10 and 1:60. The optimum extraction ratio needs to be determined when setting up the assay for a particular tissue for the first time.

In many cases it is then possible to use this crude extract directly in the radioimmunoassay. It is also possible to use xylem, phloem or cell sap collected directly from the plant in the radioimmunoassay. It may be necessary, particularly when undertaking ABA determination in a new species, to assess the possibility that contaminants within the sample may possess significant affinity for the ABA antibody binding sites. This can be assessed using thin layer chromatography (TLC) (see below).

PRINCIPLE OF THE TECHNIQUE

The assay utilises the competition between added radiolabelled ABA (labelled DL-*cis/trans* [^{3}H] ABA) and the ABA in the sample, for the binding sites of an ABA antibody. Separation of bound and unbound ABA (radiolabelled and non-radiolabelled) is achieved by flocculation of the antibody protein/ABA complex with ammonium sulphate solution. This precipitate is then counted to reveal the proportion of bound radioactivity. As the concentration of sample ABA increases, the amount of radioactively labelled ABA bound decreases. Free antibody and excess radioactive and non radioactive ABA are removed from the assay before the amount of radiolabelled ABA is determined.

Several antibodies exist which are highly specific for the free acid (+)-ABA, the physiologically active form of the acid. Using the AFRC MAC 252 antibody, Quarrie (*personal communication*) has shown that a 125pg (+)-ABA in 50μl standard gave the same cpm in the RIA using MAC252 as did a 250pg (±)-ABA in 50μl standard, demonstrating the antibody's affinity for the biologically active form of the acid. Table 1 details the cross-reactivity of the MAC252 monoclonal antibody with ABA and some of its derivatives.

Table 1. Cross-reactivity of the MAC252 monoclonal antibody with ABA and some common derivatives.

Compound	Percentage cross-reactivity
(+)-2-*cis*-abscisic acid	100
(+)-2-*trans*-abscisic acid	<0.1
(±)-2-*cis*-abscisic acid	49
(+)-2-*cis*-abscisic acid methyl ester	0.4
(+)-2-*cis*-abscisic acid glucose ester	<0.1
phaseic acid	<0.1
dihydrophaseic acid	<0.1
xanthoxin	<0.1

Procedure

1. For each sample add 200µl of 50% phosphate buffered saline (50mM sodium dihydrogen phosphate, 50mM di-sodium hydrogen phosphate and 100mM NaCl adjusted to pH 6) (Sigma, Poole, Dorest, U.K.) to 1.5ml microfuge tubes (Starstdet Ltd. Beaumont Leys, Leicester, UK).

2. Add 50µl of the sample to be analysed into the phosphate buffered saline, together with, 100µl of H^3 ABA (*cis-trans*-[G-3H]-ABA, Amersham International Plc. Aylesbury, Buckinghamshire, UK) dissolved in buffer (phosphate buffered saline and 5mg γ-globulin per ml) (Sigma, Poole, Dorset, UK).

3. Finally add 100µl of MAC 262 in buffer (phosphate buffered saline with 5mg bovine serum albumin and 4mg of polyvinylpyrrolidone (PVP) per ml) (Sigma, Poole, Dorset, UK) to each of the microfuge tubes.

4. The assay also requires that suitable standards be run along side samples in which ABA concentration is to be determined. Prepare ABA standards in the range of 125 - 2000 pg $50\mu l^{-1}$ (125, 250, 500, 1000 and 2000 pg $50\mu l^{-1}$ prepared from synthetic ABA (Sigma, Poole, Dorset, U.K.). These 50µl aliquots of these standards should be alongside the samples for ABA determination.

5. To parameterise the calibration (see below) run the assay together with two extra tubes containing either 50µl of concentrated ABA (10^{-3} M) or 50µl of deionised distilled water. These two extra samples provide the counts per minute (cpm) obtained with minimum binding of radioactive ABA to the antibody (see below) or the cpm obtained with maximum binding, respectively (see below).

6. Lids should then be placed on the microfuge tubes and each one thoroughly mixed using a vortexer.

7. The reaction should be allowed to proceed for 1 hour at 4 °C in the dark. The reaction is then stopped flocculating out the antibody/ABA complexes with 500µl of saturated ammonium sulphate solution (Sigma, Poole, Dorset).

8. Tubes should then be vortexed and left for 1 hour, then centrifuged at 5000g for 5 minutes. The supernatant should then be discarded.

9. Add 1ml of 50% saturated ammonium sulphate to re-dissolve the pellet with the aid of vortexing.

10. Re-centrifuge the tubes, add 1ml of 50% saturated ammonium sulphate solution and re-suspend the pellet. Centrifuge again to recover the pellet. Discard the supernatant.

11. Add 100μl of deionised distilled water to each tube to dissolve the pellet.

12. Add 1.5ml of scintillant (e.g. Ecoscint H, National diagnostic, NJ, USA supplied by Mensura Technology, Wigan, Lancashire, UK) to each tube. Samples should then be counted in a scintillation counter for six minutes each (e.g. Packard tri-carb 300, meriden, CT, USA).

Calibration of the assay and determination of ABA concentrations

A standard curve of counts per minute (cpm) versus ABA concentration can be produced from the cpm determined for the known standards (Fig. 2A). In order to obtain a linear calibration curve (Fig. 2B) the cpm from the standards is logit transformed to allow a linear log/logit plot to be constructed. The logit transformation of a variable B is given by:

Equation 1

$$LogitB = Ln\left\{\frac{\dfrac{B - B_{min}}{B_{max} - B_{min}}}{1 - \dfrac{B - B_{min}}{B_{max} - B_{min}}}\right\}$$

Where, B_{max} is the cpm obtained from the assay tube containing 50μl of deionised distilled water instead of sample, B_{min} is obtained from the cpm of the assay tube contained and B is the cpm from the tubes containing sample for [ABA] determination.

Assessing the immunoactivity of contamination in crude extracts

In many cases specific immunoreactive contamination may exist in crude extracts. Quantification of such contamination is relatively simple. A possible protocol is included here.

1. Extract a crude sample as above, ideally at a 1:10 ratio of sample to water.
2. Apply 100µl of this extract via repeated application 5cm from the base of a pre-prepared silica gel TCL plate (Aldrich, Gillingham, Dorest , England) by washing overnight in 20:80 methanol/ethyl acetate mixture and drying thoroughly.
3. Alongside the sample, apply 5µl of 10^{-3} M ABA. To prevent contamination between the two samples, remove a thin strip of silica along the entire length of the plate to isolate the lanes of the two samples.
4. To concentrate the samples, place the plate in a tank of ethyl acetate/water and allow the solvent edge to run to the edge of the application zone twice. Then run the plates twice until the solvent edge reaches a specific point near the top of the plate.
5. Briefly expose the plate to a UV-A source to locate the position of the concentrated ABA sample.

The profile of immunoactivity along the TLC plate can then be determined by removing silica from ten bands on the plate between where the samples were applied and the front of the solvent after the plates had been run. The edges of the plate can be discarded to remove any possible edge effects during the running of the plate.

The immunoreactive components within the silica from each band can then be extracted in deionised distilled water at 4°C and aliquots subjected to radioimmunoassay (above). By reference to the concentrated ABA sample, the concentration of ABA in the same region of the plate deposited from the test sample and the apparent concentrations of ABA determined elsewhere on the plate, the significance of any contamination can be assessed. The vast majority of immunoreactivity should be found in the same band as that in which the ABA sample is visualised under UV.

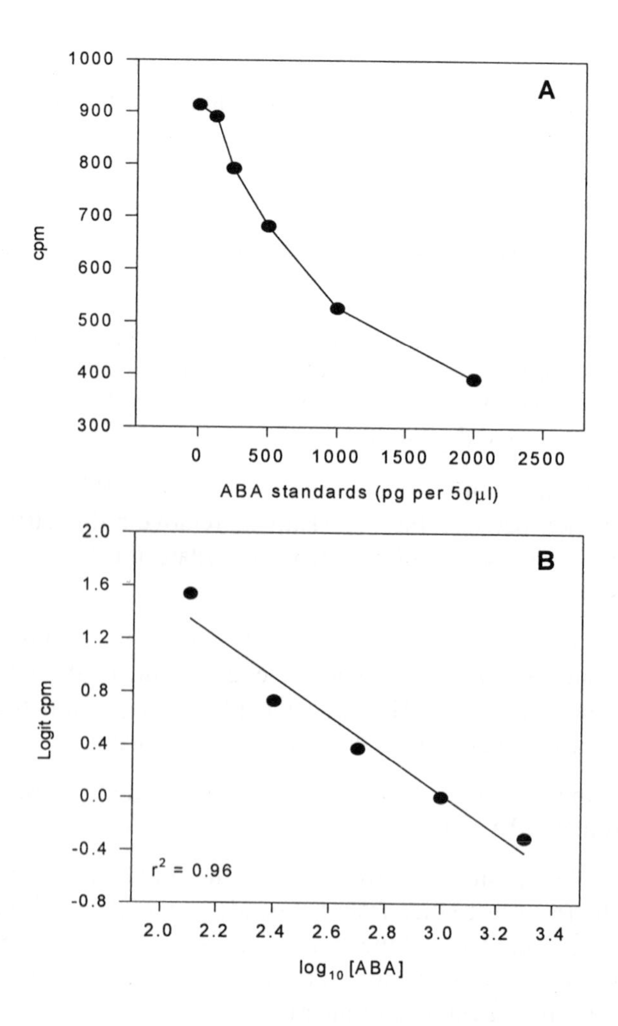

Figure 2. A plot of counts per minute (cpm) obtained during scintillation counting with different added standard concentrations of ABA (0, 125, 250, 500, 1000 and 2000 pg $50\mu l^{-1}$) (A) and a Log/Logit transformation of the data to yield a linear calibration curve for use in converting the cpm of samples into concentrations of ABA (B).

It is also possible to assess the level of non-specific interference within the assay that may be caused by contamination within a crude extract using a 'spike-dilution' test. This test involves spiking a series of serial dilutions of tissue extract with known amounts of ABA, to generate a series of lines (Fig. 3). In the absence of interference the series of lines should be parallel.

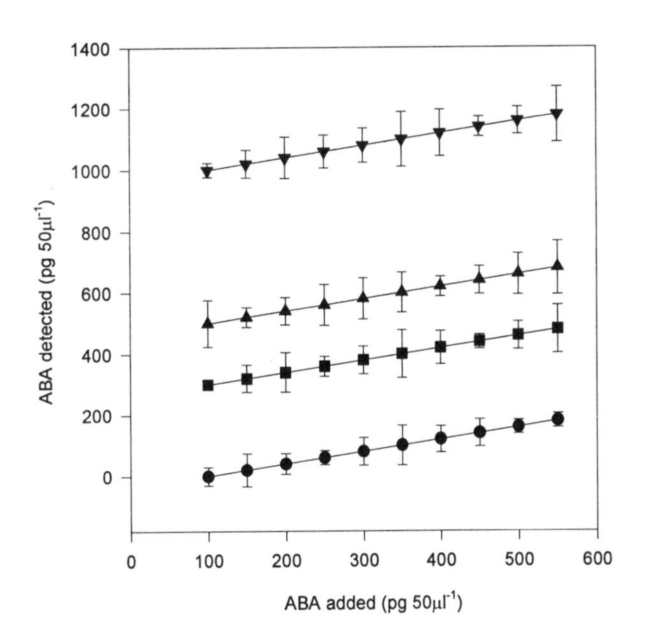

Figure 3. A typical spike-dilution test of a sample showing no interference (see text).

Both of these analyses should be undertaken when any new species or tissue is to be analysed for ABA concentrations using radioimmunoassay. It may also be important to undertake such analysis when the effects of a stress on concentrations of ABA are to be considered. Using thin layer chromatography, the appearance of significant levels of immunoactivity in the absence of any apparent interference in the sample (determined by the spike-dilution test), may suggest the existence of ABA conjugates within the crude sample. The significance of such compounds in mediating known ABA-dependent processes can be assessed by removing the bands of the TLC plate in which they reside, extracting the silica and applying the extract into a stomatal bioassay system as described by Weyers & Meidner (1990).

If significant interference and /or contamination is present, samples may need to be purified. Several methods may be appropriate. One procedure worth considering is adding HPLC grade methanol and AnalaR grade acetic acid to the aqueous extract to give a ratio of 80:18:2

(sample:methanol:acetic acid). 1ml samples can then be passed through a C18 column (3m 200mg sorbent, SepPak, waters). The column can then be washed with water/methanol/acetic acid (80:18:2) and the ABA subsequently eluted with 80% methanol, 20% water buffered to pH 6-8 with ammonium acetate. The elute can then be dried in a stream of dry air, re-dissolved in 0.5 ml of deionised distilled water, acidified to pH 1-3 with hydrochloric acid and partitioned against ether. Samples can then be re-dissolved in 250µl of distilled, deionised water and applied to TLC plate to allow immunoactivity to be re-profiled against a known ABA standard. Palmer (1996) has shown this cleaning procedure has been shown to remove all significant immunoactivity not associated with free ABA in an aqueous extract of sunflower. Such extraction procedures may recover *c*.90% of the ABA in the original sample and show no change in efficiency as a result of changes in the concentration of contaminant or sample ABA (Palmer, 1996).

Future developments

Radioimmunoassay is likely to remain a robust technique for ecophysiologists for many years. The development of methods to isolate specific compartments and extract the ABA within them without the risk of creating significant artifacts will prove crucial in the continuing research effort to understand the role of ABA in controlling gas exchange and growth in plants. While ELISA and GC-MS remain expensive and in the case of GC-MS, unattractive for determining ABA concentrations in a large number of samples, technical advances and potential changes in the way samples are collected, will undoubtedly enhance the suitability of these alternative methods.

REFERENCES

Addicott F.T. *Abscisic Acid.* New York, USA: Praeger, 1983

Bacon M.A., Wilkinson S., Davies W.J. pH-regulated leaf cell expansion in droughted plants is abscisic acid dependent. Plant Physiol 1998; 118:1507-1515

Davies W.J., Jones H.G., eds. *Abscisic Acid.* Oxford, UK: BIOS Scientific Publishers, 1991

Hartung W., Sauter A., Turner N.C., Fillery I., Heilmeier H. Abscisic acid in soils: What is its function and which factors and mechanisms regulate its concentration. Plant Soil 1996; 184:105-110

Munns R., Cramer G.R. Is co-ordination of leaf and root growth mediated by abscisic acid? Plant Soil 1996; 185:33-49

Munns R., King R.W. Abscisic acid is not the only stomatal inhibitor in the transpiration stream of wheat plants. Plant Physiol 1988; 88:703-708

Palmer S. *Leaf Expansion – its Inhibition at Low Nitrogen Availability*. PhD Thesis. Lancaster, UK: University of Lancaster, 1996

Quarrie S.A., Whitford P.N., Appleford N.E.J., Wang T.L., Cook S.K., Henesen I.E., Loveys B.R. A monoclonal antibody to (S)-abscisic acid: its characterisation and use in a radioimmunoassay for measuring abscisic acid in crude extracts of cereal and lupin leaves. Planta 1988; 173:330-339

Saab I.N., Sharp R.E., Prtichard J. Effect of inhibition of abscisic acid accumulation on the spatial-distribution of elongation in the primary root and mesocotyl of maize at low water potentials. Plant Physiol 1992; 99:26-33

Schurr U. Xylem sap sampling - new approaches to an old topic. Trends Plant Sci 1998; 8:293-298

Sharp R.E., Hsiao T.C., Silk W.C. Growth of the maize primary root at low water potentials 2. Role of growth and deposition of hexose and potassium in osmotic adjustment. Plant Physiol 1989; 93:1337-1346

Spollen W.G., LeNoble M.E., Samuels T.D., Bernstein N., Sharp R.E. Abscisic acid accumulation maintains maize primary root elongation at low water potentials by restricting ethylene production. Plant Physiol 2000; 122:967-976

Spollen W.G., Sharp R.E., Saab I.N., Wu Y. "Regulation of Cell Expansion in Roots and Shoots at Low Water Potentials." In *Water Deficits*. Oxford, UK: BIOS Scientific Publishers, 1993

Tejo C. T., Davies W.J., Ruiz L. Sensitivity of stomata to abscisic acid – an effect of the mesophyll. Plant Physiol 1993; 102:497-502

Voetberg G.S., Sharp R.E. Growth of the maize primary root at low water potentials 3. Role of increased proline deposition in osmotic adjustment. Plant Physiol 1991; 96:1125-1130

Weyers J., Meidner H. *Methods in Stomatal Research*. Harlow Longman Scientific and Technical, 1990

Wilkinson S. pH as a stress signal. Plant Growth Regul 1999; 29:87-99

Wilkinson S., Corlett J.E., Oger L., Davies, W.J. Effects of xylem sap pH on transpiration from wildtype and flacca mutant tomato leaves: a vital role for abscisic acid in preventing excessive water loss from well-watered plants. Plant Physiol 1998; 117:303-309

Wilkinson S., Davies W.J. Xylem sap pH increase: a drought signal received at the apoplastic face of the guard cell that involves the suppression of saturable abscisic acid uptake by the epidermal symplast. Plant Physiol 1997; 113:559-73

Wu Y.J., Sharp R.E., Durachko D.M., Cosgrove D.J. Growth maintenance of the maize primary root at low water potentials involves increases in cell-wall extension properties, expansion activity, and cell wall susceptibility to expansion. Plant Physiol 1996; 111:765-772

Wu Y.J., Spollen W.G., Sharp R.E., Hetherington P.R., Fry S.C. Root-growth maintenance at low water potentials - increased activity of xyloglucan endotransglycosylase and its possible regulation by ABA. Plant Physiol 1994; 106:607-615

Zhang J., Davies W.J. Antitranspirant activity in the xylem sap of maize plants. J Exp Bot 1991; 42:317-321

CHAPTER 27

RADIOCHEMICAL TECHNIQUES TO QUANTIFY ALLELOCHEMICALS IN PLANTS

Geneviève Chiapusio and François Pellissier
Laboratoire de Dynamique des Ecosystemes d'altitude. LDEA. Université de Savoie. France

INTRODUCTION

Several steps are necessary to demonstrate that one plant is phytotoxic upon another one. The first step is to identify and quantify chemical compounds (terpens, steroids, phenols...) released by donor plants. The second one is to study their becoming into the soil. Finally, these allelochemicals must be absorbed by the target plant in order to induce (or not) phytotoxic effects.

The latter step is usually demonstrated with bioassay experiments in order to determine allelochemicals phytotoxicity upon germination, growth, photosynthesis...of target plants. Such noticed biological effects are supposed to be due to the allelochemicals. Nevertheless, the proof of allelochemicals absorption in target plant is not made! Radiochemical techniques can then reveal that allelochemicals are really absorbed (or not) by target plants. As allelochemicals are mainly secondary metabolites, it is difficult to distinguish between the natural allelochemical occurring in plants and the applied one. The distinction becomes easier when the applied one is radiolabelled. Quantifying the radiolabelled compound in different plant organs is also a way to identify a possible allelochemical sink organ. Making hypothesis upon physiological target becomes then easier.

M.J. Reigosa Roger, Handbook of Plant Ecophysiology Techniques, 443–452.

Radiochemical techniques are widely used in agronomy to understand herbicide translocation into plants. In this chapter, no general information about radioactivity will be given neither about necessary precautions to work with radioactivity (^{14}C).

Objective of this chapter is to present basic aspects of widely used quantitative techniques. Based upon a practical example, we will focus successively on i) the need for an adequate sample preparation; ii) the advantages and disadvantages of three methods to quantify the radioactivity in samples and, iii) a general discussion on the need to use such techniques to improve allelopathy studies.

SAMPLE PREPARATION

The way to obtain target plants with incorporated radiolabelled allelochemicals will influence the selection of the technique to detect the radioactivity. Thus, this section will explain the way to prepare radioactive solution, will present plant growth conditions and will indicate how to express results.

Preparation of the radiochemical: the 'mother solution'

Most of the time, radiochemicals are sold in a powder form. Adding solvent is then necessary to obtain a solution. The adequate quantity will depend on the radiochemical specific activity. In general, the solubilisation is made in order to obtain a convenient dilution to prepare the further radiochemical solutions (e.g. Figure 1, step 2). The choice of the solvent will depend on the chemical to dissolve, but not on the radioactivity. For example, ethanol is commonly used with radiolabelled phenols. This solution, the 'mother solution', has to be made very carefully because it will be the base of preparation of all further radiochemical solutions. This solution has to be kept in freezer until further utilisation.

Warning: Take all precautions manipulating radioactive powder; it is easy to inhale!

• [Ring-U^{14}C] *p*-hydroxybenzoic acid
(radiochemical purity > 97%, specific activity 12 mCi

Mother solution :
radioactive powder dissolved into 1ml of ethanol 99.9%
concentration: 8.7 10^{-3} mmol ml^{-1}

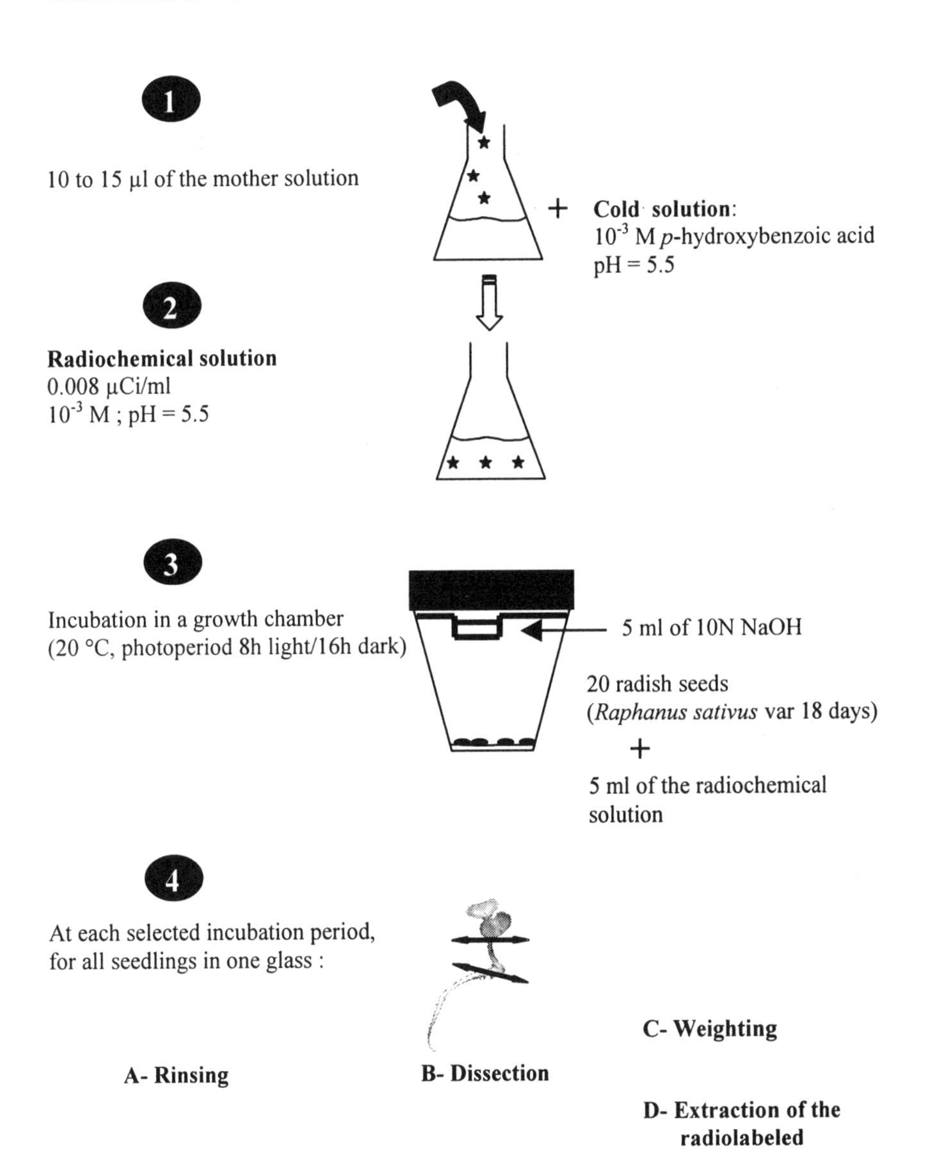

Figure 1. *p*-hydroxybenzoic acid (POH) translocation in radish seedlings (non-sterile culture)

Radiochemical solution

This solution is a mixed solution of a classical solution of the allelochemical, 'the cold solution' and, the 'mother solution'. The 'cold solution' is prepared at the tested concentration and an adequate quantity of the mother solution is added. Sufficient radioactivity has to be added to the cold solution in order to remain upper the threshold measurement of the radioactivity detector.

Warning: Before making the radiochemical solution, check how the addition of the mother solution changes the 'cold solution' concentration.

Plant growth conditions

No special culture conditions are required. The number of seeds, photoperiod, incubation temperature... are depending on the used species. The choice of sterile or non-sterile culture depends on the objective of the work. In non-sterile condition, a release of $^{14}CO_2$ could occurr. The $^{14}CO_2$ has then to be trapped in a NaOH solution.

Warning: Glass beakers are better than plastic ones to recover the entire radioactivity (it is difficult to wash plastic as carefully as glass).

Samples preparation

At the end of each selected incubation period, seedlings from each glass are washed three times in the solvent (e.g. ethanol) to remove adsorbed compounds. Then, seedling parts are separated. Each organ type removed from the same glass is weighed and frozen together (-25°C). These constitute one sample of one organ type.

Expression of results

Allelochemical having penetrated in seedlings should be either expressed as concentration (for example in μmol g^{-1} FW) or as quantity (% of applied ^{14}C allelochemical) for one beaker.

Quantity is the ratio between the Disintegration Per Minute (DPM) of the initial ^{14}C allelochemical solution which was added to each glass and the recovery obtained from separated organs of seedlings. Allelochemical concentration is calculated at the equivalence of the cold solution moles, represented by x DPM detected in organs.

Concentration pattern reflects the distribution of allelochemical in plant organs. This unit is widely used to obtain a ratio between compound content and plant biomass, which allows then comparisons between different studies. However, using only concentration could lead to erroneous conclusion because seedlings biomass *per se* is necessary to understand distribution of allelochemical in organs of target plants (Table 1). Thus, allelochemical content is used in order to detect preferential accumulation organ.

Warning: The total DPM counted in one glass (seedlings + culture medium + rinsing seedlings and beaker + $^{14}CO_2$) should be at minimum 80 % of the initial radioactive solution added to the culture medium. It is the only way to validate the experimentation (under 80%, experiments are considered as non-valuable).

Table 1. Biomass (mg Fresh Weight ± standard deviation) of dissected radish seedlings grown with 10^{-3}M mixed solution (POH + ^{14}POH) in non-sterile conditions. Data represent means of at least 3 replicates per incubation time and per culture condition. * indicates statistically significant higher biomass compared to other organs biomass, according to Mann-Whitney *U* non-parametric test ($P<0.05$). Biomass variation of each radish organs: cotyledon biomass is about three times more than roots or hypocotyls (statistically significant for $P<0.05$).

Incubation period (days)	***Medium***	**Roots**	**Hypocotyls**	**Cotyledons**
D3	*Non-Sterile*	76 ± 21	77 ±9	329 ± 32 *
D4	*Non-Sterile*	168 ± 26	147 ± 9	458 ± 19 *

Statistical analyses.

Use statistics as usually, depending on the number of replicates, the distribution and normality of data...

RADIOACTIVITY QUANTIFICATION

Extraction by grinding and quantification by Liquid Scintillation (example: Table 2)

Principle

Samples are ground with a mortar and pestle and soluble radiolabelled allelochemical is extracted with adequate solvent.

Homogenates are centrifuged (3000 rpm) for 3 min at 20 °C. The supernatant is set aside for further analysis after each extraction. The ground residue is extracted twice in the mortar and pestle, supernatants being combined.

Table 2. POH translocation in radish seedlings in non-sterile condition of incubation. POH concentration (expressed as µmol g^{-1} Fresh Weight ± standard deviation) and quantity (expressed as % of applied ^{14}POH for 20 seedlings ± standard deviation) were measured after grinding extraction method.
Data represent means of at least 3 replicates per incubation time. * indicates inside a same incubation time, POH content of the organ is statistically different from the others according to Mann-Whitney *U* non-parametric test ($P<0.05$). POH concentration in each organ is depending of the time and the type of organ. Focusing on % of applied, cotyledons are the POH sink organ at anytime.

Incubation period (days)	*POH quantification*	Roots	Hypocotyls	Cotyledons	Total in seedlings
D3	*µmol g⁻¹FW*	5.9 ± 1.4	5.8 ± 0.9	10.3 ± 1.8 *	22.0 ± 4
	% of applied	8 ± 1	5 ± 0.4	35 ± 3 *	47 ± 4
D4	*µmol g⁻¹FW*	2.8 ± 1.4	2.7 ± 1.3	5.5 ± 1.9	10.9 ± 4.5
	% of applied	5 ± 2	4 ± 1	23 ± 8 *	32 ± 11

Counting

Supernatants and pellets are mixed with liquid scintillation cocktail. The choice and the quantity of the added liquid scintillation cocktail are made according to manufacturer recommendations.Radioactivity within samples is counted using a Liquid Scintillation Counter. Results appear as DPM.

Advantages / Disadvantages

Principle of extraction is very simple. Time to make the extraction is long. Care is needed during grinding due to the radioactivity risk and during addition of liquid scintillation cocktail (chemical risk).

Extraction by oxidising and quantification by Liquid Scintillation (example: Table 3)

Principle

Frozen radioactive samples are wrapped in a paper (Germaflor) for oven drying at 80°C for 48 hours. Each sample is then combusted in a biological oxidiser at 900°C for about 3 min. As samples are directly burned, $^{14}CO_2$ is directly trapped into a vial containing liquid scintillation cocktail.

Table 3. POH translocation in radish seedlings in sterile condition of incubation. POH concentration (expressed as µmol g^{-1} Fresh Weight ± standard deviation) and quantity (expressed as % of applied ^{14}POH for 20 seedlings ± standard deviation) were measured after oxidiser extraction method.
Data represent means of at least 3 replicates per incubation time. * indicates inside a same incubation time, POH content of the organ is statistically different from the others according to Mann-Whitney *U* non-parametric test ($P<0.05$). Cotyledons and roots are radish organ having highest POH concentrations. Focusing on % of applied, cotyledons are the POH sink organs.

Incubation period (days)	*POH quantification*	Roots	Hypocotyls	Cotyledons	Total in seedlings
D3	*µmol g^{-1}FW*	15.7 ± 4.1	7.1 ± 0.9 *	12.4 ± 2.4	35.1 ± 5.9
	% of applied	5 ± 1	5 ± 1	30 ± 2 *	40 ± 4
D4	*µmol g^{-1}FW*	14.8 ± 3.1	5.5 ± 0.9 *	12.4 ± 1.0	33.2 ± 2.4
	% of applied	11 ± 5	6 ± 1	38 ± 3 *	53 ± 7

Counting

Radioactivity within samples is counted using a Liquid Scintillation Counter and results are expressed as DPM.

Advantages / Disadvantages

Very simple and fast in manipulating. Few radioactive waste.

Extraction and quantification

Principle

After extraction by grinding (as above), the supernatants are analysed by means of TLC (Thin Layer Chromatography, see Chapter 21) coupled to an Imaging scanner or/and HPLC (High Performance Liquid Chromatography, see Chapter 18) coupled to a radio-analyser.

Initial separation of the radioactive components can be performed by TLC to find out what classes of compound are involved. The distribution of radioactivity on chromatogram is analysed with an imaging scanner. HPLC technique can provide higher resolution for the separation of closely related molecular species.

Counting

Radioactivity of radiolabelled compounds is detected and quantification is made with area of peaks, using HPLC or TLC.

Advantages / Disadvantages

Identifying all radioactive molecules by comparison with standard (co-chromatography). Unknown molecules may be also identified with the help of mass spectrometry.

IS THE USE OF RADIOLABELLED COMPOUNDS USEFUL FOR ALLELOPATHY RESEARCH?

The three previously described techniques are complementary.

The highest concentration of radiolabelled chemical recovered in seedlings is obtained by oxidising extraction method. When samples are oxidised, every ^{14}C is recovered whereas it is quite impossible with grinding extraction. Difference between concentrations obtained by grinding and by oxidising extraction provides indication on two forms - soluble and bound- for the allelochemical in cells of target plants.

These two techniques do not allow identifying the radiolabelled counted molecule. Every detected ^{14}C is considered to belong to the same molecule. Actually, several modifications could occur to the radiolabelled allelochemical such as degradation in the culture medium or in the plant itself. The only technique that could identify each radiolabelled molecule is the combination of grinding and quantification by TLC or HPLC.

Radiolabelled techniques provide the proof of what really occurs to allelochemicals in target plants. These techniques are easy to set up. Care is necessary during manipulation because of the radioactivity. The cost of the experiments is quite high when including the purchase of the radiolabelled molecule and all radioactive wastes issued from experiments.

Thus, the reply of the heading question of this section is: **yes!**

FURTHER READINGS

Barz W., Weltring K.M. "Biodegradation of Aromatic Extractives of Wood." In *Biosynthesis and Biodegradation of Wood Components*. T. Higuchi, ed. San Diego, California: Academic Press, 1985

Chao J.F., Hsiao A.I., Quick W.A., Hume J.A. Effect of decapitation on absorption, translocation, and phytotoxicity of imazamethabenz in wild oat (*Avena sativa* L.). J Plant Growth Regul 1994; 13:153-158

Delrot S., Bonnemain J.L. "Le Transport à Longue Distance des Herbicides dans la Plante." In *Les Herbicides: Mode d'Action et Principes d'Utilisation* R. Scalla, ed. Paris, France: INRA, 1991

Dayan F.E., Duke S.O., Weete J.D., Hancock H.G. Selectivity and mode of action of carfentrazone-ethyl, a novel phenyl triazolinone herbicide. Pestic Sci 1997; 51:65-73

Martin J.P., Haider K. "Microbial Degradation and Stabilization of ^{14}C-labeled Lignins, Phenols, and Phenolic Polymers in relation to Soil Humus Formation." In *Lignin Biodegradation: Microbiology, Chemistry and Potential Applications*. T.K. Kirk, T. Higuchi, H. Chang, eds. Boca Raton, Florida: CRC Press, 1980

Simmen U.R.S., Gisi U. Uptake and distribution in germinating wheat of [14C]SAN 789F and [14C] cyproconazole applied as seed treatment. Crop Prot 1996; 15:275-281